RECENT APPLICATIONS IN SOL-GEL SYNTHESIS

Edited by **Usha Chandra**

Recent Applications in Sol-Gel Synthesis
http://dx.doi.org/10.5772/65824
Edited by Usha Chandra

Contributors

Bekir Salih, H. Mehmet Kayili, Mehmet Atakay, Nurhan Onar Camlibel, Buket Arik, Fei Li, Xiao Huang, Guo-Jun Zhang, Helena Cristina Vasconcelos, Afonso Silva Pinto, Ana Gómez-Varela, María Aymerich, Daniel Nieto, María Teresa Flores-Arias, Carmen Bao-Varela, Yolanda Castro, Pieter A.A. De Beule, Ezequiel Álvarez, Guadalupe Valverde Aguilar, Mónica Jaime Fonseca, Angeles Mantilla Ramírez, Antonio Gustavo Juárez Gracia, Dhanasekaran Vikraman, Shrividhya Thiagarajan, Anandhavelu Sanmugam, Ahmad Azmin Mohamad, Muhamad Zamri Yahaya, Mohd Asyadi Azam, Mohd Asri Mat Teridi, Pramod Kumar Singh, Susana Devesa, Manuel Pedro Graça, Luís Costa, Maria Zielecka, Elzbieta Bujnowska, Katarzyna Suwala, Magdalena Wenda, Mohamad Hafiz Mamat, Marmeezee Mohd. Yusoff, Mohamad Rusop Mahmood, Robabeh Bashiri, Norani Muti Mohamed, Chong Fai Kait, Abdul Naeem, Asad Mahmood, Franco Ferrero, Monica Periolatto

Published by InTech

Janeza Trdine 9, 51000 Rijeka, Croatia

Publishing Process Manager Martina Blecic
Technical Editor SPi Global
Cover InTech Design team

First published June, 2017
Printed in Croatia
Legal deposit, Croatia: National and University Library in Zagreb

Additional hard copies can be obtained from orders@intechopen.com

Recent Applications in Sol-Gel Synthesis, Edited by Usha Chandra
p. cm.
Print ISBN 978-953-51-3245-5
Online ISBN 978-953-51-3246-2

For EU product safety concerns:
IN TECH d.o.o., Prolaz Marije Krucifikse Kozulić 3, 51000 Rijeka, Croatia,
info@intechopen.com or visit our website at intechopen.com.

Meet the editor

Dr. Usha Chandra, retired professor in Physics, emeritus fellow (UGC), and emeritus scientist (CSIR), University of Rajasthan, Jaipur, India, is an experimentalist carrying out research on nanomaterials, minerals, and meteorites under high pressure using Mössbauer spectroscopy. Nanocrystalline perovskites (LaSrMnO) and pyrites (CoFeS2) synthesized using sol-gel process and studied under high pressure using various techniques are the achievements worth mentioning. As an expert in X-ray absorption spectroscopic (XANES, EXAFS) and photoemission (XPS) techniques, she has published several research articles in international journals. She is a member of various professional bodies in this arena. As a member of the American Physical Society (USA.), the Indian Physics Association (India), and the International Advisory Board—Mössbauer Effect Data Center— her contribution to the field is exclusive.

Contents

Preface

In 2000, during my endeavor to synthesize nanocrystalline rare-earth perovskite systems, I got fascinated by "sol-gel process" producing homogeneous nanoparticles. Fortunately, discussions and interactions with Prof. R. C. Mehrotra (national professor, Department of Chemistry, University of Rajasthan, Jaipur, India, and vice-chancellor, Delhi University, Delhi), an expert and pioneer in synthesizing alkoxide precursors using sol-gel wet chemical process with lots of research publications on the subject, gave me confidence to venture into this new field. Thus, I started my remarkable journey of high-pressure studies on nanocrystalline systems synthesized using the sol-gel wet chemical process.

Among the various synthesizing techniques, sol-gel is one of the attractive, versatile techniques to obtain both the matrix and nanocomposites with varying structures and particle sizes to define varying properties. The technique widely used in the fields of material sciences and ceramic engineering has grown to cover wide range of fields like thin films, fibers, electronics, biosensors, etc.

The goal of the book *Recent Applications in Sol-Gel Synthesis* is to expose researchers to the capabilities of sol-gel technique in enhancing physical and chemical properties of the materials. The requirement of low temperature in the process enhances the utility (application) of the sol-gel technique in various fields like optics, biosystems, ultrahigh-temperature ceramics (UHTC), etc.

In recent years, the field has grown up enormously. To keep track with recent developments, I have entrusted collaboration of experts in their individual fields. Consequence of such collaborative effort is reflected in the variety of the writing style and methods of approach to a given topic. However, my role as an editor was not to fit everyone in the same mold but to retain the imprints of the individual authors.

Suggestions and criticisms are welcome for future endeavors. I am grateful to authors for their valuable time and sincere efforts toward the preparation of the respective contributions.

Prof. Usha Chandra
Department of Physics, University of Rajasthan,
Jaipur, India

Chapter 1

Facile Methodology of Sol-Gel Synthesis for Metal Oxide Nanostructures

Shrividhya Thiagarajan,
Anandhavelu Sanmugam and
Dhanasekaran Vikraman

Additional information is available at the end of the chapter

http://dx.doi.org/10.5772/intechopen.68708

Abstract

Sol-gel is a low temperature, highly controllable and cost effective method for production of homogeneous, highly stoichiometric and high quality ultrafine nanostructures. Sol-gel route is adoptable way to choose desired shape of the metal oxide (MO) nanostructures such as nanospheres, nanorods, nanoflakes, nanotubes, nanoribbons, nanospheres and nanofibers for shape-dependent applications and comparative accessibility. Biomedical applications involving drug deliveries, mimicking of natural bone and teeth, antimicrobial activities and pharmaceuticals employ sol-gel prepared MO nanostructures because of their low temperature synthesis, homogeneity and purity. Apart from this, sol-gel route is preferred for synthesis of MO-based nanostructures with several ranges of applications such as magnetic applications, energy generation, conversion and storage devices, electronic device applications and sensors and actuators materials. In this chapter, we have discussed about the comprehensive ideas of sol-gel technique to synthesis metal oxide nanostructures.

Keywords: sol-gel, metal oxides, nanostructures

1. Introduction

Sol-gel process have been extensively explored for producing metal oxide nanostructures in the field of engineering and technological applications probably due to the controlled shape and size exhibited by the obtained products. Since the synthesis of silica gel by Ebelman in 1846, this method has been developed progressively and sol-gel synthesized materials have been implemented in several applications with excellent optical, magnetic, electrical, thermal and mechanical properties [1]. Several forms of materials such as thin films, nanoparticles, glass and ceramics can be achieved using sol-gel method in a cost-effective way [2]. Low temperature chemistry, reproducibility and high surface to volume ratio of obtained products are other features that add merit to this technology [3]. Apart from this, sol-gel process have opened up some new avenues in bioengineering fields including drug delivery, organ implantation, pharmaceuticals and biomaterial synthesis due to the purity and quality of the yields from this process. These advantages have attracted researchers and industrialists to utilize this method widely for past few decades [4]. Metal oxides are class of functional materials with numerous applications and can be synthesized using sol-gel process [5]. Sol-gel synthesis of metal oxide can be done at relatively low temperature compared to the solid-state reactions. In general, sol-gel process involves formation of sol from homogeneously mixed solution, converting them into gel by polycondensation process and finally heat treating the product according to the material required [6]. The formation of crystalline materials such as nanoparticles or thin films and non-crystalline materials like ceramics, xerosol, aerosol and glasses depends upon the final heat treatment steps [7]. This chapter provides the fundamental steps involved in sol-gel process, various controlling parameters, the size- and shape-controlled synthesis of various metal oxides and application of sol-gel synthesized metal oxide nanoparticles in various fields.

2. Feasibility of sol-gel method

Figure 1 shows the schematic diagram of sol-gel method. The experimental set up is very simple. Sol is obtained by either hydrolysis or polymerization reactions by adding suitable reagents in the precursor solution. The sol can be deposited onto preferred substrates as thin films using two techniques, viz. (1) spin coating and (2) dip coating. The gelation process done through condensation of the sol or addition of polymers converts this sol to gel. This gel can be used to form materials of different types such as nanoparticles, xerogel, glass or ceramics depending upon the further processing steps involved. Nanoparticles and xerogels can be obtained by simple evaporation of solvent. The obtained xerogel can be formed as ceramics by heat treatment and glassy nature can be induced by melting techniques. Thus, sol-gel method can be used to obtain different forms of materials, controlled phase and shape and size of the derived materials [6].

The parameters that could be controlled in sol-gel method includes (1) concentration of precursor used, (2) nature of solvent used, (3) pH of the solution, (4) type of additives added and their concentration, (5) pre and post heat treatment of the materials, (6) aging of the solution and (7) nature of polymer used for condensation [8]. The particles formed in gel matrix possess uniform shape and size that enhances the optical, electrical, magnetic and other intrinsic nature of the materials. Low cost, low temperature chemistry, simple experimental set up and highly controllable synthesis are the major advantage of this method over other synthesis procedure. The large surface to volume of sol-gel-derived material makes it suitable

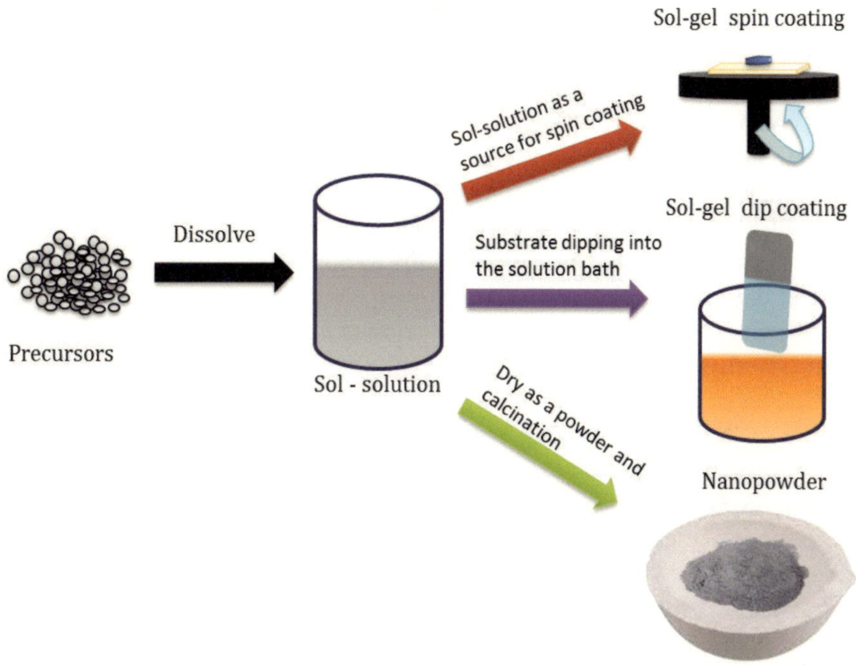

Figure 1. Schematic diagram of sol-gel processing.

for catalytic applications. Apart from this, low temperature chemistry ensures less defects to be induced in the materials formed. The purity and high quality yield with good physical properties makes researchers in biomedical and biotechnical field to adapt this method mostly for biomedical applications [9].

3. MO nanostructures using sol-gel

The synthesis of solid materials often involves wet chemistry reactions and sol-gel chemistry based on the transformation of molecular precursors into an oxide network by hydrolysis and condensation reaction [9–11]. Morphology plays a vital role in improving the properties of any material by enhancing high surface to volume ratio. The control of shape, size and packing structure of the particles in materials plays a crucial role for building next-generation devices and therapeutic materials [1]. Sol-gel method is an excellent tool to deploy a controlled architecture in material chemistry to fabricate metal oxide nanostructures (MONSs). Sol-gel prepared metal oxides have shown to exhibit excellent optical and electrical properties. The review of important MONSs derived from the sol-gel method helps us to understand the factors that could be taken into account for controlling the shape and size of the particles. In general, the solvents, additives, aging time and post heat treatment are few important factors that determine the shape and size of the building blocks of materials being synthesized [6, 12].

ZnO is a multifunctional material, which is widely being explored by many researchers for several years. Sol-gel method is preferably a low-cost simple method for preparation of numerous ZnO nanostructures with implementation in abundant scientific and industrial applications. As an example, XRD pattern of sol-gel dip coating route prepared ZnO thin film is given in **Figure 2**. The polycrystalline nature hexagonal structure of ZnO is exhibited with (002) preferential orientation at $2\theta = 34.48°$. Also, low intensity peaks are exhibited with (100), (101), (102), (110), and (103) lattice orientations as shown in **Figure 2**. Li et al. [13] illustrated that the aging time of sol influenced on the morphology of nanoparticles. They found that the longer time aged sol had better morphology than as synthesized particles. Also, they suggested that the possible reason for this trend may be due to the uneven distribution of colloidal particle size. Previously, the effect of solvent on the growth of ZnO nanorods was explained by Foo et al., where they used ZnO seeds obtained from hydrothermal method as seeds for growth of nanorods [14]. With the advent of technology, flexible electronics is becoming popular in recent past. Low temperature synthesis is mandatory for materials to be coated onto the flexible substrates as the thermal stability of substrate materials has to be taken into account. Sol-gel method opens up a new avenue for low temperature deposition of MONS onto flexible substrates for construction of flexible electronic devices. In this view point, Sun et al. successfully reported uniform sol-gel-derived ZnO thin films at relatively low temperature (≤200°C) that can function as efficient transporting layer in inverted solar cells [15]. Kim et al. fabricated sol-gel-derived ZnO on plastic substrates and demonstrated potential for Al/Sol-gel ZnO/Al for low cost flexible memory devices [16]. Xu et al. showed the control of thickness and packing density of the thin films by number of dips and post heat treatment of ZnO sols and their influence on the optical properties of the thin films [17]. ZnO nanowires synthesized by sol-gel route were investigated for photoconductivity studies and origin of the photoresponse was also analyzed by Ahn et al. [18]. Qu et al. demonstrated deposition of ZnO nanorod arrays as an antireflective layer in the polycrystalline Si solar cells [19]. Ghosh et al. adapted a cooling for the sol-gel deposited ZnO films at different cooling rates and uniform nanospheres were attained as a result. They showed that the films exhibited tuning of the optical emission and ultraviolet photosensing properties [20]. Moreover, metal doping is possible by sol-gel dip coating technique [21, 22]. As an example, Chandramohan et al. [22] have prepared Al-doped ZnO thin films and plausibly explained its variation of optical properties by increase of doping composition (**Table 1**). They demonstrated that estimation of various optical constants like band gap (E_g), refractive indices (n, k), dielectric constant (e), optical conductivity (σ), average excitation energy (E_0), oscillator strength (E_d), effective mass (m^*), plasma frequency (x_p), static dielectric constant (ε_0) and carrier concentration (N). **Figure 3** represented that the SEM micrographs and EDX patterns of undoped and Al-doped ZnO thin films. The highly transparent thin films showed nanowires protruding from stacked nanorods on SEM inspection that signifies the suitability of these thin films for gas sensors.

Iron oxide is a versatile material, which can be used in many applications such as magnetic storage, drug delivery, sensor and electrochromic applications. Sol-gel process ascertains iron oxide with improved physical and chemical properties due to uniform particle size distribution, controlled phase synthesis and morphology. Dhanasekaran et al. have shown the influence of precursor concentration on the morphology of highly dispersed, single phase Fe_2O_3

nanostructure formed using sol-gel method. The nucleation and growth kinetics affected by the shape and size of the material was elaborated. **Figure 4** shows SEM micrographs of the Fe_2O_3 nanostructures at different $FeCl_3$:$FeCl_2$ concentrations ratio. They have obtained saturation magnetization (M_s) and coercivity values (H_c) at 35 T and 58 A/m, respectively, for Fe_2O_3 nanostructures prepared at $FeCl_3$:$FeCl_2$ molar ratio of 1:1.5 [23]. This observation shows that ferrous oxide nanostructures with low coercivity and high saturation magnetization value could be used as core materials in recording media.

Sol-gel-mediated synthesis of Fe_2O_3 nanorods was described by Woo et al. [24]. They showed that the diameter and length of the rods could be controlled by addition of (H_2O/Oleic acid) and the phase could be controlled by the temperature, atmosphere and hydrous state of the

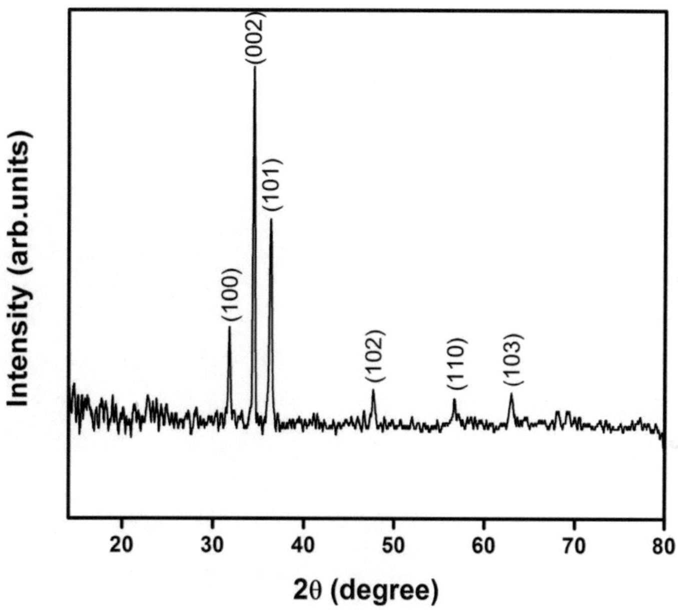

Figure 2. XRD pattern for sol-gel dip coating prepared ZnO thin film.

Sample	E_0	E_d	M_{-1}	M_{-3}	E_g	ε_∞	ω_p×10¹⁵ s⁻¹	m^*	N×10¹² cm⁻²
Pure-ZnO	1.866	4.694	2.515	0.722	3.21	6.97	2.054	0.184	1.594
Al doping—0.01 mM	1.910	3.939	2.062	0.565	3.23	5.8	1.855	0.185	1.635
Al doping—0.03 mM	2.062	3.808	1.846	0.434	3.29	4.91	1.848	0.187	1.137
Al doping—0.05 mM	2.224	3.858	1.732	0.350	3.33	4.598	1.830	0.189	1.055

E_0, average excitation energy; E_d, oscillator strength; E_g, band gap; ε_∞, static dielectric constant; ω_p, plasma frequency; m^*, effective mass and N, sheet carrier concentration.

Table 1. Calculated optical and dispersion parameters of undoped and Al-doped ZnO films [21].

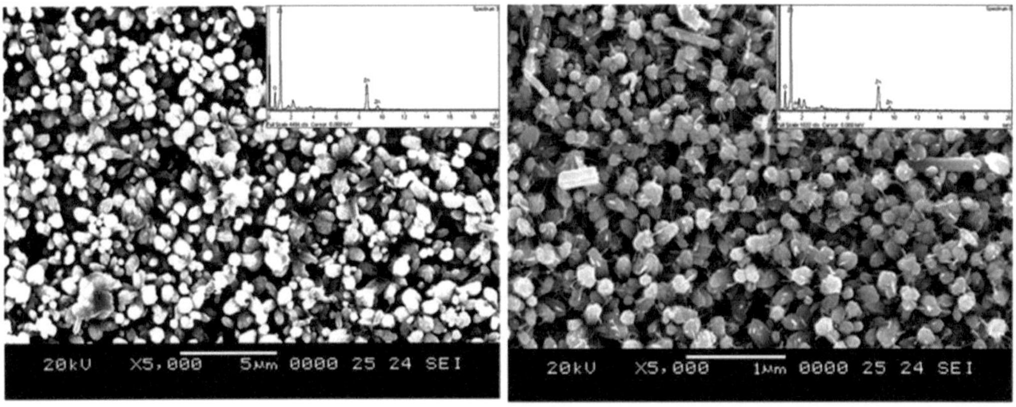

Figure 3. (a) SEM micrograph of ZnO thin film with inset EDX pattern and (b) SEM micrograph of Al-doped ZnO thin film with inset EDX pattern [21].

Figure 4. SEM micrographs of sol-gel route synthesized Fe_2O_3 nanostructures using different $FeCl_3$:$FeCl_2$ molar ratios such as (a) 1:1.0, (b) 1:1.5 and (c) 1:2.0.

gel during the crystallization. Liu et al. described α-Fe_2O_3 ultra-fine powder with an average particle size of 6–26 nm preparation by a sol-gel process. The temperature dependence of the electric conductance of the elements made of nanocrystalline α-Fe_2O_3 shows that the gas-sensing properties are strongly related to its surface. The elements exhibited good sensitivity and selectivity to ethyl alcohol, indicating that it is a promising alcohol-sensing material [25]. Leonardi et al. prepared iron oxide nanoparticles prepared by electrospinning and sol-gel method for ethanol sensing properties. Highly uniform shaped spherical rods were obtained using sol-gel method. They concluded that both the methods exhibited a good sensing response. This report also attributes the advantage of low temperature simple sol-gel method over other expensive techniques [26]. Sanchez et al. reported simple one pot sol-gel synthesis of epsilon polymorph of iron oxide stabilized in a silica gel. These reports provide an insight that sol-gel not only controls the shape and size of the material synthesis but also provides an excellent phase control and formation of different polymorphs in an orderly trend [27].

Tin oxide is an important transition metal oxide widely being studied for electronic and sensor devices. The additives also influence the morphology of materials in sol-gel process. The SnO_2 nanostructure was grown on Al_2O_3 substrates by a sol-gel spin coating method by Khadhim et al. A novel H_2 gas sensor based on a SnO_2 nanostructure was operated at room temperature (25°C). They observed that the addition of glycerine as additive to the sol solution increased the porosity of the SnO_2 nanostructure surface, which increased the adsorption/desorption of gas molecules leading to the high sensitivity of the sensor [28]. Seval et al. fabricated homogeneous nanostructured SnO_2 films using sol-gel spin coating method from tin (II) acetate solutions. A p-n heterojunction diode was proposed by depositing SnO_2 film formed at 600°C on p-Si substrates by them. They determined that the interface states played an important role in the conduction mechanism of the diode [29]. The role of polymerization agent was described by Bagherian et al., in their report where SnO_2 nanosheets were synthesized through sol-gel method using gelatin as the natural polymerization agent. The calcination temperature variables such as 650, 700 and 750°C were found to influence the resultant optical and structural properties. Morphological studies manifested that wide area SnO_2 nanosheets were obtained by using gelatin at the chosen calcination temperatures. Being a conductive semiconductor, this unique structure will have prospective applications as a conductive transparent substrate [30].

CuO is one of the most broadly studied metal oxides in semiconductor processing-based applications. Dhanasekaran et al. have demonstrated the influence of bath temperature on dip coating of CuO sol onto glass substrates on the structural and morphological properties of the thin films obtained. Homogeneous distributions of ellipsoidal granular-shaped particles were observed on the surface of the films. These films were exploited to have single phase structure and have excellent optical and magnetic properties. Shrividhya et al. [31] have demonstrated the effect of precursor concentration on the morphological properties of CuO thin films using sol-gel dip method (**Figure 5**).

The copper oxide nanostructures were synthesized using 0.2 and 0.6 M copper nitrate concentrations by a sol-gel method [32]. The X-ray diffraction pattern revealed that the prepared structures are polycrystalline in nature. All the reflections in these patterns could be indexed to standard diffraction patterns of copper oxide (JCPDS card no. # 89-5899). The typical X-ray

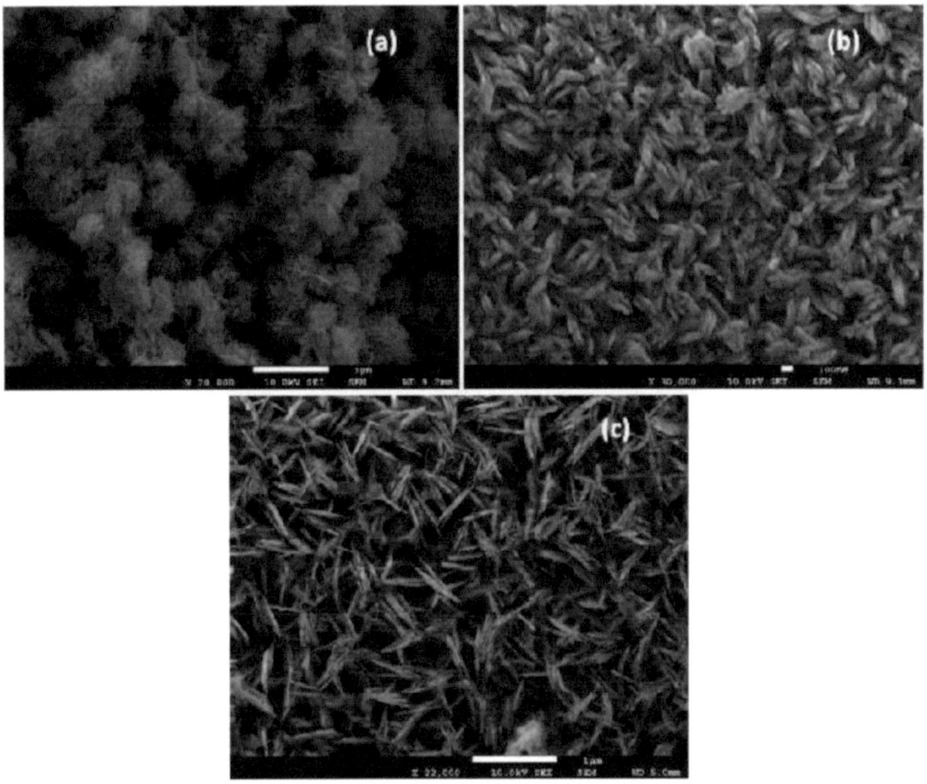

Figure 5. CuO nanostructures prepared by sol-gel dip coating method at (a) 5 mM, (b) 7 mM and (c) 9 mM CuSO4 salt concentrations [31].

diffraction pattern of CuO nanostructures prepared at 0.2 M copper nitrate concentration is shown in **Figure 6(a)**. The refined unit cell parameters (a = 4.650 Å, b = 3.518 Å and c = 5.225 Å) conformed well to the reported values of a = 4.689 Å, b = 3.420 Å and c = 5.130 Å. The highly diffracted peak is observed at an angle 2θ = 35.511 corresponding to the (−111) lattice orientation. The various diffraction peaks such as (−111), (111), (−202), (112), (−113) and (−311) are observed for copper oxide nanostructures prepared at 0.2 M copper nitrate concentration. The full width at half maximum (FWHM) value of a predominant orientation peak is 0.43°. Also the novel reflection plane (110) of the Cu_2O phase emerged on the CuO nanostructures prepared at 0.6 M copper nitrate concentration (**Figure 6(b)**). This may be due to the increase of copper content in the nanostructures. The prepared thin films exhibited good optical properties [32]. Bibi et al. explained the varying annealing temperature on the CuO nanostructures. The morphological changes in turn greatly altered the optical, dielectric and electrical properties [33]. Moreover, NiO thin films prepared by sol-gel technique using different concentrations of nickel sulfate ($NiSO_4$) and their XRD patterns are presented in **Figure 7(a)**. XRD patterns revealed face-centered cubic polycrystalline nature of NiO thin films oriented along (200) lattice preferential orientations (JCPDS – 89-7130). Moreover, (111) and (220) lattice-oriented XRD reflections are exhibited with low intensity for NiO. The higher concentration of 0.40 M $NiSO_4$ using prepared NiO optical transmission and absorption spectrum provided in **Figure 7(b,c)** [34].

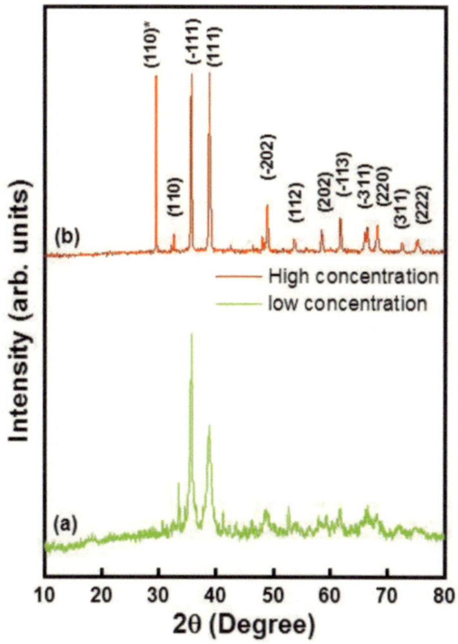

Figure 6. XRD patterns of sol-gel route prepared CuO nanoparticles using (a) low (0.2 M) and (b) high (0.6 M) copper nitrate concentrations [32].

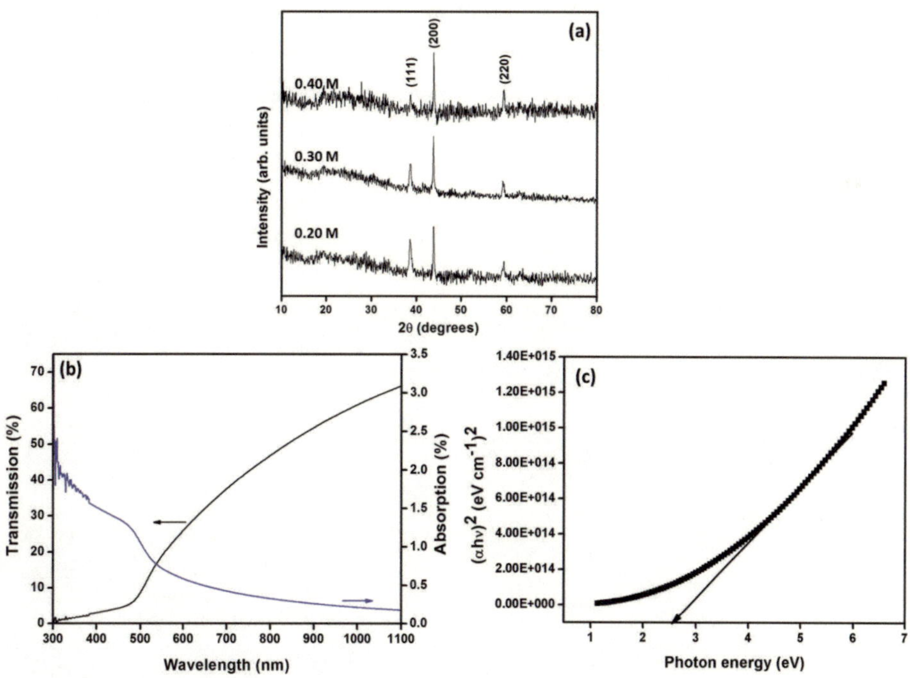

Figure 7. (a) X-ray diffraction patterns of different $NiSO_4$ concentrations using sol-gel deposited NiO thin films such as 0.20, 0.30 and 0.40 M. (b) Optical transmission and (c) optical absorption spectrum of sol-gel-coated NiO thin film prepared at 0.40 M $NiSO_4$.

4. Size-controlled synthesis for various applications

Nanomaterials exhibit a vast shape and size-dependent properties and have found applications in diverse fields, including optics, electronics, mechanics, drug delivery and catalysis. Nanostructures as building block for bottom approaches to material integration encourage functional architecture. Such materials receive considerable interest due to their interesting application-specific properties. Researchers have given extensive reports on such size-controlled applications. As we have already discussed in detail, sol-gel method provides a fine control over the size and shape of the MONS especially for the applications using surface interface chemistry like catalytic activity, optics and electronic devices. SEM images of dip-coated ZnO ($ZnSO_4$ as a source) thin films which assembled by nanoflower- and nanoplate-like grains are shown in **Figure 8**.

NiO is another functional metal oxides mainly explored for energy-related applications. Dhanasekaran et al. [34] have demonstrated that shape selective synthesis of NiO nanostructures for hydrazine oxidation as amperometric sensor. They synthesized NiO NPs, having pellet, rod, dot and cuboid shapes, using a variety of reducing agents via a simple and low-cost sol-gel approach (**Figure 9**). Also, NiO-silica core-shell ($NiO@SiO_2$) NPs were prepared using tetraethyl orthosilicate (TEOS) as a source of the porous silica. They concluded nano-pellet shape NiO exhibited the better electrocatalytic performance of 953 mAg^{-1} for hydrazine oxidation. NiO and nickel manganite composite nanostructures have been synthesized by sol-gel route based on sorbitol as a chelating agent. Magnetic properties of the synthesized nanoparticles were analyzed [35]. Mutkule et al. deposited NiO nanostructures by sol-gel synthesis and their ability to sense LPG gas was analyzed by them [36].

Recently, Dhanasekaran et al. [37] demonstrated magnetic properties induced due to the size-controlled preparation in CuO nanoleaf structures (**Figure 10**). The reported saturation magnetization and coercivity values were at 0.24 T and 355.5 Oe, respectively, for nanoleaf structure of CuO (bath temperature at 95°C). Kato et al. illustrated the dependency of physical and chemical properties on size and shape of metal oxides [38]. Verduraz

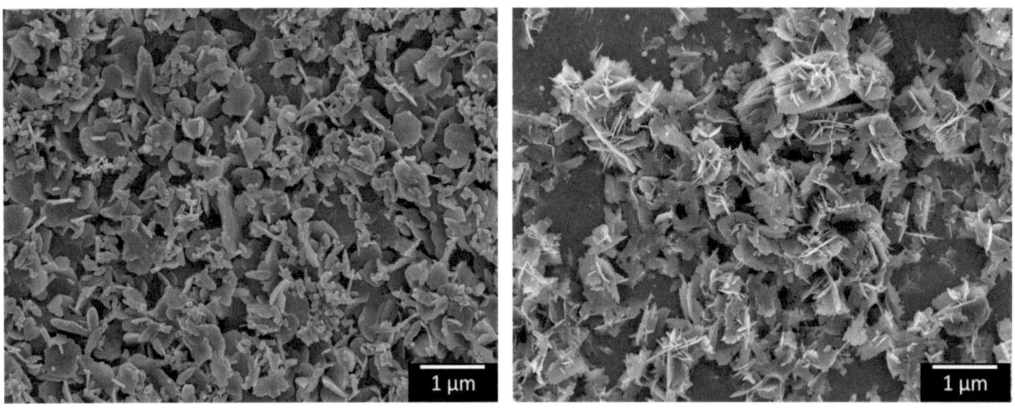

Figure 8. Nanoflower- and nanoplate-like morphological properties of ZnO nanostructures prepared by sol-gel dip coating.

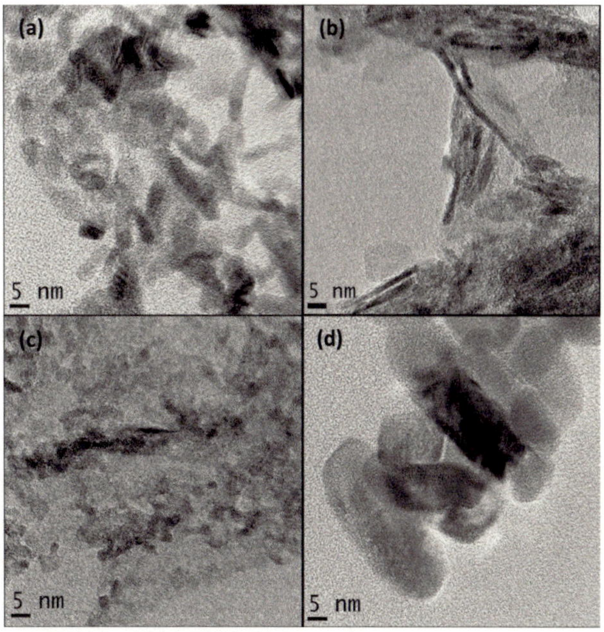

Figure 9. TEM images of NiO nanostructures prepared using different reducing agents (a) NaOH with PEG, (b) NaOH without PEG, (c) NH_3 and (d) Na_2CO_3 [34].

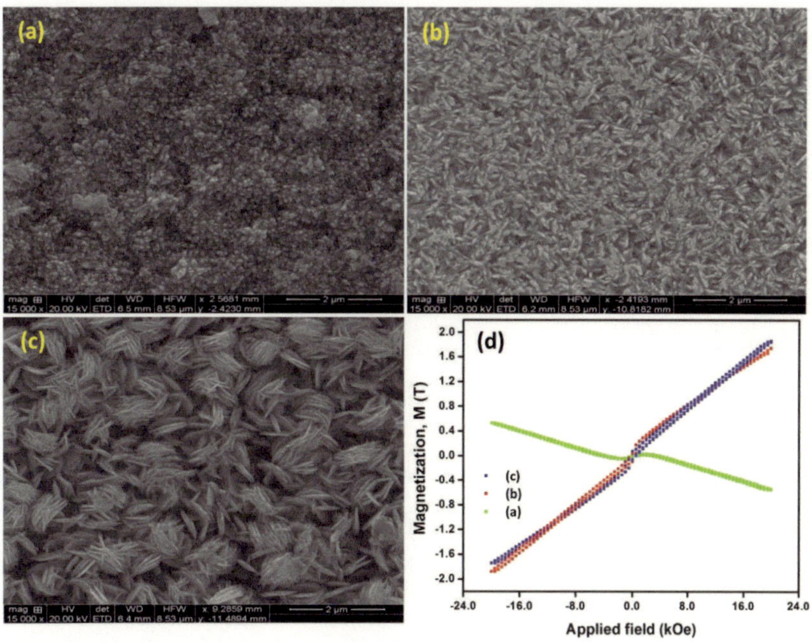

Figure 10. Morphological variations by adjusting bath temperatures for dip-coated CuO nanostructures (a) uneven and rough surface properties at 75°C, (b) elongated ellipsoidal-like grains at 85°C and (c) nanosheet-like morphology with regularly arranged surface at 95°C and (d) M-H hysteresis spectra for different bath temperatures (75, 85 and 95°C) using prepared CuO nanostructures [37].

et al. have reviewed the effect of shape and size control of MONS for catalytic applications [39]. Anandan et al. have presented a report on size and shape effect of NiO nanoparticles on their optical properties [40]. These fundamental understanding of size and shape effect on the properties of materials and knowledge about sol-gel processing would provide researcher a deep insight toward improving the existing properties of materials in a cost-effective and proficient way.

5. Conclusion

In summary, sol-gel method can be used for synthesize of MONSs in a bottom-up approach with perfect control over the shape and size of the derived products. The various steps involved in sol-gel material processing are elaborated. Also, a good level of understanding on the mechanism of shape and size control due to the controlling parameters like aging period, precursor concentration, solution pH, bath temperature and post heat treatment is achieved from the existing literature and also from our work. Uncomplicated and powerful ideas can be executed through simple cost-effective sol-gel processing to advance the technology to the next level.

Author details

Shrividhya Thiagarajan[1], Anandhavelu Sanmugam[2] and Dhanasekaran Vikraman[3*]

*Address all correspondence to: v.j.dhanasekaran@gmail.com

1 Department of Physics, Kalasalingam University, Krishnan Koil, Tamilnadu, India

2 Department of Chemistry (S & H), Vel Tech Multi Tech., Chennai, India

3 Division of Electronics and Electrical Engineering, Dongguk University-Seoul, Seoul, South Korea

References

[1] Pierre AC. Introduction to Sol-Gel Processing. New York: Springer; 2002. DOI: 10.1007/978-1-4615-5659-6

[2] Klein LC. Sol-Gel Technology for Thin Films, Fibers, Preforms, Electronics and Specialty Shapes. New Jersey: Noyes Publications; 1989. DOI: 10.1002/adma.19890010816

[3] Klein LC. Sol-Gel Optics Processing and Applications. New York: Springer; 1994. DOI: 10.1007/978-1-4615-2750-3

[4] Attia Y. Sol-Gel Processing and Applications. New York: Springer US; 1994. DOI: 10.1007/978-1-4615-2570-7

[5] Reisfeld R , Jorgensen CK. Chemistry, Spectroscopy and Applications of Sol-Gel Glasses. Berlin: Springer-Verlag; 1992. DOI: 10.1007/BFb0036963

[6] Levy D, Zayat M. The Sol-Gel Handbook: Synthesis, Characterization and Applications. Germany: Wiley-VCH Verlag GmbH & Co; 2015. DOI: 10.1002/9783527670819

[7] Lopez TM, Avnir D, Aegerter MA. Emerging Fields in Sol-Gel Science and Technology. New York: Springer US; 2003. DOI: 10.1007/978-1-4615-0449-8

[8] Livage J, Ganguli D. Sol–gel electrochromic coatings and devices: A review. Solar Energy Materials and Solar Cells. 2001;**68**:365–381. DOI: 10.1016/S0927-0248(00)00369-X

[9] Aegerter M, Mennig M. Sol-Gel Technologies for Glass Producers and Users. New York: Springer Science; 2004. DOI: 10.1007/978-0-387-88953-5

[10] Livage J. Sol-gel processes. Current Opinion in Solid State and Materials Science. 1997;**2**:132–138. DOI: 10.1016/S1359-0286(97)80057-5

[11] Guglielmi M, Carturan G. Precursors for sol-gel preparations. Journal of Non-Crystalline Solids. 1988;**100**:16–30. DOI: 10.1016/0022-3093(88)90004-X

[12] Guglielmi M, Kickelbick G, Martucci A. Sol-Gel Nanocomposites: Advances in Sol-Gel Derived Materials and Technologies. New York: Springer-Verlag; 2014. DOI: 10.1007/978-1-4939-1209-4

[13] Li Y, Xu L, Li X, Shen X, Wang A. Effect of aging time of ZnO sol on the structural and optical properties of ZnO thin films prepared by sol–gel method. Applied Surface Science. 2010;**256**:4543–4547. DOI: 10.1016/j.apsusc.2010.02.044

[14] Foo KL, Hashim U, Muhammad K, Voon CH. Sol–gel synthesized zinc oxide nanorods and their structural and optical investigation for optoelectronic application. Nanoscale Research Letters. 2014;**9**:429. DOI: 10.1186/1556-276X-9-429

[15] Sun Y, Seo JH, Takacs CJ, Seifter J, Heeger AJ. Inverted polymer solar cells integrated with a low-temperature-annealed sol-gel-derived ZnO film as an electron transport layer. Advanced Materials. 2011;**23**:1679–1683. DOI: 10.1002/adma.201004301

[16] Kim S, Moon H, Gupta D, Yoo S, Choi Y-K. Resistive switching characteristics of sol–gel zinc oxide films for flexible memory applications. IEEE Transactions on Electron Devices. 2009;**56**:696–699. DOI: 10.1109/TED.2009.2012522

[17] Xu L, Li X, Chen Y, Xu F. Structural and optical properties of ZnO thin films prepared by sol–gel method with different thickness. Applied Surface Science. 2011;**257**:4031–4037. DOI: 10.1016/j.apsusc.2010.11.170

[18] Ahn SE, Ji HJ, Kim K, Kim GT, Bae CH, Park SM, Kim YK, Ha JS. Origin of the slow photoresponse in an individual sol-gel synthesized ZnO nanowire. Applied Physics Letters. 2007;**90**:153106. DOI: 10.1063/1.2721289.

[19] Qu Y, Huang X, Li Y, Lin G, Guo B, Song D, Cheng Q. Chemical bath deposition produced ZnO nanorod arrays as an antireflective layer in the polycrystalline Si solar cells. Journal of Alloys and Compounds. 2017;**698**:719–724. DOI: 10.1016/j.jallcom.2016.12.265

[20] Ghosh S, Basak D. A simple process step for tuning the optical emission and ultraviolet photosensing properties of sol–gel ZnO film. RSC Advances. 2017;**7**:694–703. DOI: 10.1039/C6RA25921A

[21] Chandramohan R, Dhanasekaran V, Ezhilvizhian S, Vijayan TA, Thirumalai J, John Peter A, Mahalingam T. Reprinted from Journal of Materials Science: Materials in Electronics, Spectral properties of aluminium doped zinc oxide thin films prepared by SILAR method. With permission of Springer. 2012;**23**:390–397. DOI: 10.1007/s10854-011-0439-1

[22] Chandramohan R, Vijayan TA, Arumugam S, Ramalingam HB, Dhanasekaran V, Sundaram K, Mahalingam T. Effect of heat treatment on microstructural and optical properties of CBD grown Al-doped ZnO thin films. Materials Science and Engineering: B. 2011;**176**:152–156. DOI: 10.1016/j.mseb.2010.10.017

[23] Dhanasekaran V, Anandhavelu S, Polychroniadis EK, Mahalingam T. Microstructural properties evaluation of Fe_2O_3 nanostructures. Materials Letters. 2014;**126**:288–290. DOI: 10.1016/j.matlet.2014.04.065

[24] Woo K, Lee HJ, Ahn JP, Park YS. Sol–gel mediated synthesis of Fe_2O_3 nanorods. Advanced Materials. 2003;**15**:1761–1764. DOI: 10.1002/adma.200305561

[25] Liu XQ, Tao W, Shen YS. Preparation and characterization of nanocrystalline α-Fe_2O_3 by a sol-gel process. Sensors and Actuators B: Chemical. 1997;**40**:161–165. DOI: 10.1016/S0925-4005(97)80256-0

[26] Leonardi SG, Mirzaei A, Bonavita A, Santangelo S, Frontera P, Panto F, Antonucci PL, Neri G. A comparison of the ethanol sensing properties of α-iron oxide nanostructures prepared via the sol–gel and electrospinning techniques. Nanotechnology. 2016;**27**:075502. DOI: 10.1088/0957-4484/27/7/075502

[27] Lopez-Sanchez J, Serrano A, Del Campo A, Abuin M, Rodriguez de la Fuente O, Carmona N. Sol–gel synthesis and micro-Raman characterization of ε-Fe_2O_3 micro and nanoparticles. Chemistry of Materials. 2016;**28**:511–518. DOI: 10.1021/acs.chemmater.5b03566

[28] Kadhim IH, Hassan HA. Hydrogen gas sensing based on SnO_2 nanostructure prepared by sol–gel spin coating method. Journal of Electronic Materials. 2017;**46**:1419–1426. DOI: 10.1007/s11664-016-5166-1

[29] Aksoy S, Caglar Y, Caglar M, Ilican S. Influence of annealing temperature on the structural and optical characteristics of nanostructure SnO_2 films and their applications in heterojunction diode. Journal of Nanoelectronics and Optoelectronics. 2016;**11**:115–121. DOI: 10.1166/jno.2016.1885

[30] Bagherian S, Zak AK. X-ray peak broadening and optical properties analysis of SnO_2 nanosheets prepared by sol-gel method. Materials Science in Semiconductor Processing. 2016;**56**:52–58. DOI: 10.1016/j.mssp.2016.07.021

[31] Shrividhya T, Ravi G, Hayakawa Y, Mahalingam T. Determination of structural and optical parameters of CuO thin films prepared by double dip technique. Journal of Materials Science: Materials in Electronics. 2014;**25**:3885–3894. DOI: 10.1007/s10854-014-2103-z

[32] Dhanasekaran V, Soundaram N, Seong-Il Kim, Chandramohan R, Srinivas Mantha, Saravanakumar S, Mahalingam T. Optical, electrical and microstructural studies of monoclinic CuO nanostructures synthesized by a sol–gel route. Centre National de la Recherche Scientifique (CNRS) and The Royal Society of Chemistry. 2014;**38**:2327–2333. DOI: 10.1039/c4nj00084f

[33] Bibi M, Javed QA, Abbas H, Baqi S. Outcome of temperature variation on sol-gel prepared CuO nanostructure properties (optical and dielectric). Materials Chemistry and Physics. 2017;**192**:67–71. DOI: 10.1016/j.matchemphys.2017.01.074

[34] Dhanasekaran V, Park HJ. Shape-selective synthesis of NiO nanostructures for hydrazine oxidation as a nonenzymatic amperometric sensor. The Royal Society of Chemistry Advances. 2016;**6**:86101–86107. DOI: 10.1039/c6ra12805j

[35] Freitas Cabral AJ, Remedios CMR, Ospina CA, Carvalho AMG, Morelhao SL. Structure of antiferromagnetic NiO/ferrimagnetic $NiMn_2O_4$ composite prepared by sorbitol-assisted sol-gel method. Journal of Alloys and Compounds. 2017; **696**:304–309. DOI: 10.1016/j.jallcom.2016.11.272

[36] Mutkule SU, Navale ST, Jadhav VV, Ambade SB, Naushad M, Sagar AD, Patil VB, Stadler FJ, Mane RS. Solution-processed nickel oxide films and their liquefied petroleum gas sensing activity. Journal of Alloys and Compounds. 2017;**695**:2008–2015. DOI: 10.1016/j.jallcom.2016.11.037

[37] Dhanasekaran V, Park HJ, Kim SI, Thaiyan M. Magnetic, structural and optical behavior of cupric oxide layers for solar cells. Journal of Alloys and Compounds. Elsevier. 2016;**686**:616–627. DOI: 10.1016/j.jallcom.2016.06.032

[38] Kato K, Dang F, Mimura KI, Kinemuchi Y, Imai H, Wada S, Osada M, Haneda H, Kuwabara M. Nano-sized cube shaped single crystalline oxides and their potentials; composition, assembly and functions. Advanced Powder Technology. 2014;**25**:1401–1414. DOI: 10.1016/j.apt.2014.02.006

[39] Bozon Verduraz F, Fievet F, Piquemal JY, Brayner R, El Kabouss K, Soumare Y, Viau G, Shafeev G. Nanoparticles of metal and metal oxides: Some peculiar synthesis methods, size and shape control, application to catalysts preparation. Brazilian Journal of Physics. 2009;**39**:134–140. DOI: 10.1590/S0103-97332009000200002

[40] Anandan K, Rajendran V. Morphological and size effects of NiO nanoparticles via solvothermal process and their optical properties. Materials Science in Semiconductor Processing. 2011;**14**:43–47. DOI: 10.1016/j.mssp.2011.01.001

Chapter 2

Sol-Gel-Derived Silicon-Containing Hybrids

Maria Zielecka, Elzbieta Bujnowska,
Katarzyna Suwala and Magdalena Wenda

Additional information is available at the end of the chapter

http://dx.doi.org/10.5772/67824

Abstract

Silicone-containing hybrid systems including functionalized and modified silica nanospheres as well as silica aerogels were synthesized using sol-gel process. The most important advantage of the developed silicon-containing hybrids lies in their high application potential, e.g. for the production of façade paints, polymer nanocomposites, highly efficient insulating materials and other high-tech products for construction industry. The possibility of precise steering of sol-gel process and synergism of very rich silicon compound chemistry allows to obtain materials with strategically designed architecture. Original methods of manufacturing following silicon-containing hybrids such as functionalized and non-functionalized silica nanospheres, silica nanospheres containing immobilized silver and/or copper nanoparticles and silica aerogels were developed. Homogeneous structural properties of synthesized materials were achieved due to carefully selected parameters of sol-gel process. The technology offers a possibility to control the grain size and uniformity of nanospheres by process parameters. The smallest nanosphere size is 10 nm and the largest 600 nm. The properties of aerogels obtained according to the original technology developed in the Industrial Chemistry Research Institute (ICRI) are extremely good and repetitive particularly taking into account their thermal conductivity. The main advantages of the developed aerogels are very good mechanical properties, very low dust formation and significantly decreased costs when compared to traditional silica aerogels.

Keywords: silicon-containing hybrids, silica nanospheres, aerogels

1. Introduction

Synergism resulting from the possibilities of precise steering of sol-gel process and very rich silicon compound chemistry allows to obtain silicon-containing hybrids, especially spherical nanosilica and aerogels with strategically designed architecture. The key factor in the chemistry of sol-gel process is the complex equilibrium state, which should be achieved for the simultaneous reactions of hydrolysis and condensation according to the following reaction schemes (see **Figure 1**).

A number of parameters affect the sol-gel process [1]. The most important are the ratio of alkoxysilane precursors to water, reaction mixture pH and temperature and mixing speed and time. The successful control of this process needs a lot of practical experience as well as the knowledge of the correlation of the influence of particular factors. Careful adjustment of reaction condition and substrate ratio in sol-gel process allows to obtain several kinds of materials with different physical properties and structures. The structural properties are strongly influenced by the values of water to silicon ratio designated as r. Depending on the value of r, materials with different properties can be obtained, e.g. r = 3.8 or 5.1 amorphous xerogel [2], r = 6.2 aerogel [3], r = 10.9 advanced coating materials, r = 1/2 fibres [4] and r = 20/50 monodisperse nanospheres (see **Figure 2**) [5].

Reaction mixture pH strongly influences the structure and porosity of these materials [6]. Longer siloxane chains formed in pH <7 are less tightly packed giving the structure of 'cross-linked spaghetti' having long and narrow pores. Highly branched siloxane chains produced in pH >7 can form so-called Eden clusters with further condensation to 'raisin buns' having cylindrical pores (see **Figure 3**) [7].

Moreover, the parameters of drying process and the subsidiary substances added to the reaction mixture during sol-gel process strongly influence the pore size and their distribution. Medium porosity materials, so-called xerogels, are formed during the evaporation of solvents in normal conditions, and drying in supercritical conditions gives very porous aerogel.

The innovative, original methods of manufacturing of silica-containing nanospheres and silica aerogels were developed in the Industrial Chemistry Research Institute during recent years. Sol-gel-derived functionalized and non-functionalized silica and silica-titania nanospheres of the designed architecture tailored to the needs of the potential application were developed [8, 9].

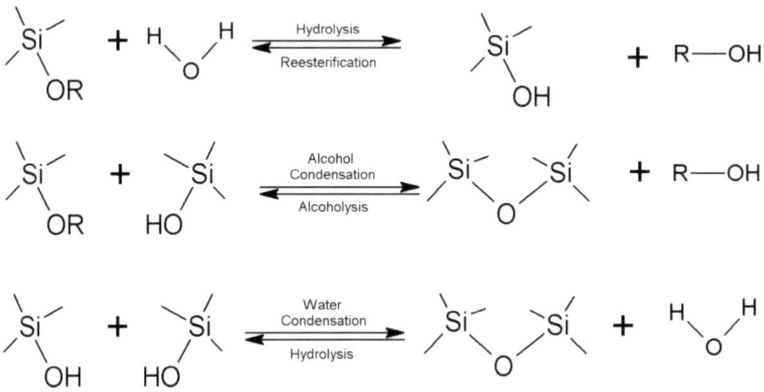

Figure 1. Simultaneous reactions of hydrolysis and condensation occurring in sol-gel process, where R is alkyl.

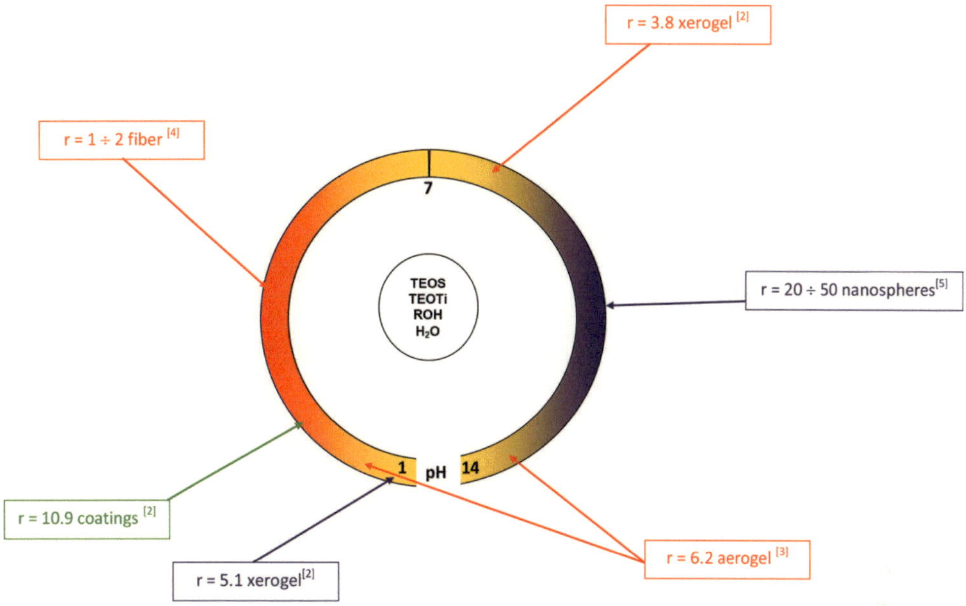

Figure 2. The effect of pH and r on physical properties of sol-gel-derived products.

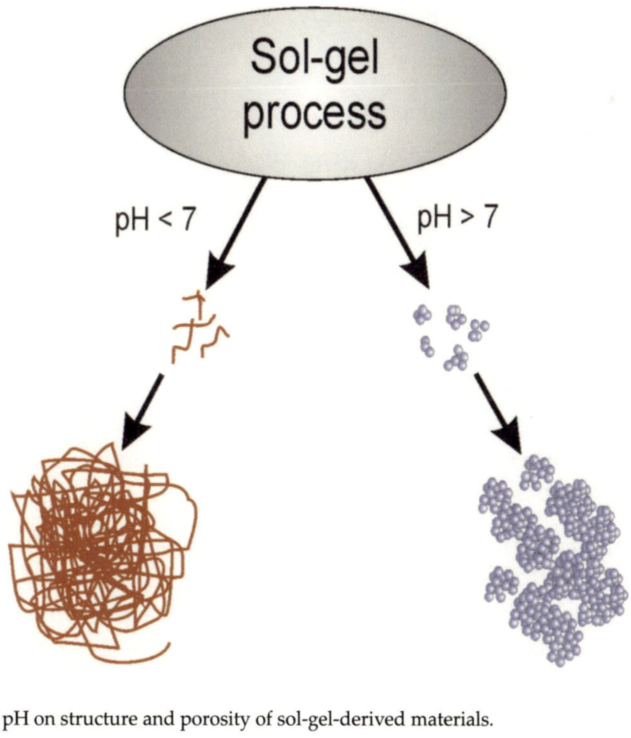

Figure 3. The effect of pH on structure and porosity of sol-gel-derived materials.

Moreover, the technologies of silica nanospheres modified by durable immobilization of silver and/or copper nanoparticles of narrow particle size distribution were also elaborated. Sol-gel process was also applied in the innovative technology of silica aerogels obtained without using supercritical drying process. The most important advantage of the developed silicon-containing hybrids lies in their high application potential, e.g. for the production of façade paints [10], polymer nanocomposites [11–13], highly efficient insulating materials and other high-tech products for construction industry.

2. Experimental part

2.1. Materials

The following chemicals were applied in the sol-gel process:

-tetraethoxysilane (TEOS, technical grade), Wacker Chemie, Germany

-aqueous ammonia (reagent grade, 25 wt%, d = 0.91 g/cm^3, POCH S.A., Poland

-ethyl alcohol (absolute, reagent grade), POCH S.A., Poland

-silver nitrate, POCH S.A., Poland

-copper acetate, POCH S.A., Poland

-distilled water (prepared in the laboratory)

-hexane (reagent grade), POCH S.A., Poland

-trimethylchlorosilane (TMCS, technical grade), Wacker Chemie, Germany

-hexamethyldisilazane (HDMS, technical grade), Wacker Chemie, Germany

2.2. Methods

2.2.1. Methods of synthesis

Functionalized and non-functionalized silica nanospheres as well as silica-titania nanospheres were obtained according to the earlier procedure [8–10]. The detailed procedure of manufacturing sol-gel-derived silica nanospheres containing immobilized silver or copper nanoparticles was presented in our earlier studies in Ref. [14].

The process for preparing aerogel was carried out at a temperature of 25–60°C with a stirrer speed of 250 rpm. Tetraethoxysilane (TEOS) was used as a silica precursor. Synthesis was carried out in an aqueous-ethanolic reaction mixture using weight ratio of TEOS:H_2O:EtOH as 1:3.5:3.9. In the first phase, the pH of the reaction mixture was kept ~3. In order to start the gelation process, the pH was increased to 8 by adding liquid ammonia. The obtained silica-based sol was poured into the prepared mould and allowed to gel. An advantage of the invention is to carry out the modification process by silanization of gel obtained in the sol-gel process. The resulting gel is subjected to age by carrying out in two stages at atmospheric pressure initially at 60–80°C for 12–24 h and then at 180–400°C. Transparent, water-repellent, ultralight aerogel granules were obtained [15].

2.2.2. *Methods of testing*

FT-IR spectra in the mid-IR range were recorded on PerkinElmer System 2000 spectrometer using KBr pellets or on KRS crystals in the purpose to observe characteristic bands of obtained materials as described earlier [10]. Particle size (Z-average diameter) and particle size distribution in resulting sols of non-functionalized, functionalized and modified silica nanospheres were measured using photon correlation spectroscopy (PCS) and Malvern apparatus (Zetasizer Nano ZS) according to the presented earlier method [14]. Scanning electron microscopy (SEM) was performed using a JEOL JSM-6490LV operating in the high vacuum mode at an accelerating voltage of 15 kV to compare the morphology of silica nanoparticles [10]. The oxidation level of silver and copper immobilized on silica nanospheres was confirmed by powder X-ray diffraction (XRD) using Siemens D500 X-ray powder diffractometer with high-resolution solid-state detector Si[Li] (operating parameters U = 40 kV and I = 30 mA) and radiation CuK (λ = 0.1541837 nm). Silver or copper content was determined by absorption atomic spectroscopy using 5100 PC, Perkin Elmer AAS Spectrometer.

Analysis of the specific surface area by the Brunauer-Emmett-Teller (BET) method was performed using a TriStar II 3020 V1.03, Micromeritics. The principle of measurement of the specific surface area is based on the measurement of the low-temperature (77 K) adsorption of inert gas (usually nitrogen or argon) at various adsorbate pressures. Based on the determined isotherms total pore volume, distribution volume, surface area and the average radius were determined by Barrett-Joyner-Halenda (BJH) method.

The bulk aerogel density was also calculated, on the basis of their mass-to-volume ratio. Hydrophobic properties of aerogels were tested by placing a small sample (approx. 1 cm^2) in a vessel filled with distilled water at a temperature of 22°C. The sample that floated on the water surface for a minimum time of 30 min and immediately come to the surface after the forced immersion was classified as hydrophobic.

The thermal conductivity of aerogels was measured using FOX200 apparatus, TA I Instruments. FOX plate apparatus 200 is equipped with two sensors of heat flow density (upper and lower). It meets the requirements of ISO standards and PN-EN for this type of devices. The volume of the measured sample was determined by the properly prepared frame of the size 10 × 10 × 0.5 cm (length, width, height).

Emission of dust releasing during the procedure of the thermal conductivity of silica aerogels was measured. Measurements were done by Central Institute for Labour Protection-National Research Institute. Quantitative concentration of particles (number of particles/cm^3) over time (minutes) was measured by using DiscMini apparatus, Matter Aerosol.

2.3. Results and discussion

2.3.1. Non-functionalized silica and silica-titania nanospheres

Silica and silica-titania nanospheres were obtained by sol-gel process using tetraethoxysilane and tetraethoxytitanate as alkoxysilane and alkoxytitanate precursors at a temperature of 25°C in various process conditions at pH >7, according to the reaction (see **Figure 4**).

Pure silica or silica-titania nanospheres were obtained after drying of resulting homogeneous, opaque sols. The preliminary results confirmed the completion of the hydrolysis reaction of Si—O—C bonds in alkoxysilane precursor what was justified by IR spectrum not revealing any content of organic material, in particular in the region of stretching vibrations C—H at 2900 cm^{-1} what has been thoroughly discussed previously [10].

Homogeneous structural properties of synthesized materials were achieved due to carefully selected parameters of sol-gel process (pH, temperature, dosing rate). The technology offers a possibility to control the grain size and uniformity of nanospheres by process parameters. The smallest nanosphere size is 10 nm and the largest 600 nm. The monomodal particle size distribution and very low dispersion of particle size were observed for homogeneous sols obtained by sol-gel process (see **Figure 5**).

The preliminary results concerning the effect of temperature and stirring rate confirmed that different conditions of sol-gel process play an important role in the growth, polydispersity and stability of nanoparticles [14]. The optimization of silicon-containing hybrid synthesis using sol-gel process was carried out for silica-titania nanospheres by using the statistically designed factorial experiment which was considered to be the best method of finding the optimal formulation and process conditions. The scope of the optimal recipes of the sol-gel system was determined based on a factorial experiment designed as a 2^2 array with additional 'star' experimental points to obtain greater accuracy of the resulting regression equations. The software, "Statistica" 5.5—StatSoft Poland Ltd., was applied for the mathematical calculations and for the visualization of the results as contour plots. One possible approach that can be applied to the selection of the optimal formulation is presented in **Figure 6**, where the area of the optimal silica-titania properties is hatched.

Factorial design is the most suitable tool for this optimization, since it allows simultaneous determination of the effect of the parameters characterizing the properties of the sol and the obtained silica-titania nanoparticles, especially their catalytic activity in the reaction of cyclohexanone

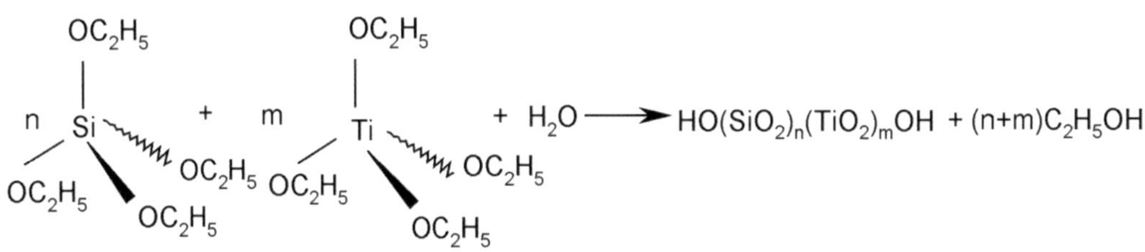

Figure 4. The reaction scheme for the preparation of silica-titania nanospheres.

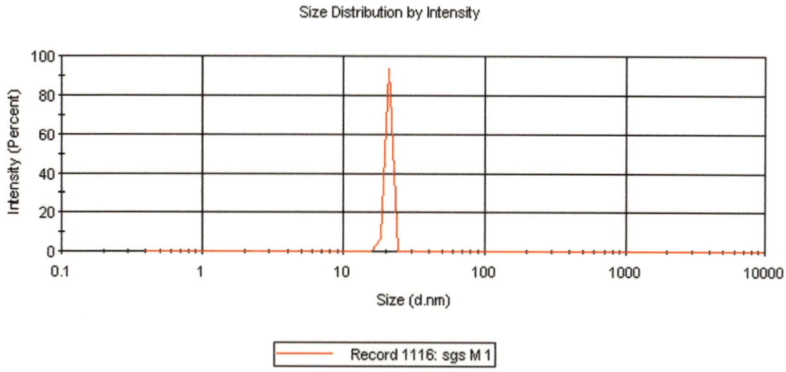

Figure 5. Monomodal particle size distribution of non-functionalized silica nanospheres of 20 nm.

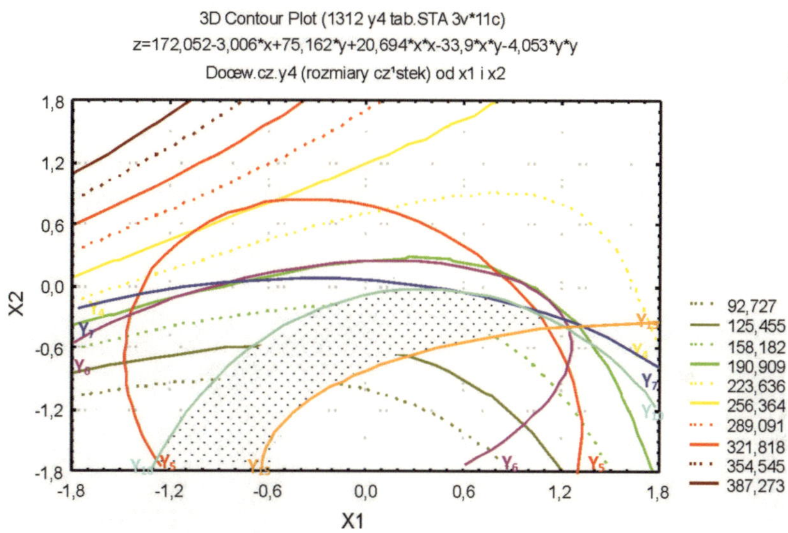

Figure 6. Determination of the area of optimal recipes (hatched).

ammoxidation. Testing of catalytic activity of obtained silica-titania nanoparticles was an integral part of the designed factorial experiment in the purpose to optimize the composition and catalytic properties of these materials. The determined parameters of the catalytic activity of nanoparticles were taken as dependent variables in the designing of factorial experiment. The area of optimal sol recipes and properties of silica-titania nanospheres was obtained by overlapping, on the contour plot of the selected dependent variable, the contour lines obtained for the assumed limiting values of the other critical dependent variables. The following dependent variables y_i were considered: y_4 - particle size [nm]; y_5 - particle size distribution [nm]; y_6 - polydispersity; y_7 - sol appearance; y_{10} - cyclohexanone oxime yield (mol%); and y_{15} - consumption of H_2O_2 on cyclohexanone oxidation. The limiting values of variables x_1 and x_2 in the area of optimal properties are presented in **Table 1**.

Real values of the independent variable	From	To
X_1-Ti/Si ratio (mol%)	1.217	8.201
X_2 content of NH_{3aq} in sol (wt%)	0.660	0.875

Table 1. The limiting values of variables x_1 and x_2 in the area of optimal properties.

In order to synthesize silica-titania nanoparticles with optimal properties, the values of x_1 and x_2 variables corresponding to the selected value of the variable y_i should be selected within the hatched area. Silica-titania nanospheres prepared according to the recipes from the area of optimal properties were characterized by homogeneous particle sizes, as illustrated by SEM micrograph (see **Figure 7**).

Supramolecular structure of received Si-containing nanospheres depends considerably on the method of drying. Drying in the oven drier allows to obtain separate nanoparticles, while after drying in spray drier, spherical microstructures can be observed (see **Figure 8a**, **b**).

The differences in the structure of the microaggregates composed of silica or silica-titania nanospheres can be easily observed. Silica nanosphere microaggregates are composed of single nanoparticles, whereas silica-titania aggregates have solid structure. The observed differences may have a significant impact on the redispersibility of the received nanomaterials used both as nanofillers and catalysts.

Figure 7. SEM micrograph illustrating homogeneous particle size distribution of silica-titania nanospheres (sample dried in oven drier).

http://dx.doi.org/10.5772/67824

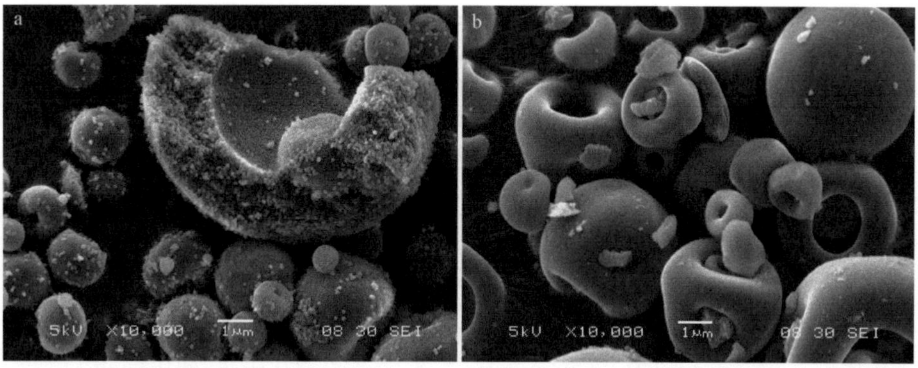

Figure 8. SEM micrographs of nanospheres dried in spray drier (a) silica nanospheres and (b) SEM silica-titania nanospheres.

Very promising results confirming the use of these nanomaterials as nanofillers were received. Thus, the sol of non-functionalized silica nanospheres having a particle size of about 50 nm was applied as poly(vinyl chloride) (PVC) nanofiller, which was introduced during the suspension polymerization process. The resulting PVC composite showing the homogeneous distribution of silica nanospheres was characterized by significantly improved mechanical properties, already at a content of 0.5 wt% of these nanoparticles [16–18].

2.3.2. Functionalized silica nanospheres

Functionalized silica nanospheres were developed in the purpose of permanent embedding to cross-linked polymer matrix, which should allow obtaining nanocomposites with significantly improved properties. The following reactive groups, vinyl, amino, methacryloxy and glycidyloxy, were selected and introduced through the use of suitable carbofunctional silanes according to the following reaction (see **Figure 9**).

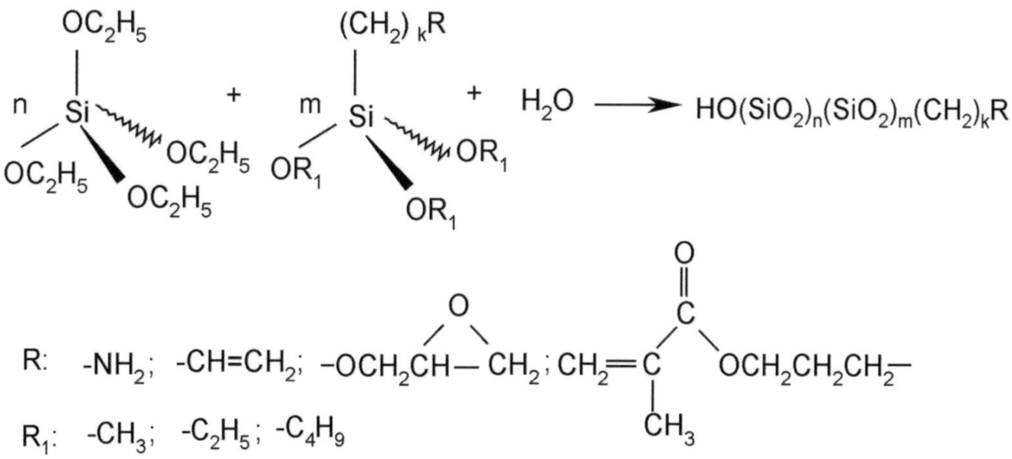

Figure 9. The reaction scheme for the preparation of functionalized silica nanospheres.

The presence of the reactive groups in functionalized silica nanospheres was confirmed based on the intensity of the following characteristic bands in the IR spectrum for vinyl group, 1640–3070 cm^{-1}; for amine group, 1710–3300 cm^{-1}; for methacryloxy group, 1640–1650 cm^{-1} and for glycidyloxy group, 1000–1100 cm^{-1}. The applicability of silica nanospheres functionalized with vinyl groups was tested in polymer composites based on silicone rubber cross-linked by hydrosilylation reaction according to the following reaction (see **Figure 10**).

The substantial effect of the content and the average particle size of embedded silica nanospheres on the tensile strength of silicone rubber nanocomposites can be observed (see **Figure 11**).

The best results were achieved for nanoparticles characterized by a particle size of 82.4 nm added in the content range from 1.0 to 1.5 wt%. A visible decrease of tensile strength can be observed for both larger and smaller content of nanoparticles introduced in the amount of 2 and 0.5 wt%, respectively. Silica nanospheres functionalized with amine, glicydoxyl and methacryloxyl groups were successfully applied for manufacturing of polymer nanocomposites based on polyamide, polyolefins and other engineering polymers [11, 13].

2.3.3. Modified silica nanospheres by immobilization of silver or copper nanoparticles

Modification of sol-gel-derived silica nanospheres by durable immobilization of silver or copper nanoparticles was studied as the effective stabilization method of metal nanoparticles

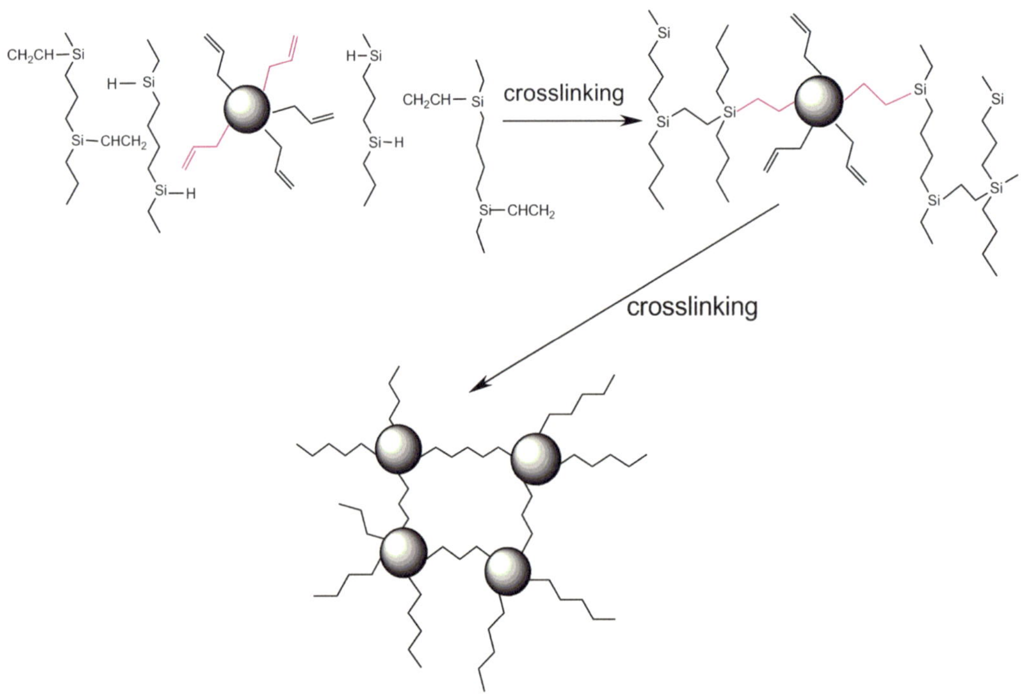

Figure 10. Scheme of formation of silicone rubber matrix with durably built-in silica nanospheres.

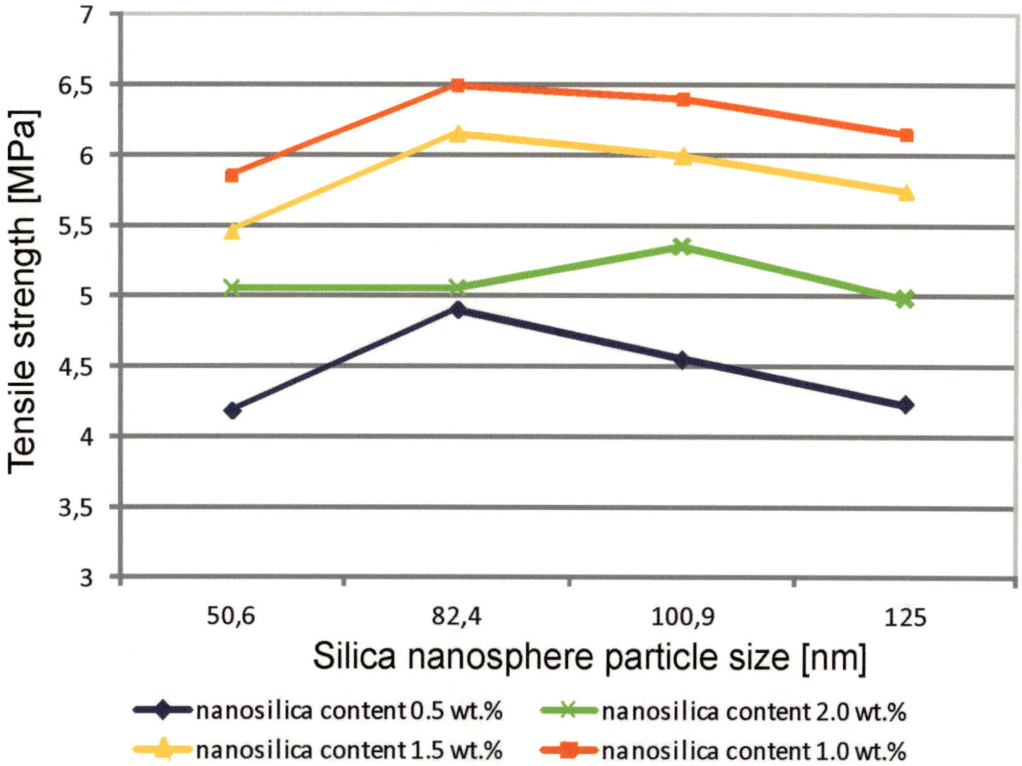

Figure 11. The effect of the content and the average particle size of embedded silica nanospheres on the tensile strength of silicone rubber nanocomposites.

against their agglomeration. However, the antimicrobial activity of silver and copper, especially as nanoparticles, has long been recognized [19, 20]; the successful application of these metals as biocides was limited due to the low stability of colloids of these nanometals. The agglomeration of silver or copper nanoparticles in colloids can result in the decrease of their biocidal effect. Nanometal particles can be stabilized by the protective colloids [21]. The application of such stabilized colloids is limited due to the negative impact of protective stabilizers on the properties of the obtained materials. Studies on the microbial activity of silver and copper have shown that both the nanoparticles of metals and their ions or colloidal solutions demonstrate biocidal activity [22]. Nanoparticles of these metals are sufficiently small to damage the bacterial cell membrane and significantly affect the function and operation of the enzymes contained in the cell [23–25].

One of the most interesting examples of protection of metal nanoparticles before agglomeration is their immobilization in the films of silica [26]. The synthesis is conducted using a sputter deposition silica film containing metal particles.

One method of Cu nanoparticle absorption into silica nanoparticles is the synthesis by ion implantation using a laser of appropriate wavelength (ultraviolet) [27, 28]. The obtained modified nanosilica can be used as active nanofiller for polymer nanocomposites.

The use of the sol-gel method is an easier way to obtain nanosilica modified by metal nanoparticles characterized by homogeneous distribution of metal nanoparticles on the silica surface [29, 30]. The exemplary SEM micrograph reveals a uniform distribution of copper nanoparticles on the surface of silica nanospheres (see **Figure 12**).

The oxidation degree of copper or silver was examined by XRD showing the diffraction peaks 110, 200, 220 and 311 that confirm the presence of Ag° or Cu° [14]. Metallic silver or copper form the cubic F type cells, the angular positions of the reflections are characteristic for silver: 38.15; 44.32; 64.49; 77.55; 6.90 or copper: 43.33; 50.48; 74.20; 90.30. Copper or silver nanoparticles incorporated in the silica structure are stable and do not undergo coagulation during storage, which ensures durable biocidal activity and solves the problem of the loss of properties due to the agglomeration of silver or copper nanoparticles.

Several results of antimicrobial activity of the obtained silica nanospheres containing immobilized silver or copper nanoparticles [10] as well as of polymer nanocomposites containing these fillers [12] were reported earlier. Recently, interesting application results of anti-algae and antifungal activity of silica nanospheres containing immobilized copper nanoparticles used as silicone-acrylic paint nanofillers were obtained (see **Figure 13a–c**).

It can be easily observed that the growth of algae on the surface covered by silicone-acrylic paint containing 0.5 ppm of copper nanoparticles immobilized on silica nanospheres having a particle size of 80 nm was completely stopped (**Figure 13a**), while the untreated control sample was almost completely overgrown with algae (**Figure 13b**, **c**)

Summarizing the results of the performed research on silica nanospheres, it should be underlined that this is a very promising area of research that allows the use of the developed silica-based nanomaterials for innovative and original applications.

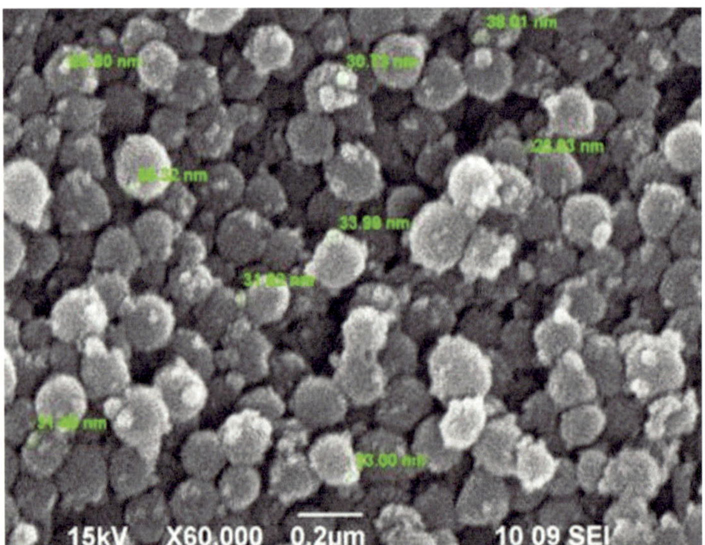

Figure 12. SEM micrograph of silica nanospheres containing immobilized silver nanoparticles.

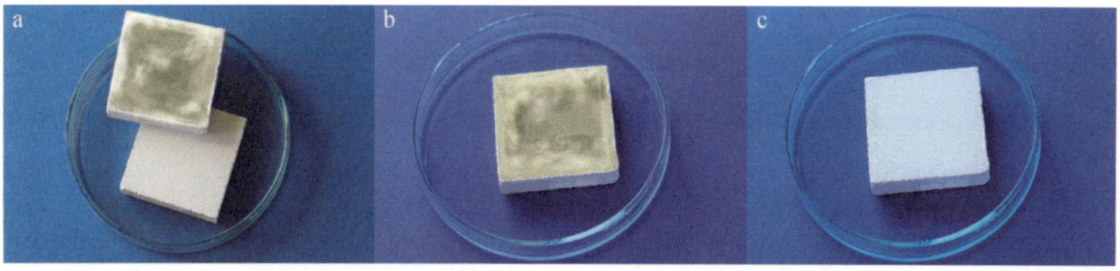

Figure 13. Anti-algae activity of copper nanoparticles immobilized on silica nanospheres applied as filler of silicone-acrylic paint: (a) sample protected with silicone-acrylic paint showing no algae growth and control sample overgrown with algae, (b) control sample after 4 weeks of weathering before testing anti-algae activity and (c) tested sample after 4 weeks of weathering before testing anti-algae activity.

2.3.4. Sol-gel-derived silica aerogels

Aerogels belong to the class of the advanced materials. Specific properties of the aerogels are resulting from their unique composition in which 90–98% is air and the rest is silica forming the three-dimensional structure. The air entrapped in the aerogel is the key agent that gives unique characteristics giving the enormous potential for various applications. The unique properties include low density, very high specific surface area, porosity and low thermal conductivity. Aerogels are non-toxic, environmentally friendly and fire resistant (melting point 1200°C) what make them ideal insulating material both thermal and acoustic [31]. Insulation - mainly thermal) is an important portion of the building materials market [32, 33]. Therefore, aerogels were in the spotlight of researchers and become a medium of inspiring all who see their unique properties used in the practical application.

However, the main problem associated with the extensive use of silica aerogels is their high price and complex production process. Moreover, a serious disadvantage of silica aerogels is their dustiness. The development of cheaper methods of production undoubtedly will increase their area of application [34]. Formation of a wet gel, which is a product of hydrolysis and polymerization (condensation) and drying as the second important process, which is a critical step in the creation of the spatial structure of aerogels, see **Figure 14**.

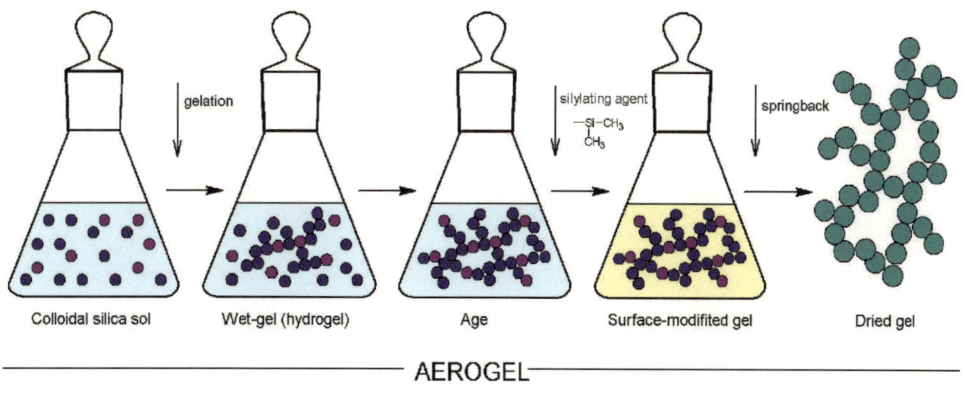

Figure 14. Aerogel manufacturing scheme.

During typical procedure gel should be aged and modified (if needed) and then dried with total removal of all liquids from the pores.

Gel ageing is needed to strengthen its structural and mechanical properties. This results in the increase of the three-dimensional network cohesion by creating new bridge bonds ≡Si–O–Si≡ which results in the shrinkage of the gel matrix. This is connected with water and alcohol release from the pores (syneresis). This process has a major effect on the porosity of the final product; therefore, it is important to control the conditions under which the ageing takes place, i.e. temperature and time. The next step is immersion of gel in a suitable anhydrous organic solvent, e.g. hexane, which inhibits the formation of hydrogen bonds and also leads to a reduction of capillary forces helping in porosity increase. Thus, prepared gel is subjected to a modification by using silylating agent in the purpose of hydrophobic group introduction [35]. The modification process is necessary to replace the reactive OH groups with non-reactive CH_3 groups to avoid the formation of siloxane bridges at the pore walls (see **Figure 15**).

The drying step is the greatest technological challenge. During the drying process, the residual solvents are removed that causes weight loss of the product with a simultaneous increase in volume. Incorrect parameters of the drying can result in the reduction of gel porosity.

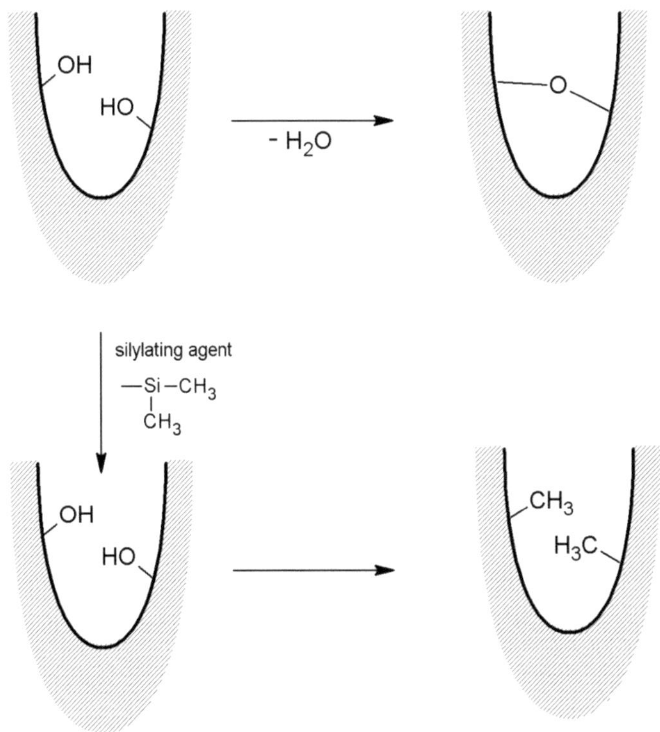

Figure 15. The replacement of the reactive OH groups with non-reactive CH_3 groups.

Several methods of aerogel drying are used. The main known from the literature is the drying under supercritical conditions [36, 37], but this method is expensive and complicated and requires specialized equipment.

In our studies, we were focused on the development of original aerogel technology. The key point of this technology is the process of drying at atmospheric pressure, which has simple experimental equipment requirements and easy to control test parameters [15].

Several samples of silica aerogel using tetraethoxysilane as a precursor at a fixed ratio of the main reactants in the sol-gel process were prepared (see Section 2.2.1). For all described silica aerogel samples, the same procedure of sol manufacturing was applied. Differences in the properties of aerogel samples result from the application of different gel treatment procedures. Technological parameters of gel treatment procedure are presented in **Table 2**.

Samples A4 and A5 obtained in a pilot scale, the remaining samples were obtained at a laboratory scale.

The obtained silica aerogels are transparent, water repellent and very light. Dustiness to a large extent had been reduced, which is important from the point of view of their application. Satisfactory results of the thermal conductivity in the range of 0.024–0.038 W/(mK) were obtained. As a control sample commercial silica aerogel SJ1800Series (CHKAM) was used. The results obtained for the selected samples are shown in **Table 3**.

The lowest thermal conductivity was measured for the sample A-1. In order to confirm the obtained result, two other samples (A-8 and A-9) were prepared according to the same

Sample	Ageing time (h)	Silylating agent	Drying method and temperature (°C)
A-1, A-8, A-9	48	TMCS	180
A-2	48	TMCS	Muffle furnace 400
A-3	336	TMCS	180
A-4	48	TMCS	180
A-5	48	HDMS	180
A-6	48	TMCS	Oven 60
A-7	48	TMCS	Oven 60
A-10	48	TMCS	Oven 200
A-11	48	TMCS	Oven 200
A-10	48	TMCS	Oven 250
A-12	48	TMCS	Oven 300
A-13	48	TMCS	Muffle furnace 400

TMCS, trimethylchlorosilane; HDMS, hexamethyldisilazane

Table 2. Technological parameters of gel treatment procedure applied for samples A-1–A-13.

Sample	Specific surface S_{BET} (m^2/g)	Total pore surface (m^2/g)	Pore size (nm)	Total pore volume (cm^3/g)	Thermal conductivity (W/(mK))	Bulk density (g/cm^3)
Control sample CHKAM	507	123.78	22	2.87	0.030	0.030
A-1	565	185.16	21	3.07	0.025	0.074
A-2	978	989.44	13	3.31	0.029	0.031
A-3	868	194.06	9	2.05	0.038	0.056
A-4	689	292.56	15	2.62	0.030	0.0843
A-5	667	498.92	10	1.07	0.033	0.1735

Table 3. Characteristics of the selected sample.

procedure. The values of thermal conductivity obtained for samples A-8 and A-9 confirm the repeatability of the synthesis procedure.

The key indicator of silica aerogel properties is also their porous structure. For this purpose, a measurement of the specific surface area by the BET method was applied for selected samples. Nitrogen adsorption isotherms reflect the structure of the porosity of studied adsorbents in Ref. [14].

The observed BET specific surface area for the silica aerogel was in the range from 565 to 978 m^2/g, and the total pore volume was in the range of 1.07–3.31 cm^3/g. The measured bulk density values indicate the lightness of the obtained sample that is in good accordance with the published data [34, 38].

Isotherms obtained by low-temperature nitrogen adsorption indicated that the obtained materials are mesoporous (see **Figure 16**).

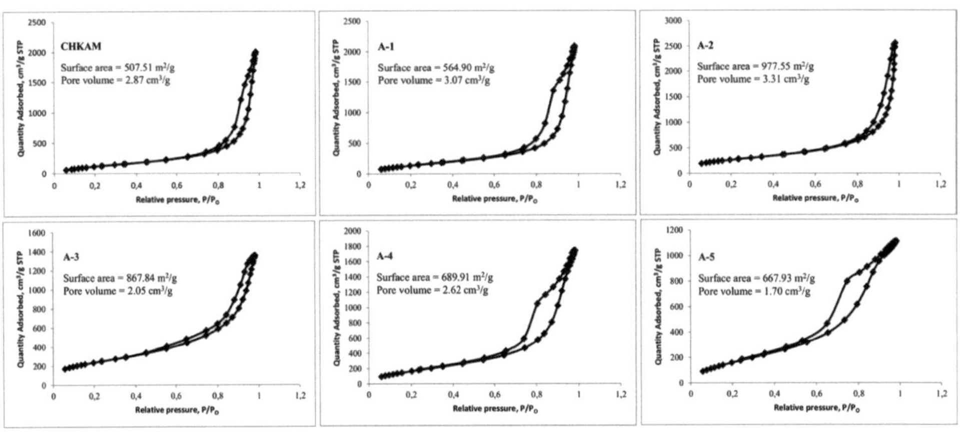

Figure 16. N_2 adsorption-desorption isotherms at 77 K for the silica aerogel.

http://dx.doi.org/10.5772/67824

According to the IUPAC classification [39, 40], the shape of the isotherms can be described as type IV with H2 hysteresis loop (samples A-1, A-4, A-5) or H3 (samples CHKAM, A-2, A-3), which indicates the presence of pores in the shape of 'ink bottle' or cylindrical.

More complete information about the texture of the tested adsorbent materials can be obtained based on an analysis of the distribution curve of the pore volume (see **Figure 17**).

On this basis, the structure of the adsorbent can be concluded. All curves are shown in one scale; the pore diameter distribution is in the range from 2 to 100 nm, confirming the high proportion of mesopores in the structure of adsorbents (according to the IUPAC classification) [39, 40].

The narrow size of the peaks in the samples A-4 and A-5 indicates the high homogeneity of pore size. A wider range of sizes of the peaks in the sample A-2 are due to the presence of pores of different diameters. This affects a very large surface area of the adsorbent (**Table 1**). The pore size distribution in the samples CHKAM, A-2 and A-3 is heterogeneous, where we find more than one peak and a broad range of pore sizes.

The effect of drying temperature under atmospheric pressure conditions on the thermal conductivity coefficient of obtained aerogels, was studied (see **Table 4**).

The samples were dried in a convection plate oven or in a muffle furnace. The best results of thermal conductivity were obtained for the samples dried in the convection plate oven at 180°C.

Drying under atmospheric conditions, it is a promising technique [41], which could be used on an industrial scale, as demonstrated in the verification process carried out on a prototype, large scale laboratory line for the production of aerogels.

As mentioned earlier, one of the major disadvantages of the silica aerogels is their dustiness which is the tendency of a material to release the dust into the atmosphere during manufacturing and use of the material, defined as the amount of dust emitted during the standard test procedure

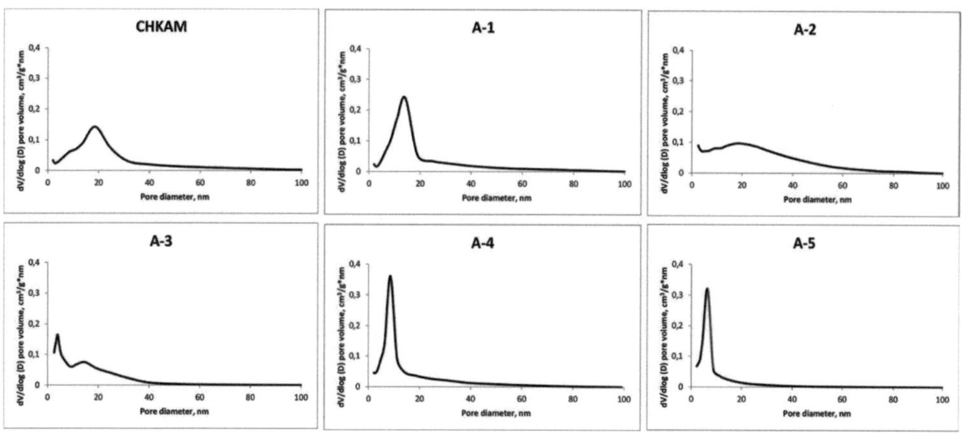

Figure 17. BJH pore size distribution of silica aerogel.

Sample	Thermal conductivity (W/(mK))	Drying method and temperature (°C)
A-6	0.035	Oven 60
A-7	0.030	Oven 60
A-8	0.025	Oven 180
A-9	0.024	Oven 180
A-10	0.033	Oven 200
A-11	0.031	Oven 200
A-10	0.034	Oven 250
A-12	0.031	Oven 300
A-13	0.028	Muffle furnace 400
A-2	0.029	Muffle furnace **400**

Table 4. Thermal conductivity of silica aerogel terms of drying method.

application. This parameter depends not only on the properties of the material but also on the testing method [42]. In the present case, dust emission measurements during the test procedure related to measurement of thermal conductivity of the obtained silica aerogel and comparative sample were realized. These studies confirmed the significantly lower emissions of the obtained silica aerogel when compared to the control sample (see **Figure 18**).

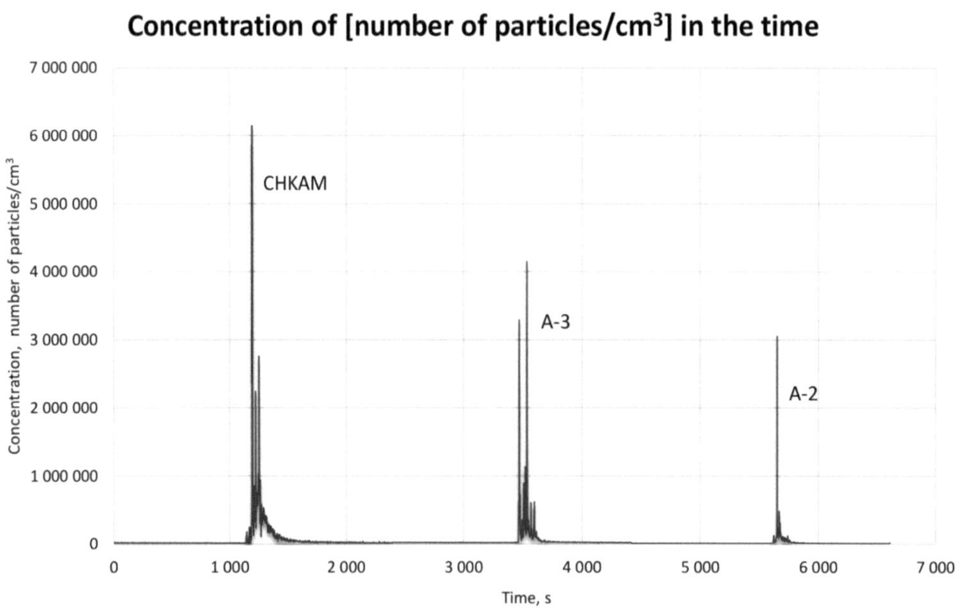

Figure 18. Emissions of dust from aerogels defined as the concentration of particles in 1 cm^3.

The main advantage of aerogel drying at atmospheric pressure is connected with the fact that it does not require high pressure and can be carried out continuously. This has a significant impact on the economy and ecology of the process of aerogel manufacturing. Such aerogel is a perfect component of construction products improving their thermal insulation properties. Moreover, such aerogel can be used in all applications where its lightness, non-toxicity and low coefficient of thermal conductivity as well as its hydrophobicity are needed.

2.3.5. Concluding remarks

The optimization of the synthesis of sol-gel-derived silicon-containing hybrids was carried out under different conditions of temperature and stirring rate confirming that the stirring rate plays an important role in the growth, polydispersity and stability of nanoparticles. Supramolecular structure examination proved homogeneous structure of the obtained silica nanopowders and aerogels. The most important advantage of the developed silicon-containing hybrids lies in their high application potential, e.g. for the production of façade paints, polymer nanocomposites, highly efficient insulating materials and other high-tech products for construction industry. The application of the developed additives allows to achieve higher hygienic level with the significant decrease of the applied amount of aggressive biocides, which are harmful to human health.

Acknowledgements

This work has been financially supported by projects No UDA-POIG.01.03.01-00-073/08-00 and No UDA-POIG.01.03.02-14-074/10-00 co-financed by the EU (European Regional Development Fund) and project GEKON1/O2/213318/14/2014 co-financed by the National Centre for Research and Development and National Fund for Environmental Protection and Water Management in Poland.
The authors thank Dr. Eng. Elzbieta Jankowska from the Central Institute for Labour Protection - National Research Institute for the measurements of dust emission, Dr. Eng. Osazuwa Osawaru from the Industrial Chemistry Research Institute for the measurements of surface area and Dr. Aneta Łukomska of the Industrial Chemistry Research Institute for preparation of SEM micrographs and Mr. Marek Panasiuk of the Industrial Chemistry Research Institute for measurements of particle size, thermal conductivity and all technical assistance.

Author details

Maria Zielecka*, Elzbieta Bujnowska, Katarzyna Suwala and Magdalena Wenda

*Address all correspondence to: maria.zielecka@ichp.pl

Industrial Chemistry Research Institute, Warsaw, Poland

References

[1] Levy D, Zayat M, editors. The Sol-Gel Handbook. 1st ed. Weinheim: Willey-VCH Verlag; 2015. 1508 p.

[2] Tanaka S, Mizukami F, Niwa S, Toba M, Maeda K, Shimada H, Kunimori K. Preparation of highly dispersed silica-supported palladium catalysts by a complexing agent-assisted sol-gel method and their characteristics. Applied Catalysis A: General. 2002;**229**:165–174. DOI:10.1016/S0926-860X(02)00025-X

[3] Brinker CJ, Scherer GW. Sol–Gel Science. The Physics and Chemistry of Sol–Gel Processing. 1st ed. London: Academic Press; 2013. 912 p.

[4] Sakka S, Kamiya J. Hydrolysis and polycondensation of dimethyldiethoxysilane and methyltriethoxysilane. Journal of Non-Crystalline Solids. 1982;**48**:31–46. DOI:10.1016/0022-3093(86)90106-7

[5] Stober W, Fink A, Bohn E. Controlled growth of monodisperse silica spheres in the micron size range. Journal of Colloid and Interface Science. 1968;**26**:62–69. DOI:10.1016/0021-9797(68)90272-5

[6] Murakata T, Sato S, Ohgawara T, Watanabe T, Suzuki T. Control of pore size distribution of silica gel through sol-gel process using inorganic salts and surfactants as additives. Journal of Material Science. 1992;**27**:1567–1574. DOI:10.1007/BF00542919

[7] Brook MA. Silicon in Organic, Organometallic and Polymer Chemistry. 1st ed. New York: John Wiley & Sons Inc.; 2000. 680 p.

[8] Polish patent N° 198188, 2008.

[9] Polish patent N° 204519, 2010.

[10] Zielecka M, Bujnowska E, Kepska B, Wenda M, Piotrowska M. Antimicrobial additives for architectural paints and impregnates. Progress in Organic Coatings. 2011;**72**:193–201. DOI:10.1016/j.porgcoat.2011.01.012

[11] Jeziorska R, Swierz-Motysia B, Zielecka M, Szadkowska A, Studzinski M. Structure and mechanical properties of low-density polyethylene/spherical silica nanocomposites prepared by melt mixing: the joint action of silica's size, functionality, and compatibilizer. Journal of Applied Polymer Science. 2012;**125**:4326–4337. DOI:10.1002/app.36579.

[12] Jeziorska R, Zielecka M, Gutarowska B, Zakowska Z. High-density polyethylene composites filled with nanosilica containing immobilized nanosilver or nanocopper: thermal, mechanical, and bactericidal properties and morphology and interphase characterization. International Journal of Polymer Science. Volume 2014; Article ID 183724, 13 p. DOI:10.1155/2014/183724

[13] Jeziorska R, Swierz-Motysia B, Zielecka M, Studzinski M. Polyamide/spherical nanosilica nanocomposites. Polimery. 2009;**54**:647–656.

[14] Zielecka M. Characteristics of silica nanopowders and sols containing immobilized nanoparticles of copper and silver. Polimery. 2011;**56**:765–768.

[15] Polish patent application N° P-415812, 2016.

[16] Obloj-Muzaj M, Zielecka M, Kozakiewicz J, Abramowicz A, Szulc A, Domanowski W. Polymerization of vinyl chloride in the presence of nanofillers – effects on the shape and morphology of PVC grains. Polimery. 2006;**51**:133–137.

[17] Obloj-Muzaj M, Abramowicz A, Kumosinski M, Zielecka M, Kozakiewicz J, Gorska A. Properties of rigid films made of PVC nanocomposites. In: Proceedings of the VIII International Conference on "Times of Polymers and Composites"19–23 June 2016; Naples, AIP Conference proceedings; 2016. p. 1736, 020132 (2016); DOI:10.1063/1.4949707

[18] European patent EP 2 473 537, 2015.

[19] Sondi I, Salopek-Sondi B. Silver nanoparticles as antimicrobial agent: a case study on *E. Coli* as model for Gram-negative bacteria. Journal of Colloid and Interface Science. 2004;**275**:177–182. DOI:10.1016/j.jcis.2004.02.012

[20] Jong-Min L, Dae-wook K, Young-Doo J, Seong-Geun O. Preparation of silica–silver heterogeneous nanocomposite particles by one-pot preparation strategy using polyol process: size-controlled immobilization of silver nanoparticles. Materials Research Bulletin. 2006;**41**:1407–1416. DOI:10.1016/j.materresbull.2006.02.010

[21] Yoon KY, Byeon JH, Park JH, Hwang J. Susceptibility constants of *Escherichia coli* and *Bacillus subtilis* to silver and copper nanoparticles. Science of the Total Environment. 2007;**373**:572–575. DOI:10.1016/j.scitotenv.2006.11.007

[22] Choi O, Deng KK, Nam-Jung K, Ross L Jr, Rao Y, Surampalli YH, Hu Z. The inhibitory effects of silver nanoparticles, silver ions, and silver chloride colloids on microbial growth. Water Research. 2008;**42**:3066–3074. DOI:10.1016/j.watres.2008.02.021

[23] Utracki LA. Clay-Containing Polymeric Nanocomposites. Rapra Techn. Ltd., Shawbury: 2004;**9**:73–93, 260–270.

[24] Kashiwagi T, Grulke E, Hilding J, Groth K, Harris R, Butler K, Shields J, Kharchenko S, Douglas J. Thermal and flammability properties of polypropylene/carbon nanotube nanocomposites. Polymer. 2004;**45**:4227–4239. DOI:10.1016/j.polymer.2004.03.088

[25] Usuki A, Kato M, Okada A, Kurauchi T. Synthesis of polypropylene oligomer-clay intercalation compounds. Journal of Applied Polymer Science. 1997;**63**:137–139. DOI:10.1002/(SICI)1097–4628(19971128)66:9<1781::AID-APP17>3.0.CO;2-Y

[26] Joshi M, Butola BS. Isothermal crystallisation of HDPE/POSS nanocomposite: effect of POSS as nanofiller. Journal of Applied Polymer Science. 2007;**105** (2):978–985. DOI:10.1002/app.26318

[27] Jeziorska R, Swierz-Motysia B, Zielecka M, Studzinski M. Polyamide/Spherical Silica Nanocomposites. Poznan: Wyd. Uczelniane Politechniki Poznanskiej; 2007;81–86.

[28] Golebiewski J. Polymer nanocomposites. Structure, synthesis and properties. Przemysl Chemiczny. 2004;**83**:15–20.

[29] Patent USA 9,126,839, 2015 also CN102985366 and PL217463 and WO 2011139170 A8

[30] Patent USA 9,371,586, 2016 also PL 217617 and WO 2011093731 A1

[31] Gurav J, Jung IK, Park HH, Kang E S, Nadargi DY. Silica aerogel: synthesis and applications. Journal of Nanomaterials. Volume 2010 (2010); Article ID 409310, 11 p. DOI:10.1155/2010/409310

[32] Baetens R, Jelle BP, Gustavsen A. Aerogel insulation for building applications: a state-of-the-art review. Energy and Buildings. 2011;**43**:761–769. DOI:10.1016/j.enbuild.2010.12.012

[33] Hrubesh LW. Aerogel applications. Journal of Non-Crystalline Solids. 1998;**225**:335–342. DOI:10.1016/S0022-3093(98)00135-5

[34] Patel R, Purohit NS, Suthar AM. An overview of silica aerogels. International Journal of ChemTech Research. 2009;**1**(4):1052–1057.

[35] Zhou XC, Zhong L, Xu Y. Surface modification of silica aerogels with trimethylchlorosilane in the ambient pressure drying. Inorganic Materials. 2008;**44**:976–979. DOI:10.1134/S0020168508090148

[36] Dowson M, Grogan M, Birksb T, Harrison D, Craigc S. Streamlined life cycle assessment of transparent silica aerogel made by supercritical drying. Applied Energy. 2012;**97**:396–404. DOI:10.1016/j.apenergy.2011.11.047

[37] Sakka S, editor. Handbook of Sol-Gel Science and Technology, Processing, Characterization and Applications. 1st ed. New York: Kluwer Academic Publishers; 2005. 1967 p.

[38] Solemani Dorcheh A, Abbasi MH. Silica aerogel; synthesis, properties and characterization. Journal of Materials Processing Technology. 2008;**199**:10–26. DOI:10.1016/j.jmatprotec.2007.10.060

[39] Rouquerol F, Rouquerol J, Sing K, Llewellyn P, Maurin G. Adsorption by Powders and Porous Solids Principles, Methodology and Applications. 2nd ed. Amsterdam Academic Press; 2013. 646 p.

[40] Sing K, Everett D, Haul R, Moscou L, Pierotti R, Rouqerol J, Siemieniewska T. Reporting physisorption data for gas/solid systems with special reference to the determination of surface area and porosity. Pure and Applied Chemistry. 1985;**57**(4):603–619.

[41] Jung IK, Gurav JL, Ha TJ, Choi SG, Baek S, Park HH. The properties of silica aerogels hybridized with SiO_2 nanoparticles by ambient pressure drying. Ceramics International. 2012;**38S**:S105-S108. DOI:10.1016/j.ceramint.2011.04.060

[42] Jankowska E, Sobiech P. Metody badania pylistości (Dustiness testing methods). Podstawy i Metody Oceny Srodowiska Pracy. 2013;**75**:7–20.

Chapter 3

Preparation of Ultra-High Temperature Ceramics–Based Materials by Sol-Gel Routes

Fei Li, Xiao Huang and Guo-Jun Zhang

Additional information is available at the end of the chapter

http://dx.doi.org/10.5772/67783

Abstract

Ultra-high temperature ceramics (UHTCs) are a class of inorganic materials that have melting point over 3000°C and are typically borides, carbides, and nitrides of early transition metals. UHTCs are considered as the promising candidate used in the extreme environment involved with the hypersonic aviation thermal protective system. Synthesis of UHTC-based materials can be divided into solid-based and solution-based protocols according to the state of the raw materials. A sol-gel technique is one of the solution-based protocols for the preparation of UHTC-based materials, which involves the hydrolysis, condensation of the metal organic and/or metal inorganic compounds, gelation, and the posthigh temperature treatment of the dried gels. The sol-gel technique enables the synthesis of UHTC-based materials at 1300–1600°C. UHTC-based materials with desired shapes, such as nanopowders, fibers, and porous monoliths, can also be prepared via sol-gel routes.

Keywords: ultra-high temperature ceramic (UHTC), sol-gel, nanopowders, ceramic fibers, porous ceramics

1. Introduction

Temperature at 3000°C is not just hot, but it is extremely hot. It is just a baseline in the world of extreme environment engineering. Very few materials can meet this criterion due to the facts that most of the materials have already melted or decomposed below this temperature. Materials with melting temperatures above 3000°C are very limited, which are shown in **Figure 1** [1–4]. The borides, carbides, and nitrides of group IV and V elements have melting temperatures above 3000°C and are considered as ultra-high temperature ceramics (UHTCs), such as zirconium diboride (ZrB_2), hafnium diboride (HfB_2), tantalum diboride (TaB_2), zirconium carbide (ZrC), hafnium carbide (HfC), tantalum carbide (TaC), and hafnium nitride (HfN) [1].

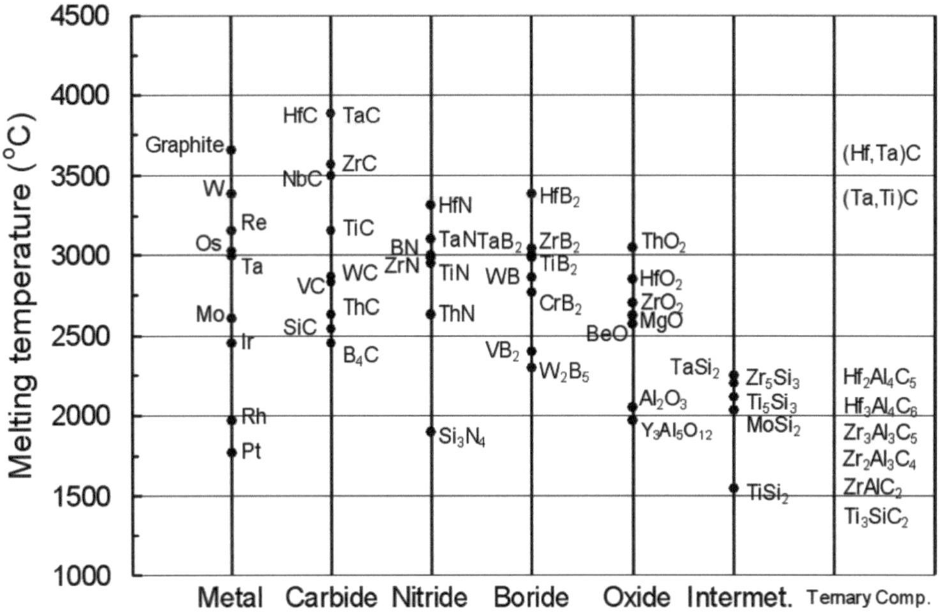

Figure 1. A comparison of the melting temperatures of the most refractory members of several classes of materials.

Recent interest in UHTCs has been motivated by the search for materials that can withstand extreme environments, such as extreme temperature, chemically aggressive environments, and rapid heating/cooling, etc. [5, 6]. Due to the combination of series of excellent physical and chemical properties, such as high hardness, good high temperature stability, and excellent solid-phase stability, UHTCs are considered as promising candidates for a variety of high-temperature structural applications, including engines, hypersonic vehicles, plasma arc electrode, cutting tools, furnace elements, and high temperature shielding. However, owing to the strong covalent bond and relatively low volume diffusion coefficient, UHTCs are difficult to get successfully sintered. Densification of these materials is typically performed by hot pressing and spark plasma sintering with or without sintering aids.

UHTCs can be synthesized by a variety of routes, which can be divided into solid phase protocols and solution based protocols based on the state of the starting materials [7, 8]. Among the solid-phase protocols, transition metal oxides, carbon black, boron oxide, and boron carbide are commonly used as the starting materials. After ball-milling and high temperature carbothermal reduction, the mixtures are converted into UHTC powders. Examples of the reactions used to synthesize borides and carbides are listed in **Table 1** [7]. These reactions are written for ZrB_2 and ZrC, but analogous processes produce other borides and carbides. These solid phase protocols are very straightforward and simple and are used to produce UHTC powders commercially. However, due to the heterogeneity of the reactants, the reactions need to be performed at high temperatures for several hours to complete. Also, those synthesized UHTC powders are micron-sized particles.

Reactions	Category	Example	
1	Carbothermal	$ZrO_2 + 3C \rightarrow ZrC + 2CO$	(1)
2	Carbothermal	$ZrO_2 + B_2O_3 + 5C \rightarrow ZrB_2 + 5CO$	(2)
3	Borothermal	$ZrO_2 + 4B \rightarrow ZrB_2 + B_2O_2$	(3)
4	Boron carbide	$7ZrO_2 + 5B_4C \rightarrow 7ZrB_2 + 3B_2O_3 + 5CO$	(4)
5	Combined	$2ZrO_2 + B_4C + 3C \rightarrow 2ZrB_2 + 4CO$	(5)

Table 1. Examples of reduction reactions that can be used to synthesize borides and carbides [7].

In solution-based protocols, soluble transition metal containing precursors, boron precursors, and carbon sources are used as the raw materials to form homogeneous sols/solutions or polymers [8–10]. These sols/solutions or polymers can be converted into UHTC materials after hydrolysis, curing/drying, and pyrolysis at high temperatures. Actually, the formation of UHTCs via solution-based protocols is still based on the carbothermal or borothermal reactions of metal oxides with carbon. However, due to the homogeneous distribution of all the reactants at the molecular level, the following high temperature reduction reactions can proceed at much lower temperature and fine UHTC powders with particle sizes down to few hundred nanometers can be synthesized. Not only granular UHTC particles but also some other UHTC materials with different morphology can be synthesized through these solution-based routes.

A sol-gel technique is one of the solution-based processes, which enables processing of ceramics at low temperature in various desirable shapes, such as monoliths, films, fibers, or nano-sized powders, as shown in **Figure 2** [11, 12]. Typically, the sol-gel technique involves the hydrolysis, condensation of the metal organic and/or metal inorganic compounds, gelation, and the posttreatment of the dried gels to obtain final ceramic products. Small molecular compounds are used as the starting materials in sol-gel routes. The homogeneity of the starting compounds at molecular level facilitates the preparation of these materials at relatively low temperatures via the sol-gel technique. The sol-gel route is an old but an efficient way to produce high quality, fine UHTC powders under moderate conditions compared with other solution-based protocols [8, 10, 13]. Combined with various shaping techniques, UHTC-based materials in various desirable shapes, such as nanopowders, nanofibers, and porous ceramics can be prepared from similar starting materials and simple reactive processes via the sol-gel technique [8, 14–16].

In this chapter, we will present recent works covering the synthesis of nanopowders, nanofibers, and porous ceramics of UHTC-based materials via sol-gel routes. The discussions mainly focus on the synthesis of ZrC- and ZrB_2-based materials, but analogous processes produce other borides and carbides. Meanwhile, numbers of publications discussing the utilization of UHTC sols to prepare fiber-reinforced ceramic composites via polymer infiltration pyrolysis routes will not be discussed here.

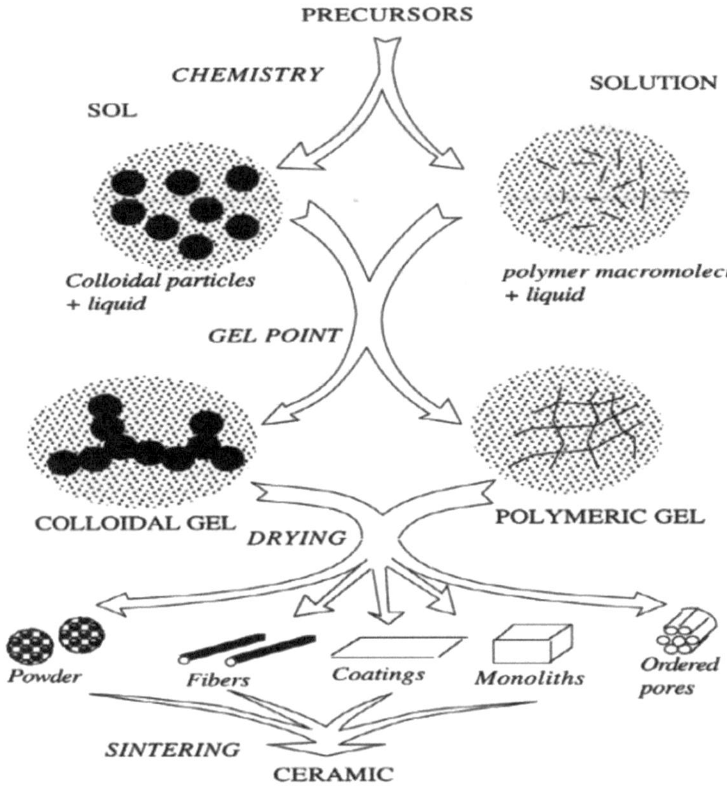

Figure 2. Simplified chart of sol-gel processes [12].

2. Synthesis of the precursors for UHTC

It is very straightforward and simple for the researchers to apply a sol-gel technique to prepare UHTC nanopowders. It seems simple to obtain the sols of the target materials by mixing the precursors of a transition metal with a carbon source and a boron source in liquid state. On drying and after high temperature ceramization, the UHTC nanopowders can be prepared [8, 10]. However, it is not always easy to get homogeneous precursor sols of UHTC materials without considering the chemical compatibility and stability of the precursors.

Generally, transition metal alkoxides, oxychloride, and chloride are used as the starting metal oxide precursors. Bidentate ligands, such as β-diketones, acetic acid, and cellulose acetate, are used to modify the high reactivity of the metal alkoxide allowing a slow-down of the system's reactivity [17, 18]. The chemically modified metal alkoxide undergoes hydrolysis and condensation in a controlled manner to form preceramics, as illustrated in Eq. (6) [17].

$$n\text{-RO}-\underset{\text{acac}}{\overset{\text{acac}}{\text{Zr}}}-\text{OR} + n\text{-HO}-\text{X}-\text{OH} \longrightarrow n\text{-RO}-\underset{\text{acac}}{\overset{\text{acac}}{\text{Zr}}}-\text{O}-\text{X}-\text{OH} + n\text{-ROH} \quad (6)$$

http://dx.doi.org/10.5772/67783

Polyzirconoxane (PZO) can be synthesized by chelation of acetyl acetone to zirconium oxychloride and is widely used as the powdery and soluble zirconia precursor [19, 20]. The synthesis process of PZO is shown in Eq. (7) [20]. It is also possible for the synthesis of powdery preceramic polymers for ZrC through reactions of zirconium oxychloride with acetyl acetone and salicyl alcohol in a one-pot reaction procedure according to Eq. (8) [21]. The preceramic polymers can be converted into ZrC nanopowders after the high temperature treatment.

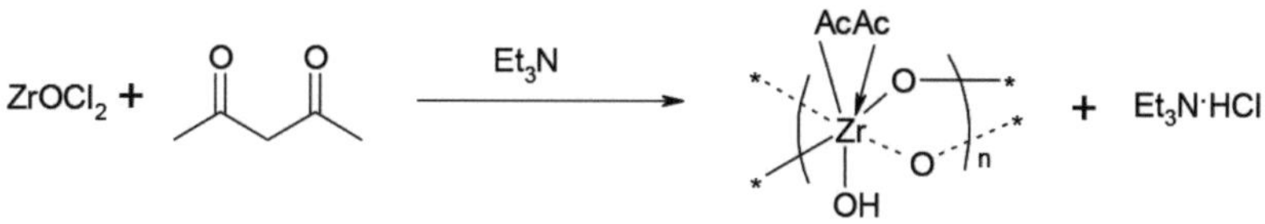

(7)

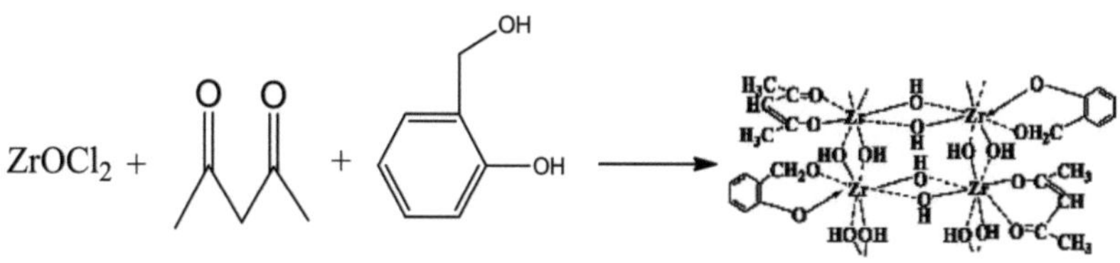

(8)

While sucrose, phenolic resin, and polyacrylonitrile are used as the carbon source; boron-containing compounds (e.g., boric acid and triethyl borate) are usually used as the boron source to synthesize metal borides. By mixing the modified metal precursors with carbon and boron sources in controlled molar ratios, the precursor sols for UHTC are prepared.

3. Synthesis of UHTC-based nanopowders

UHTC nanopowders can be synthesized by directly heating the dried gels in an inert atmosphere. By controlling the composition of the starting sols, unary UHTC nanopowders (e.g., ZrC, HfC, ZrB_2, and HfB_2) and composite nanopowders (e.g., ZrC-SiC, ZrB_2-SiC, and ZrB_2-ZrC-SiC) can be prepared.

3.1. Synthesis of unary UHTC nanopowders

3.1.1. Synthesis of metal carbide nanopowders

Sacks et al. [8] used zirconium and hafnium alkoxides as the metal precursors, acetylacetone as a chemical modifier, and phenolic resin and glycerol as the liquid carbon source. After controlled hydrolysis, condensation and drying of the precursor sol, the dried gel was obtained. The dried gel was heated at temperatures up to 1200–1600°C for carbothermal reduction reaction (Eq. (1)). During heating process, the organic components decompose and leave the fine-scale mixed metal oxide and amorphous carbon. The carbothermal reduction reaction begins at 1200°C is demonstrated by the appearance of ZrC diffraction peaks in the pyrolyzed products, as shown in **Figure 3a**. The carbothermal reduction reaction is completed after heat treatment at 1400°C for 2 h. After heated at 1475°C, the average crystallite size of

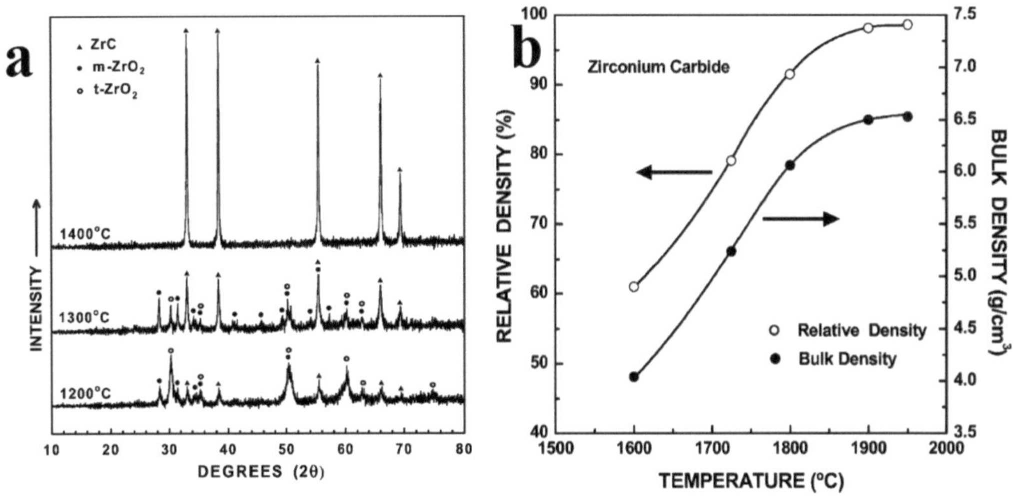

Figure 3. (a) X-ray diffraction patterns for heat treated samples prepared from a (zirconium n-propoxide/phenolic resin)-based sol. (b) Plots of relative density and bulk density versus sintering temperature for dry-pressed ZrC powder compacts [8].

the ZrC nanoparticles is 100 nm. The oxygen content within ZrC nanopowders decreases with the increasing heating temperatures, from 3.3 wt% at 1475°C to 0.1 wt% at 1800°C. The ZrC nanopowders show good sinterability and can reach 98% of relative density after sintering at 1900°C with no pressure, as shown in **Figure 3b**.

High pyrolyzing temperatures are effective to synthesize metal carbide nanopowders with low impurity content. However, it causes the increase of the average crystallite size of the products. Dollé et al. [10] developed a two-step heat treatment approach to realize the purification of ZrC nanopowders with no significant increase in the average crystallite size by using metal alkoxide, sucrose, and acetic acid as the staring materials. Only diffraction peaks of ZrC can be observed when the precursor was heated at 1400°C for 3 h in argon. Only a slight shift of the peaks to the low angles with the increase of heating temperatures indicated an increase in the lattice parameter with temperature. The average crystallite sizes increase from 93 to 150 nm with the temperatures increase from 1400 to 1600°C. During the two-step heat treatment process, the precursor was first heated at 1400°C for 150 min and then at 1800°C for 6 min maintaining the heating rate at 20°C/min. The average crystallite size came out to be 110 nm, a slight increase as compared to the one heated at 1400°C (**Figure 4**).

Tao et al. [21] developed a preceramic route to synthesize ZrC nanopowders by using zirconium oxychloride, acetylacetone, and salicyl alcohol as the raw materials (Eq. (8)). ZrC nanopowders can be prepared by directly heating the powdery precursor in argon. Only diffraction peaks of ZrC can be detected when the precursor was pyrolyzed at 1300°C for 2 h (**Figure 5a**). Such low temperature indicated a fine contact between the metal precursor and the carbon precursor. Disadvantage of the technique is the presence of the residual carbon in the final ZrC products, as can be observed in **Figure 5b**. ZrC nanoparticles seem embedded in the matrix of amorphous carbon. This might be ascribed to the reason that it is not easy to control the degree of the polymerization during the synthesis process.

http://dx.doi.org/10.5772/67783

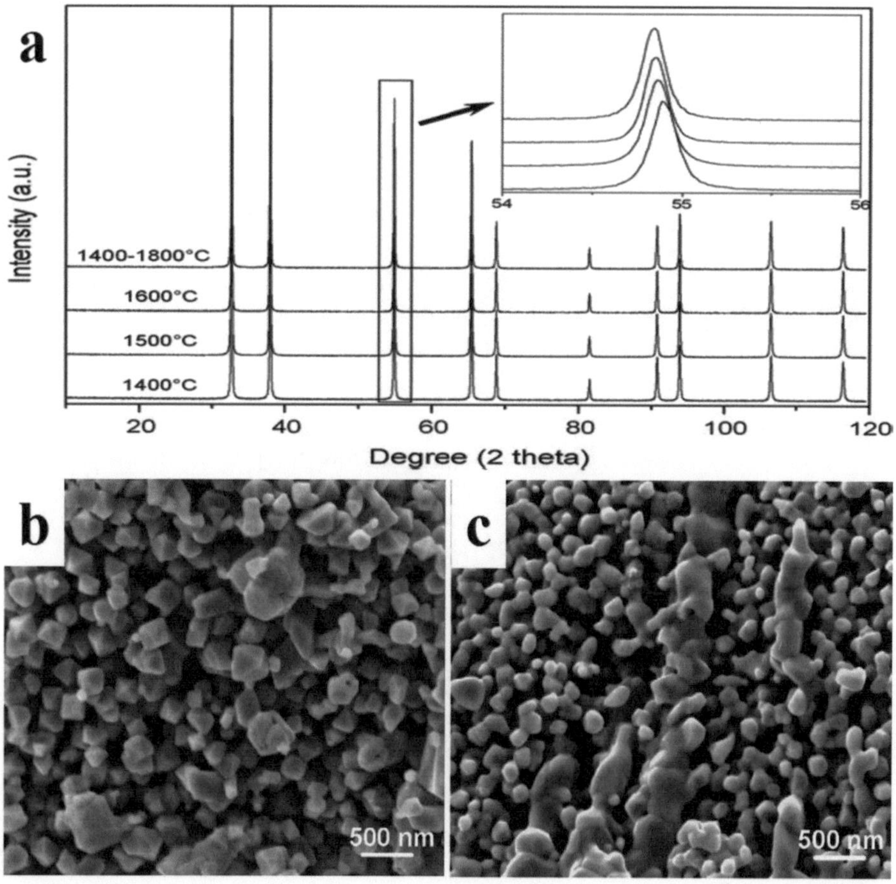

Figure 4. (a) Evolution of the XRD patterns of the different samples as a function of their respective heat treatment under argon flow. (b) SEM image of an agglomerate microstructure of a sample heat treated at 1600°C for 3 h. (c) SEM image of an agglomerate microstructure of a sample heat treated at 1400°C for 150 min then at 1800°C for 6 min [10].

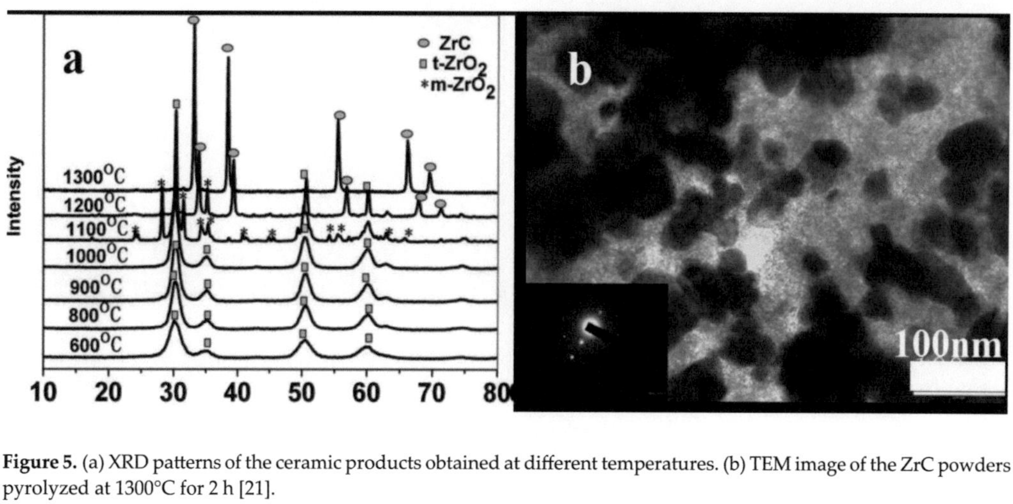

Figure 5. (a) XRD patterns of the ceramic products obtained at different temperatures. (b) TEM image of the ZrC powders pyrolyzed at 1300°C for 2 h [21].

3.1.2. Synthesis of metal diboride nanopowders

Boric acid is the commonly used boron source for the synthesis of metal diboride nanopowders via sol-gel routes. Li et al. [22, 23] prepared ZrB_2 nanopowders by using zirconium n-propoxide as the zirconia source, sucrose as the carbon source, and boric acid as the boron source. By mixing the acetylacetone-modified zirconium n-propoxide with methanol solution of boric acid and sucrose, the precursor sol was obtained. ZrB_2 nanopowder was obtained by heating the dried gels at high temperatures in an inert atmosphere (Eq. (2)). The B/Zr molar ratio plays a significant role in the formation of pure ZrB_2 during the carbothermal reduction reaction. Boron loss due to the high vapor pressure of boron oxide at high temperature must be compensated during the high temperature carbothermal reduction. Pure ZrB_2 nanopowders with particle sizes down to 100 nm can be prepared by tailoring the composition of the starting precursor sols (a B/Zr molar ratio of 2.3) and heat treatment temperature (1550°C) (**Figure 6**). Rod-like structure can also be synthesized by controlling the gelation temperature of the precursor sols.

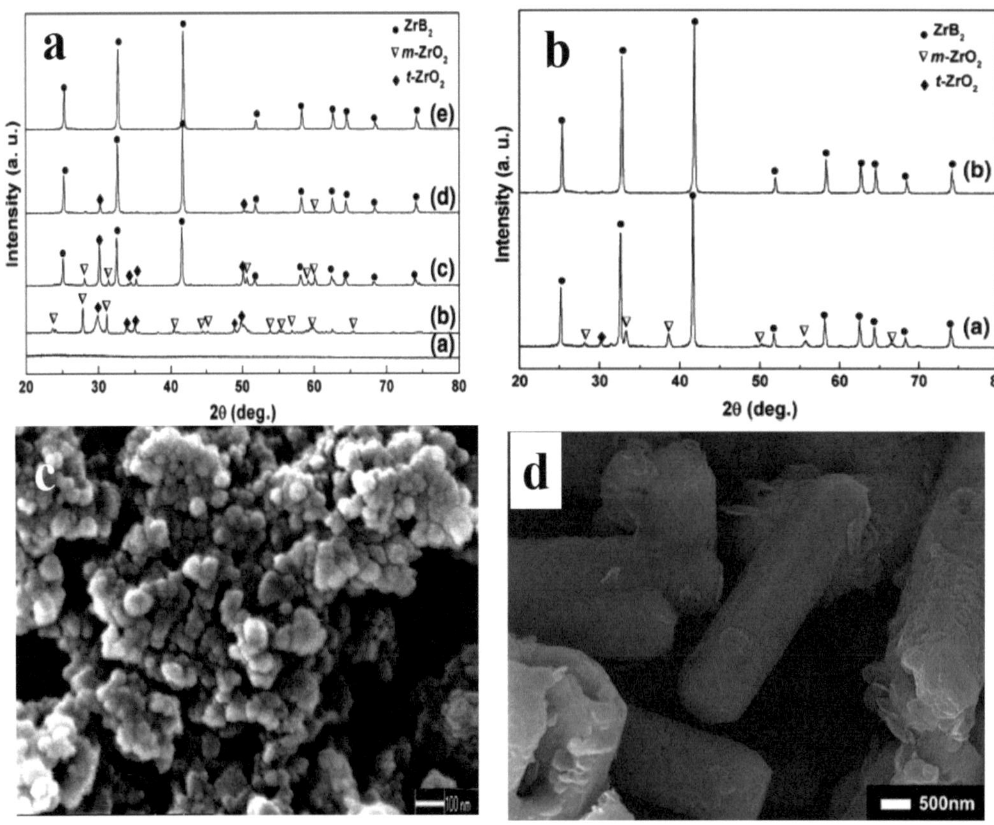

Figure 6. (a) XRD patterns of precursor and precursors pyrolyzed at various temperatures for B/Zr = 2.3. From bottom to top, precursor, 1100, 1300, 1400, 1550°C [22]. (b) XRD patterns of precursors pyrolyzed at 1550°C for 2 h with molar ratios of B/Zr 2.0 (bottom) and 2.3 (top) [22]. (c) SEM image of ZrB_2 nanoparticles pyrolyzed at 1550°C for 2 h with B/Zr = 2.3. The scale bar in the SEM is 100 nm [22]. (d) SEM image of ZrB2 nanoparticles pyrolyzed at 1550°C for 2 h with a gelation temperature of 85°C [23].

Walker and Corral [24] applied the metal oxychloride hydrates, triethyl borate, and phenolic resin as the starting materials to synthesize HfB_2 and ZrB_2, as shown in **Figure 7**. The starting materials were mixed by high-energy ultrasonication to form precipitate-free sols which turn into stable gels without any addition of catalysts. They pointed out that the conversion process to ZrB_2 involved the formation of ZrC subsequently reacting with boron oxide to form ZrB_2. Both HfB_2 and ZrB_2 prepared by optimizing the chemistry, processing, and heat treatment conditions can result in the desired final ceramic products.

3.2. Synthesis of UHTC composite nanopowders

It is not easy to sinter phase-pure UHTC materials due to their strong covalent bonds. Sintering aids (e.g., silicon carbide-SiC) are commonly introduced to the UHTC matrix to improve their sinterability and properties of the corresponding ceramic materials [7]. It is one of the advantages of sol-gel techniques to obtain homogenous distribution of all reactants at molecular level. Precursors of UHTC-SiC composite powders can be synthesized by introducing independent precursors of UHTCs and SiC into a solvent (ethanol, toluene, and tetrahydrofuran, etc.) with the liquid precursor itself acting as a mutual solvent. The precursors are dried and cured at relatively low temperatures and then heated at high temperatures to complete the carbothermal reduction reaction to form HTC-SiC composite powders, such as ZrC-SiC, ZrB_2-SiC, and HfB_2-SiC, etc. [25–30].

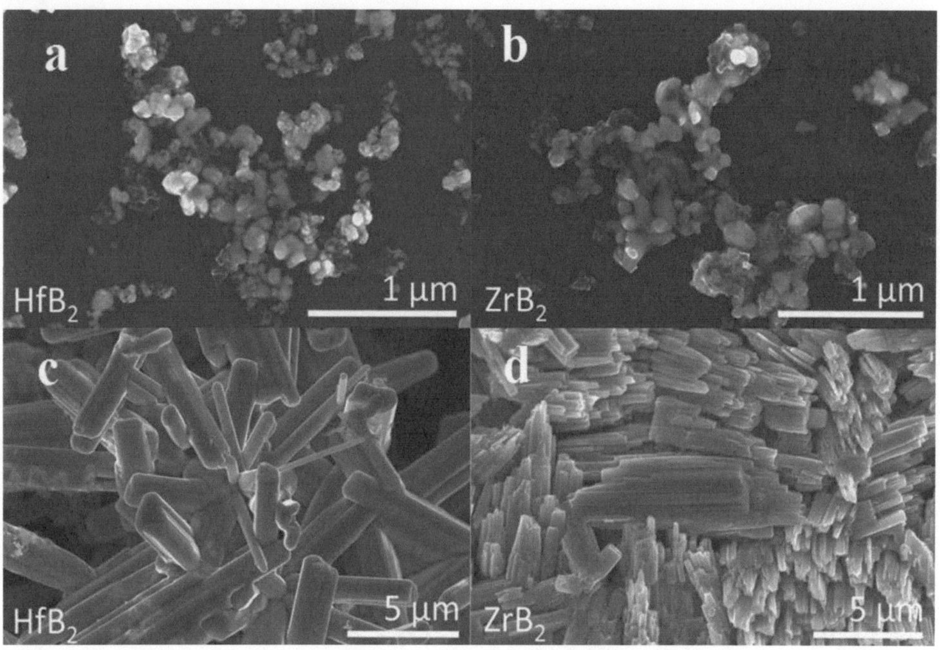

Figure 7. SEM images of heat-treated (a) HfB_2 and (b) ZrB_2 precursor with fine spherical particle morphology and extreme localized particle growth for (c) HfB_2 and (d) ZrB_2 that forms rod-like particles within the powder [24].

Cai et al. [28] used methylphenyldichlorosilane ($MePhSiCl_2$) as a SiC precursor to react with zirconcene dichloride (Cp_2ZrCl_2) to form organometallic polymers. ZrC-SiC composite powders with controlled Zr/Si molar ratios can be synthesized by pyrolyzing the polymers at 1400°C for 2 h. The particles seen in the SEM image are composed of three parts: large grains, the smaller grains, and the black matrix (**Figure 8b**). It is not easy to precisely control the residue carbon in this reaction system.

Wang et al. [31] prepared ZrC-SiC hybrid precursors by dissolving acetylacetone-modified zirconium alkoxide (PZC) into liquid polycarbosilane (LPCS). In their reaction system, LPCS simultaneously played the role of a solvent, carbon source, and silicon source. ZrC-SiC composite powders with spherical morphology and grain sizes less than 60 nm can be synthesized by heating the hybrid precursors at 1550°C in argon (**Figure 9**). The constituent of the final ZrC-SiC composite powders can be controlled by tailoring the mass ratios of LPCS/PZC. The synthesized ZrC-SiC nanopowders exhibit improved oxidation resistance compare to ZrC nanopowders.

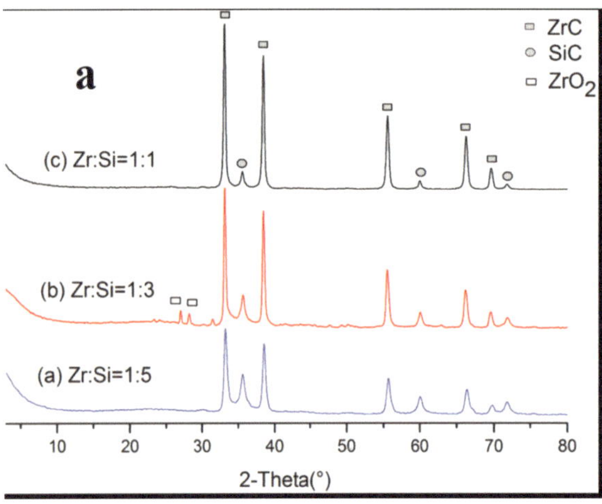

Figure 8. (a) XRD patterns of the poly-yne polymers pyrolyzed at 1400°C and (b) EDS spectra of the ceramic sample from the Zr:Si = 1:1 precursor [28].

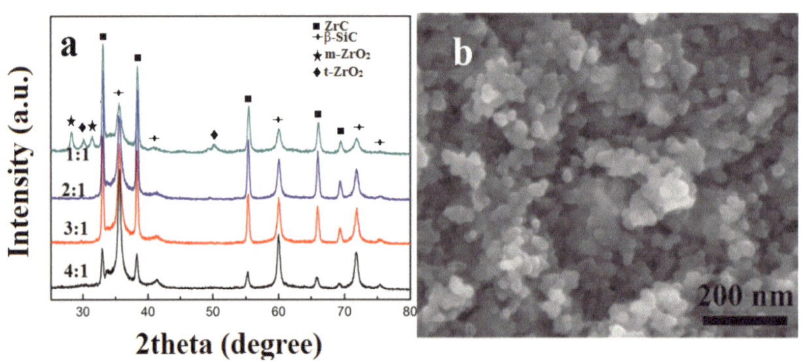

Figure 9. (a) XRD patterns of samples derived from hybrid precursors with various LPCS/PZC mass ratios after pyrolysis at 1550°C in argon and (b) SEM image of the ZrC-SiC powders derived from a hybrid precursor with LPCS/PZC = 1 in mass ratio [31].

http://dx.doi.org/10.5772/67783

Liang et al. [32] used zirconia sol, tetraethylorthosilicate (TEOS) and sucrose as the independent zirconium-, silicon- and carbon-containing source. Hybrid sol can be prepared by introducing TEOS and sucrose into zirconia sol. ZrC-SiC composite powders with controlled constituent are obtained by pyrolyzing the hybrid gels at 1500°C for 1 h in vacuum (**Figure 10**). It is possible to precisely control both the residue carbon content and SiC content in the final composite powders by using these independent carbon sources and silicon sources as the starting materials. However, no sintering behavior of the as-synthesized ZrC-SiC nanopowders is examined.

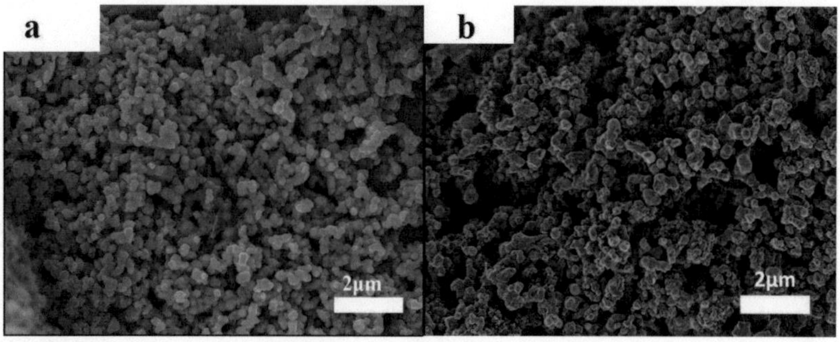

Figure 10. ZrC-SiC composite powders with different SiC volume percentage. (a) ZrC-10 vol% SiC and (b) ZrC-90 vol% SiC [32].

4. Synthesis of UHTC-based ceramic fibers

The sol-gel technique enables the processing of ceramics in various desirable shapes, such as monoliths, films, fibers, etc. [33]. The viscosity of the liquid precursors of UHTCs can be tuned by condensation polymerization of the sols themselves or by adding viscous polymers into the mixed solution. These viscous precursors can be drawn into fibrous shapes, dried, cured, and heated at high temperatures for the carbothermal reduction reaction to form UHTC ceramic fibers. Combined with different spinning techniques, such as dry spinning and electrospinning, UHTC fibers can be successfully synthesized [16, 17, 34–39].

4.1. UHTC fibers prepared by a hand-drawing method

Hand-drawing is a simple but effective method to estimate the spinnability of the spinning solution. Fibers are hand-drawn by extracting a glass rod from the viscous solutions. Spinnability can be estimated by the length of the drawn fibers. Preiss et al. [17] prepared polymeric precursors for carbothermal reactions from the chelate derivatives of zirconium alkoxides in alcohols by reaction with organic compounds having two or more reactive OH groups. These organic groups act as bridging ligands in transesterification and condensation polymerization yielding either spinnable viscous solutions or elastic gels. The spinnability of these polymeric precursors was estimated by hand-drawing fibers from the solutions. ZrC-based fibers with a diameter about 20 μm can be obtained by heating the hand-drawn precursor fibers in an argon atmosphere at 1550°C (**Figure 11a** and **b**).

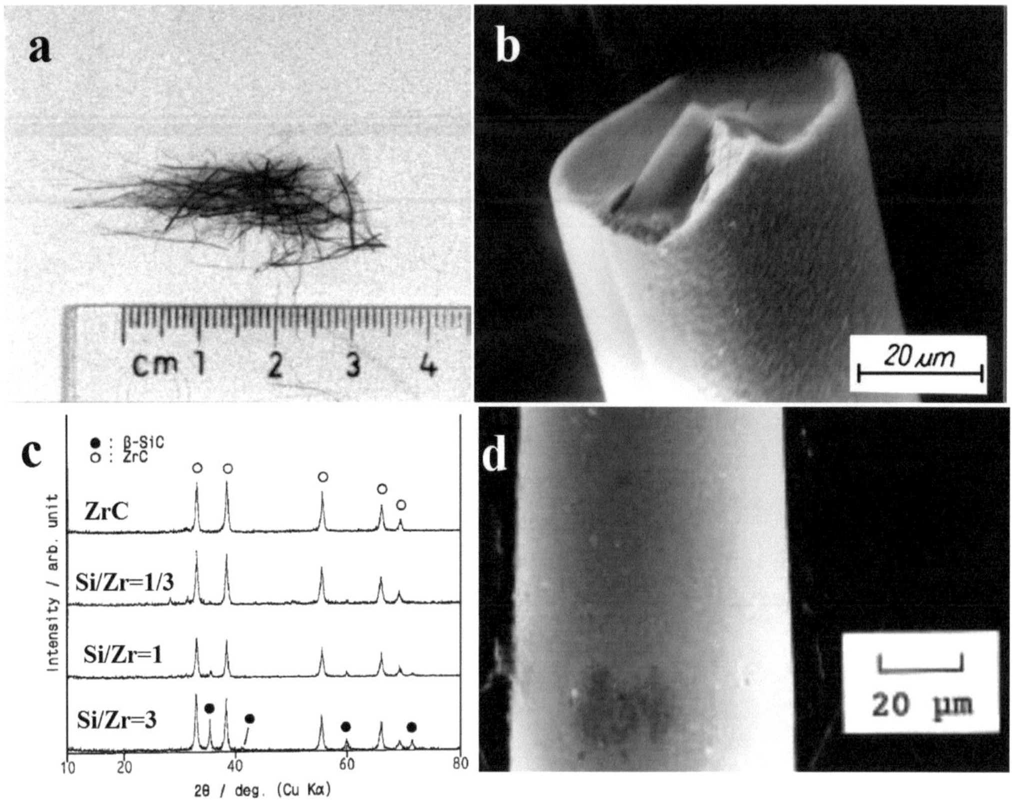

Figure 11. Photograph (a) and SEM, (b) image of ZrC fibers treated at 1550°C [17], (c) XRD patterns of fibrous products obtained by heat treatment at 1500°C in argon and(d) SEM image of ZrC-SiC fiber with a Si/Zr molar ratio of 0.92 [34].

Hasegawa et al. [34] prepared ZrC-SiC fibers by using zirconium tetrakis(2,4-pentanedionate) (ZTP), TEOS, and phenolic resin as raw materials via a hand-drawing method. The composition of the final ZrC-SiC fibers can be controlled by tailoring the Si/Zr molar ratios of the starting sols. The preceramic fibers were converted into ZrC or ZrC-SiC ceramic fibers after pyrolysis at 1500°C for 4 h under argon flow (**Figure 11c**). The diameter of the as-prepared ZrC fibers is about 40 μm (**Figure 11d**).

4.2. UHTC fibers prepared by a dry spinning method

A continuous dry spinning process was applied to prepare ZrC-based fibers by Hu et al. [14]. In their works, ZTP and phenolic resin were used as the starting precursors to form a highly viscous mixed sol. Phenolic resin played the role of a carbon source and the spinning aid. ZrC-based fibers with fine-diameter (~20 μm) and relatively smooth surfaces were obtained by dry spinning the mixed sols and heating at 1450°C. The SEM images of the as-synthesized ZrC fiber show that the fiber has a porous, extremely fine-grained microstructure (**Figure 12**).

http://dx.doi.org/10.5772/67783

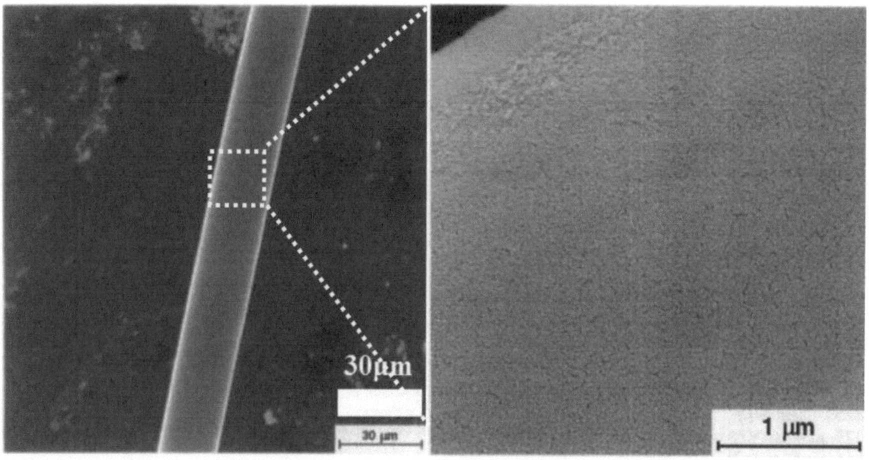

Figure 12. SEM images of a single ZrC fiber after carbothermal reduction at 1450°C [14].

However, very few studies were reported on the synthesis of UHTC fibers by the dry spinning process due to the brittleness of the ceramic fibers caused by the massive weight loss during carbothermal reduction. The state-of-the-art technique is not mature enough to apply this kind of scalable synthesis process of UHTC-based fibers.

4.3. UHTC fibers prepared by an electrospinning method

Electrospinning is a simple and versatile method for generating nanofibers of polymers and ceramics [40, 41]. The typical procedure for the preparation of ceramic nanofibers by electrospinning includes the preparation of precursor solution, the electrospinning of the precursor solution to get inorganic/polymer composite nanofibers, and the postheat treatment to get ceramic nanofibers based on the carbothermal reduction reaction. Generally, high molecular polymers such as polyvinylpyrrolidone (PVP), polyvinyl alcohol (PVA), polyacrylonitrile (PAN), and cellulose acetate are used as the spinning aids to facilitate the spinning process. The parameters including applied voltage, flow rate, and working distance influence the microstructure of the final ceramic nanofibers.

Using ZTP as the zirconia source, cellulose acetate as the carbon source and spinning aids, ZrC-based nanofibers were prepared. ZrC was initially formed at 1400°C and became the major phase at 1600°C. The fibrous structures of the electrospun green nanofibers could be well retained after high temperature ceramization.

Li et al. [39] used PZO as the zirconia source, PAN as the carbon source and spinning aid to synthesize ZrC nanofibers via electrospinning. In their reaction system, a carbon source played significant role in controlling the microstructure of the pyrolyzed ZrC nanofibers. With sucrose used as the extra carbon source in PZO/PAN spinning solution, fine ZrC nanofibers composed of ZrC nanoparticles were obtained after pyrolysis at 1400°C for 2 h (**Figure 13b**). The as-synthesized ZrC nanofibers were porous. After sintering, bamboo-like structures are formed due to Ostwald ripening.

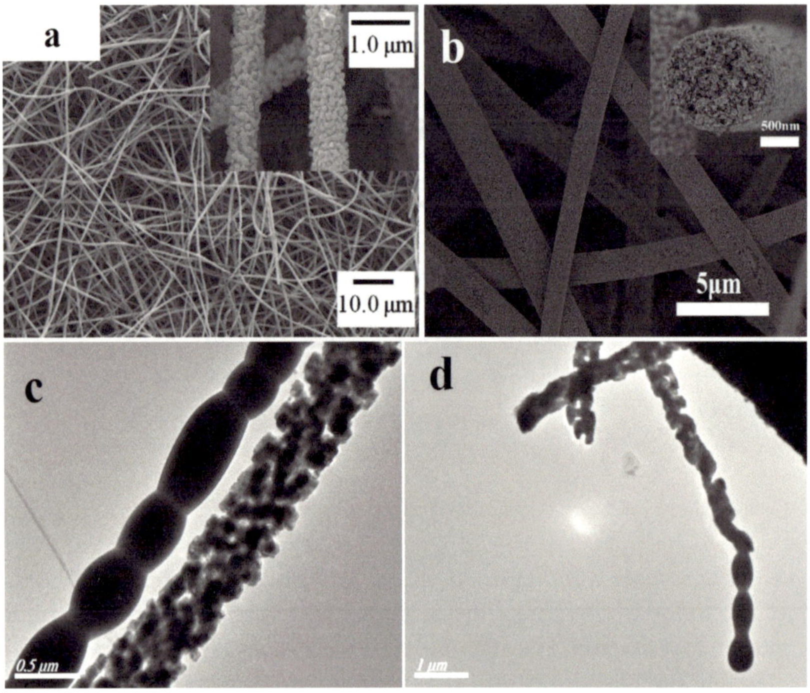

Figure 13. (a) SEM image of ZrC nanofibers heat at 1600°C for 2 h. The spinning solution was prepared from mixing ZTP and phenolic resin with a C/Zr molar ratio of 5.5 [36]. (b) SEM image of ZrC nanofibers after pyrolysis at 1400°C for 2 h under argon flow by using PZO/PAN/sucrose as the spinning solution with a C/Zr molar ratio of 3.3 [39]. (c) and (d) TEM images of ZrC nanofibers after pyrolysis at 1400°C for 2 h electrospun from PZO/PAN spinning solution.

ZrB_2 nanofibers can also be prepared by introducing a boron source into the ZrC spinning solutions [16, 37]. Formation of ZrB_2 was initiated at 1200°C with complete transformation to pure ZrB_2 at 1500°C (**Figure 14**). The as-synthesized ZrB_2 nanofibers were porous and composed of ZrB_2 nanoparticles.

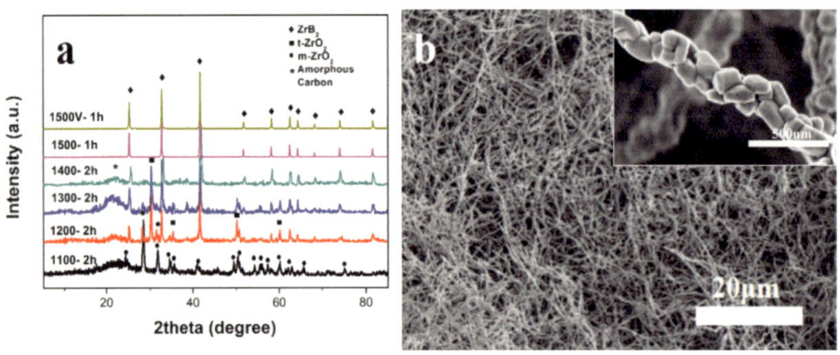

Figure 14. (a) XRD patterns of the nanofibers after pyrolysis. From bottom to top: 2 h at 1100°C, 2 h at 1200°C, 2 h at 1300°C, 2 h at 1400°C, and 1 h at 1500°C in an argon atmosphere; 1 h at 1500°C in vacuum. (b) SEM image of ZrB_2 nanofibers after pyrolysis at 1500°C for 1 h in argon. The scale bar in the inset is 500 nm [16].

http://dx.doi.org/10.5772/67783

5. UHTC-based porous ceramics via sol-gel routes

Materials containing tailored porosity exhibit special properties and features that usually cannot be achieved by their conventional dense counterparts [42, 43]. Nowadays, therefore, porous materials find many applications as end products in several technological processes. Due to the intrinsic properties of porous ceramics, such as lightweight, low thermal conductivity, high specific strength, and excellent thermal insulation, UHTC-based porous ceramics could be applied as the promising candidate materials in thermal protection systems of hypersonic vehicles.

The processing routes used for the production of porous ceramics can be classified into replica, sacrificial template, and direct-foaming methods, as schematically shown in **Figure 15** [42]. It is important to get a well-dispersed ceramic slurry or liquid precursor for the fabrication of porous ceramics with tailored microstructure and chemical composition. Sol-gel techniques have been widely used as ceramic precursors to prepare porous ceramics, such as zirconia, alumina, and silicon carbide–based porous ceramics.

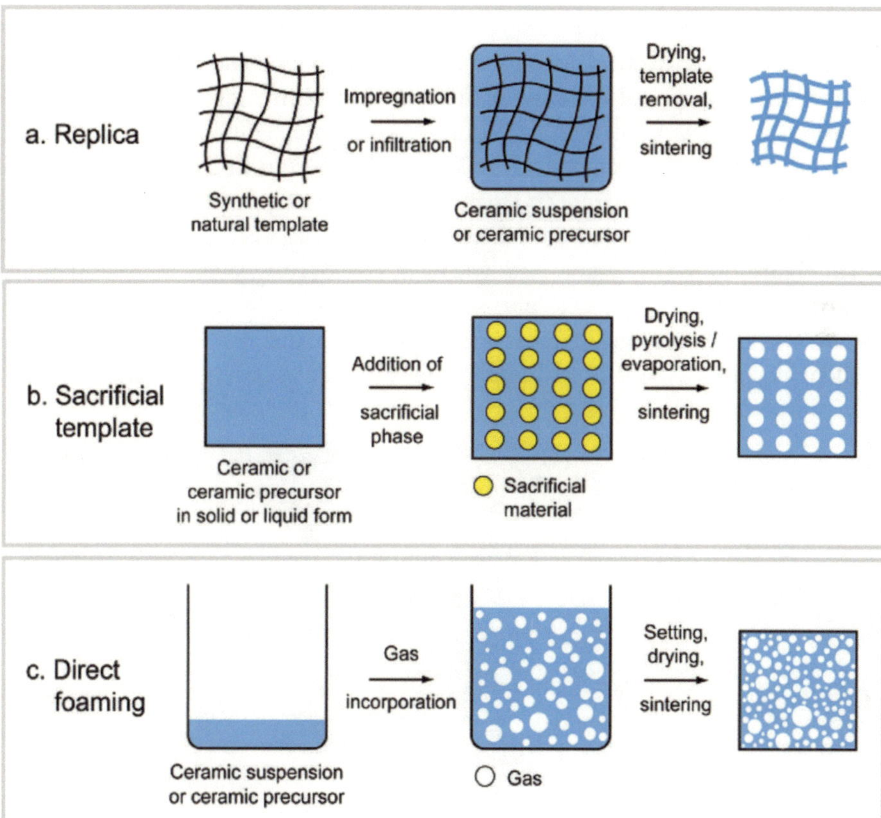

Figure 15. Scheme of possible processing routes used for the production of macroporous ceramics [42]. (a) Replica method; (b) Sacrificial template method; (c) Direct foaming method.

5.1. Porous ZrC ceramics prepared from replica routes

Though sol-gel routes have been established to synthesize UHTC nanopowders and nanofibers, focus on the preparation of UHTC-based porous ceramics is rare. The replica process was applied to prepare porous ZrC/C ceramic by Rambo et al. [44] where pine wood was used as the biological template for manufacturing of highly porous, biomorphic ZrC/C ceramics with zirconium n-propoxide as the zirconia source. Pine wood was first pyrolyzed at 800°C for 1 h under N_2 to obtain a biocarbon template. The biocarbon template was then infiltrated with zirconia sol and dried at 110°C for 1 h to obtain the ZrC/C preceramic. ZrC/C-ceramic was prepared by heating the infiltrated biocarbon template at 1600°C for 1 h under argon flow. The original pine char microstructure was reproduced in the ZrC/C-ceramic. Island-like ZrC particles were observed on the cell walls and in the channels, as shown in **Figure 16**. The porosity of the ZrC/C-ceramic was ~78%.

5.2. ZrC foams prepared from direct-foaming routes

In direct-foaming methods, porous materials are produced by incorporating air into a suspension or liquid media, which is subsequently set in order to keep the structure of air bubbles created [42]. The consolidated foams are afterwards heated at high temperatures to obtain porous ceramics. Li et al. [15] prepared ZrC foams by employing a commercial direct-foaming technique for phenolic resins. Zirconia sol was used as the zirconia precursor, while commercial foamable phenolic resin was used as the carbon source and foaming aid. By thermal-setting the wet foam and pyrolyzing the foamed green body at 1600°C, highly porous ZrC foam was prepared. The density and porosity of as-prepared ZrC foam were 0.16–0.19 g/cm^3 and 83–90%, respectively. The as-prepared ZrC foam showed closed cell structures with average cell sizes about 40 μm, as shown in **Figure 17**. The as-synthesized ZrC foam displayed excellent thermal stability up to 2400°C. However, it is not possible to precisely control the residue carbon in the ZrC foam in their reaction system. Foamability of the mixtures originated due to phenolic resin decreased as the Zr content increased. About 47 wt% amorphous carbon was observed in the final ZrC foam.

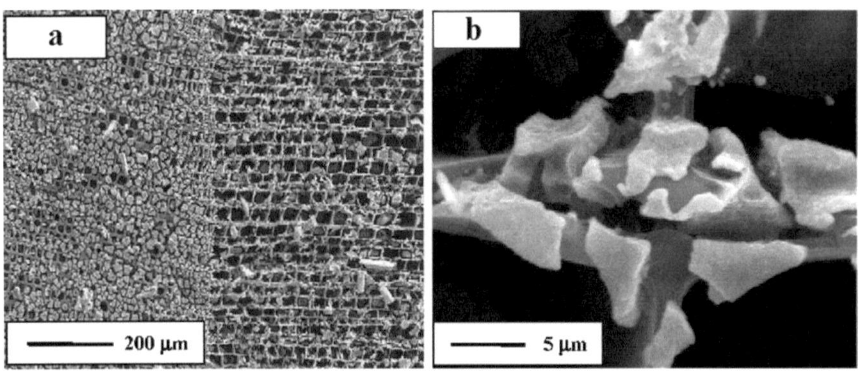

Figure 16. SEM images of the biomorphic ZrC/C-ceramic at different magnifications [44]. (a) Low magnification SEM image. (b) Higher magnification SEM image of a typical strut from (a).

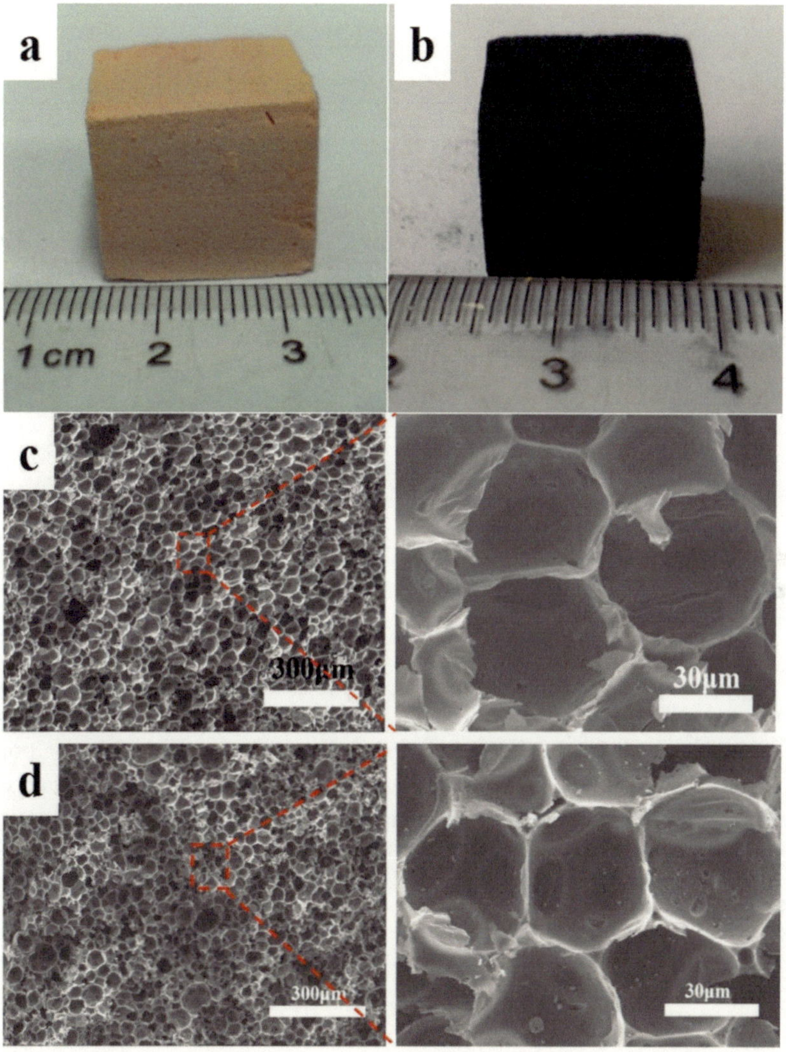

Figure 17. (a) Photographs of the foamed green body and (b) ZrC ceramic foam obtained from pyrolysis at 1600°C for 1 h. SEM images of (c) the preceramic foam and (d) ZrC foam after pyrolysis at 1600°C for 1 h in Ar [15].

Li et al. [45] developed another method for the preparation of stoichiometric ZrC foams by direct foaming of zirconia sols. Sucrose used as a carbon source was dissolved into zirconia sol to get a foamable sol. The stabilization of the wet foam was based on the gelation of the zirconia sol under thermal aging. The microstructure of the ZrC foams can be tailored by controlling the foaming parameters, such as the viscosities of the sols, the concentration of the blowing and curing agents. The ZrC foams prepared with open- or closed-cell structures are shown in **Figure 18**. The densities of the stoichiometric ZrC foams were in the range of 0.12–0.53 g/cm^3.

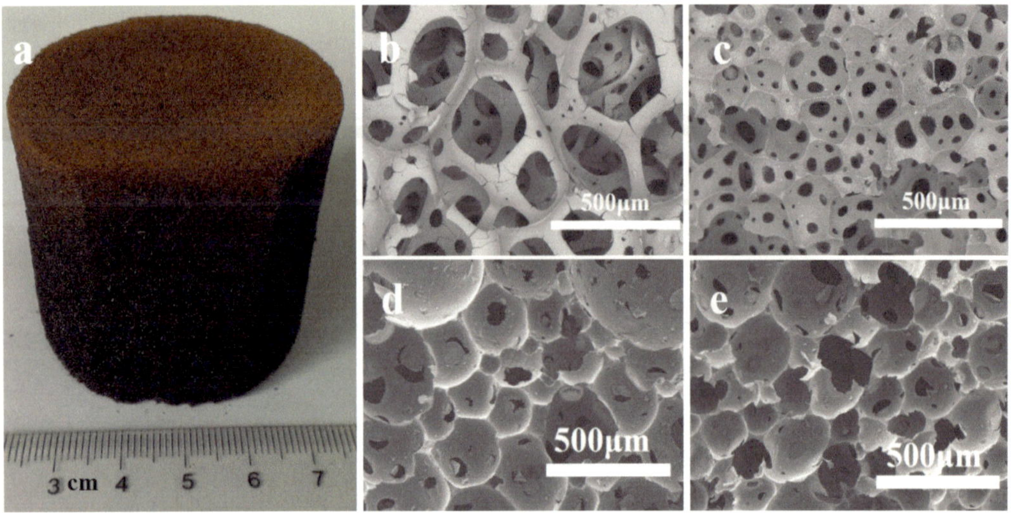

Figure 18. (a) Photograph of the preceramic foam prepared from direct foaming of zirconia sol. SEM images of the preceramic foams after being pyrolyzed at 1500°C foaming by precursor sol with various viscosities. (b), 1552 cP; (c), 2025 cP; (d), 2699 cP; (e), 2954 cP [45].

Figure 19 shows compressive stress-strain curves for ZrC foams prepared from direct foaming of zirconia sols [45]. In compression, all show a linear-elastic regime followed by a plateau of roughly constant stress, leading into a final regime of steeply rising stress. Upon compression, the foam undergoes a progressive collapse of the cells, with the lower part of the foam remains completely undamaged. In the plateau region, the stress is assumed to be independent of the strain as part of the structure collapses, while other parts of the structure remain elastic. Beyond the plateau, densification takes place and the stress rises sharply as complete densification begins.

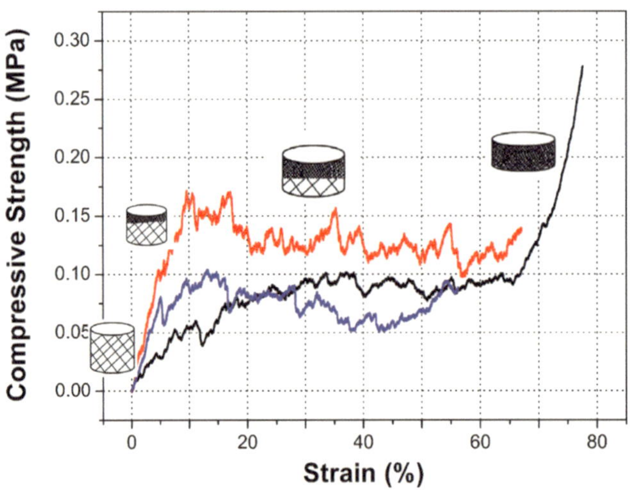

Figure 19. Typical compressive stress-strain response of ZrC foams prepared from direct foaming of zirconia sols [45].

6. Conclusions

A sol-gel technique has been applied to prepare UHTC-based materials using metal alkoxide and oxychloride as a metal precursor. First, sols of metal oxide are prepared by hydrolysis of a metal precursor with the modification of bidentate ligands (e.g., β-diketones, acetic acid, and cellulose acetate) to slow down the system's reactivity. Then soluble carbon- and boron-containing compounds, such as sucrose, phenolic resin, PAN, boric acid, and triethyl borate, are introduced to metal oxide sols. The hybrid sols are dried, cured, and heated at high temperatures to complete the conversion of metal oxides to nonoxides. UHTC nanopowders can be synthesized at 1300–1600°C under a protective atmosphere. The average size of the crystallites is controlled in the range of 60–200 nm by controlling the synthesis parameters. UHTC nanofibers can be synthesized by combining a sol-gel technique with spinning techniques, such as dry spinning and electrospinning. The as-synthesized UHTC nanofibers are porous and composed of nanoparticles. Porous UHTC materials with controllable cell structures can be prepared by the replica technique and direct-foaming methods. The as-prepared porous UHTC materials possess high porosities and can withstand 2400°C high temperature aging.

Although sol-gel routes provide great possibilities to produce UHTC-based materials with desired shapes, the high temperature thermal reduction process is still an indispensable procedure during which large mass loss occurs. It is still highly welcomed to synthesize UHTC precursors with high ceramic yield by directly converting into UHTC materials without the reduction process.

Acknowledgements

The authors express their gratitude for the financial support from the National Natural Science Foundation of China (no. 51602324).

Author details

Fei Li[1]*, Xiao Huang[1] and Guo-Jun Zhang[2]

*Address all correspondence to: lifei80380@163.com

1 Shanghai Institute of Ceramics, Chinese Academy of Sciences, Shanghai, China

2 Donghua University, Shanghai, China

References

[1] Wuchina E, Oplia E, Opeka M, Fahrenholtz W, Talmy I: UHTCs: ultra-high temperature ceramic materials for extreme environment applications. Interface. 2007;16:30–36.

[2] Paul A, Jayaseelan D, Venugopal S, Zapata-Solvas E, Binner J, Vaidhyanathan B, Heaton A, Brown P, Lee W: UHTC composites for hypersonic applications. American Ceramic Society Bulletin. 2012;91:22–28.

[3] Justin J, Jankowiak A: Ultra high temperature ceramics: densification, properties and thermal stability. Aerospacelab Journal. 2011;3:1–11.

[4] Wuchina E, Opeka M, Causey S, Buesking K, Spain J, Cull A, Routbort J, Guitierrez-Mora F: Designing for ultrahigh-temperature applications: the mechanical and thermal properties of HfB_2, HfC_x, HfN_x and αHf(N). Journal of Materials Science. 2004;39:5939–5949. DOI: 10.1023/b:jmsc.0000041690.06117.34.

[5] Fahrenholtz W, Hilmas G: Ultra-high temperature ceramics: materials for extreme environments. Scripta Materialia. 2017;129:94–99. DOI: http://dx.doi.org/10.1016/j.scriptamat.2016.10.018.

[6] Fahrenholtz W, Binner J, Zou J: Synthesis of ultra-refractory transition metal diboride compounds. Journal of Materials Research. 2016;31:2757–2772. DOI: 10.1557/jmr.2016.210.

[7] Fahrenholtz W, Hilmas G, Talmy I, Zaykoski J: Refractory diborides of zirconium and hafnium. Journal of the American Ceramic Society. 2007;90:1347–1364. DOI: 10.1111/j.1551-2916.2007.01583.x.

[8] Sacks M, Wang C, Yang Z, Jain A: Carbothermal reduction synthesis of nanocrystalline zirconium carbide and hafnium carbide powders using solution-derived precursors. Journal of Materials Science. 2004;39:6057–6066. 10.1023/B:JMSC.0000041702.76858.a7.

[9] Colombo P, Mera G, Riedel R, Sorarù G: Polymer-derived ceramics: 40 years of research and innovation in advanced ceramics. Journal of the American Ceramic Society. 2010;93:1805–1837. DOI: 10.1111/j.1551-2916.2010.03876.x.

[10] Dollé M, Gosset D, Bogicevic C, Karolak F, Simeone D, Baldinozzi G: Synthesis of nanosized zirconium carbide by a sol–gel route. Journal of the European Ceramic Society. 2007;27:2061–2067. DOI: 10.1016/j.jeurceramsoc.2006.06.005.

[11] Hench L, West J: The sol–gel process. Chemical Reviews. 1990;90:33–72. DOI: 10.1021/cr00099a003.

[12] Pierre A. Introduction to sol–gel processing. Springer Science & Business Media; Dordrecht, Netherland. 1998.

[13] Yan Y, Huang Z, Liu X, Jiang D: Carbothermal synthesis of ultra-fine zirconium carbide powders using inorganic precursors via sol–gel method. Journal of Sol–gel Science and Technology. 2007;44:81–85. DOI: 10.1007/s10971-007-1595-x.

[14] Hu Z,Sacks M, Staab G, Wang C, Jain A. Solution-based processing of nanocrystalline ZrC. in: 26th Annual Conference on Composites, Advanced Ceramics, Materials, and

Structures: B: Ceramic Engineering and Science Proceedings; John Wiley & Sons, Inc.; 2008. pp. 711–717.

[15] Li F, Kang Z, Huang X, Wang X-G, Zhang G-J: Preparation of zirconium carbide foam by direct foaming method. Journal of the European Ceramic Society. 2014;34:3513–3520. DOI: http://dx.doi.org/10.1016/j.jeurceramsoc.2014.05.029.

[16] Li F, Kang Z, Huang X, Zhang G-J: Synthesis of ZrB_2 nanofibers by carbothermal reduction via electrospinning. Chemical Engineering Journal. 2013;234:184–188. DOI: http://dx.doi.org/10.1016/j.cej.2013.08.100.

[17] Preiss H, Schierhorn E, Brzezinka K: Synthesis of polymeric titanium and zirconium precursors and preparation of carbide fibres and films. Journal of Materials Science. 1998;33:4697–4706. DOI: 10.1023/a:1004428818457.

[18] Preiss H, Berger L, Szulzewsky K: Thermal treatment of binary carbonaceous/zirconia gels and formation of Zr (C,O,N) solid solutions. Carbon. 1996;34:109–119. DOI: 10.1016/0008-6223(95)00147-6.

[19] Abe Y, Tomioka H, Gunji T, Nagao Y, Misono T: A one-pot synthesis of polyzirconoxane as a precursor for continuous zirconia fibres. Journal of Materials Science Letters. 1994;13:960–962. DOI: 10.1007/bf00701437.

[20] Abe Y, Kudo T, Tomioka H, Gunji T, Nagao Y, Misono T: Preparation of continuous zirconia fibres from polyzirconoxane synthesized by the facile one-pot reaction. Journal of Materials Science. 1998;33:1863–1870. DOI: 10.1023/a:1004357405815.

[21] Tao X, Qiu W, Li H, Zhao T: Synthesis of nanosized zirconium carbide from preceramic polymers by the facile one-pot reaction. Polymers for Advanced Technologies. 2010;21:300–304. DOI: 10.1002/pat.1664.

[22] Li R, Zhang Y, Lou H, Li J, Feng Z: Synthesis of ZrB_2 nanoparticles by sol–gel method. Journal of Sol–gel Science and Technology. 2011;58:580–585. DOI: 10.1007/s10971-011-2430-y.

[23] Zhang Y, Li R, Jiang Y, Zhao B, Duan H, Li J, Feng Z: Morphology evolution of ZrB_2 nanoparticles synthesized by sol–gel method. Journal of Solid State Chemistry. 2011;184:2047–2052. DOI: 10.1016/j.jssc.2011.05.040.

[24] Walker L, Corral E: Structural influence on the thermal conversion of self-catalyzed HfB_2/ZrB_2 sol–gel precursors by rapid ultrasonication of oxychloride hydrates. Journal of the American Ceramic Society. 2014;97:399–406. DOI: 10.1111/jace.12694.

[25] Wang M, Huang C, Wang Z: Polyzirconosilane preceramic resin as single source precursor of SiC–ZrC ceramics. Journal of Inorganic and Organometallic Polymers and Materials. 2016;26:24–31. DOI: 10.1007/s10904-015-0278-2.

[26] Xie Z, Deng X, Suo X, Zhou T, Gou Y: Synthesis and characterization of zirconium diboride precursor based on polycentric bridge bonds. Materials Chemistry and Physics. 2015;159:178–184. DOI: 10.1016/j.matchemphys.2015.03.068.

[27] Liu D, Qiu W, Cai T, Sun Y, Zhao A, Zhao T, Kleebe H: Synthesis, characterization, and microstructure of ZrC/SiC composite ceramics via liquid precursor conversion method. Journal of the American Ceramic Society. 2014;97:1242–1247. DOI: 10.1111/jace.12876.

[28] Cai T, Qiu W, Liu D, Han W, Ye L, Zhao A, Zhao T: Synthesis of soluble poly-yne polymers containing zirconium and silicon and corresponding conversion to nanosized ZrC/SiC composite ceramics. Dalton Transactions. 2013;42:4285–4290. DOI: 10.1039/c2dt32428h.

[29] Cai T, Qiu W, Liu D, Han W, Ye L, Zhao A, Zhao T: Synthesis, characterization, and microstructure of hafnium boride-based composite ceramics via preceramic method. Journal of the American Ceramic Society. 2013;96:1999–2004. DOI: 10.1111/jace.12270.

[30] Cai T, Qiu W, Liu D, Han W, Ye L, Zhao A, Zhao T: Synthesis of ZrC–SiC powders by a preceramic solution route. Journal of the American Ceramic Society. 2013;96:3023–3026. DOI: 10.1111/jace.12551.

[31] Wang X, Liu J, Hou F, Hu J, Sun X, Zhou Y, Koc R: Synthesis of ZrC-SiC powders from hybrid liquid precursors with improved oxidation resistance. Journal of the American Ceramic Society. 2015;98:197–204. DOI: 10.1111/jace.13274.

[32] Liang M, Li F, Ma X, Kang Z, Huang X, Wang X-G, Zhang G-J: Syntheses of ZrC–SiC nanopowder via sol–gel method. Ceramics International. 2016;42:1345–1351. DOI: http://dx.doi.org/10.1016/j.ceramint.2015.09.073.

[33] Clauß B: Fibers for ceramic matrix composites. Wiley-VCH Verlag GmbH & Co. KGaA; Weinheim, Germany. 2008. 1–20. DOI: 10.1002/9783527622412.ch1

[34] Hasegawa I, Fukuda Y, Kajiwara M: Inorganic–organic hybrid route to synthesis of ZrC and Si-Zr-C fibres. Ceramics International. 1999;25:523–527. DOI: 10.1016/s0272-8842(97)00089-8.

[35] Cui X, Nam Y, Lee J, Park W: Fabrication of zirconium carbide (ZrC) ultra-thin fibers by electrospinning. Materials Letters. 2008;62:1961–1964. DOI: 10.1016/j.matlet.2007.10.051.

[36] Nam Y, Cui X, Jeong L, Lee J, Park W: Fabrication and characterization of zirconium carbide (ZrC) nanofibers with thermal storage property. Thin Solid Films. 2009;517:6531–6538. DOI: 10.1016/j.tsf.2009.04.021.

[37] Ghelich R, Aghdam R, Torknik F, Jahannama M, Keyanpour-Rad M: Carbothermal reduction synthesis of ZrB_2 nanofibers via pre-oxidized electrospun zirconium n-propoxide. Ceramics International. 2015;41:6905–6911. DOI: http://dx.doi.org/10.1016/j.ceramint.2015.01.143.

[38] Ghelich R, Aghdam R, Torknik F, Jahannama M: Low temperature carbothermal reduction synthesis of ZrC nanofibers via cyclized electrospun PVP/Zr(OPr)4 hybrid. International Journal of Applied Ceramic Technology. 2016;13:352–358. DOI: 10.1111/ijac.12475.

[39] Li F, Kang Z, Huang X, Zhang G-J: Fabrication of zirconium carbide nanofibers by electrospinning. Ceramics International. 2014;40:10137–10141. DOI: 10.1016/j.ceramint.2014.02.011.

[40] Li D, Xia Y: Electrospinning of nanofibers: reinventing the wheel? Advanced Materials. 2004;16:1151–1170. DOI: 10.1002/adma.200400719.

[41] Li D, McCann J, Xia Y, Marquez M: Electrospinning: a simple and versatile technique for producing ceramic nanofibers and nanotubes. Journal of the American Ceramic Society. 2006;89:1861–1869. DOI: 10.1111/j.1551-2916.2006.00989.x.

[42] Studart A, Gonzenbach U, Tervoort E, Gauckler L: Processing routes to macroporous ceramics: a review. Journal of the American Ceramic Society. 2006;89:1771–1789. DOI: 10.1111/j.1551-2916.2006.01044.x.

[43] Colombo P, Bernardo E: Macro- and micro-cellular porous ceramics from preceramic polymers. Composites Science and Technology. 2003;63:2353–2359. DOI: http://dx.doi.org/10.1016/S0266-3538(03)00268-9.

[44] Rambo C, Cao J, Rusina O, Sieber H: Manufacturing of biomorphic (Si, Ti, Zr)-carbide ceramics by sol–gel processing. Carbon. 2005;43:1174–1183. DOI: http://dx.doi.org/10.1016/j.carbon.2004.12.009.

[45] Li F, Liang M, Ma X, Huang X, Zhang G-J: Preparation and characterization of stoichiometric zirconium carbide foams by direct foaming of zirconia sols. Journal of Porous Materials. 2015;22:493–500. DOI: 10.1007/s10934-015-9919-7.

Chapter 4

Sol-Gel Synthesis, Structural Characterization and Microwave Dielectric Properties of Bismuth Niobate Modified by Iron Inclusion

Susana Devesa, Manuel P.F. Graça and Luís C. Costa

Additional information is available at the end of the chapter

http://dx.doi.org/10.5772/intechopen.68211

Abstract

The current progress in communication technologies is leading to extensive studies on the development of miniaturized electronic devices with high electromagnetic performances, reliability, and low cost. Contributing to this purpose, the development and study of new materials, with promising electric properties in radio and microwave ranges, have been subject of our research in particular niobate-based materials. Bismuth niobate, $BiNbO_4$, is a low-firing ceramic that has been studied for a variety of applications in the microelectronic industry. In this work, the microwave dielectric characterization of $(Bi_{1-x}Fe_x)NbO_4$ ($0.00 \leq x \leq 1.00$) samples, prepared by the sol-gel method and heat treated at specific temperatures, is performed and related with their structure and morphology. The structural data were obtained by X-ray diffraction and Raman spectroscopy and the morphology by scanning electron microscopy. The dielectric characterization in the microwave region was made using the small perturbation theory, with a resonant cavity operating in TE_{105} mode, at the frequency of 2.7 GHz. The results show that the sol-gel method has the advantage of allowing the formation of α-$BiNbO_4$ phase at lower temperatures when compared with conventional preparation methods, and that the inclusion of iron inhibits the formation of low- and high-temperature β-$BiNbO_4$ phases.

Keywords: bismuth niobate, sol-gel, X-ray diffraction, microwaves, dielectric properties

1. Introduction: state of the art

In the modern communication systems, where microwave frequencies are used, wireless equipment are indispensable [1]. Therefore, the development of smaller, faster, and multifunctional equipment, with high electromagnetic performances and reliability, is an important technological challenge [2]. Since this goal can be achieved using ceramics technology, in the last decades a lot of research has been devoted to the improvement of dielectric ceramic materials that can be sintered at low temperature [3].

Bismuth oxides are recognized as low temperature-cofired ceramics (LTCC) and are considered promising materials for applications in the microelectronic industry [4]. In particular, bismuth niobate, $BiNbO_4$, has been studied to be used as resonators and filters in satellite and mobile communication systems [5].

The sintering temperatures of these microwave dielectrics cannot be high, since they need to cofire with low-loss conductors and low melting point electrodes, such as copper, silver, and gold [6]. Moreover, it is desirable that these ceramics have a high dielectric constant (ε'), a low dielectric loss (ε''), and a small coefficient of resonant frequency (τ_f) in the microwave range [7].

Bismuth niobate exists in two polymorph structures, an orthorhombic (α) phase (space group P_{nna}) and a triclinic (β) phase (space group P_1).

Figure 1 shows the crystal structures of (a) α-$BiNbO_4$ and (b) β-$BiNbO_4$. There are four formula units in the unit cell in both structures. The α structure consists of layers of vertex-sharing NbO_6 octahedra, which is normal to the *c* plane, with the Bi^{3+} positioned on the sites between these layers. The β phase presents pseudo layers of $[Bi_2O_2]$ units connected with each other and surrounded by sheets of formula $[NbO_4]n$ along the *bc* plane, in which the NbO_6 octahedra are joined at four vertices so that the two free oxygen atoms are in *c* axis. The Bi atoms hold these sheets together and are coordinated to eight oxygen atoms in the form of a much more distorted square antiprism. Summing up, the structure of α phase could be thought to be composed of two bismuth layers and two separated NbO_6 octahedral layers with different distortions. β-$BiNbO_4$ is composed of two bismuth layers and two joined octahedron layers [8, 9].

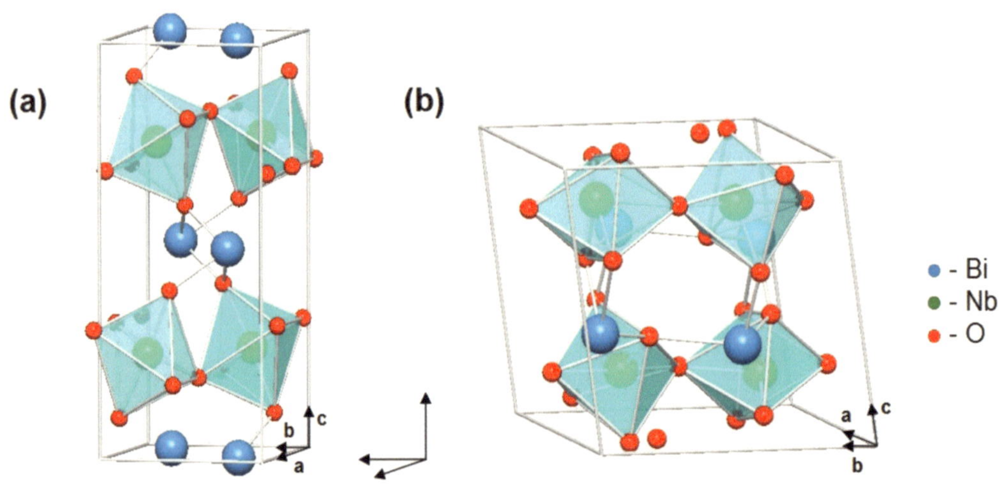

Figure 1. Schematic illustrations of crystal structures of (a) α-$BiNbO_4$ and (b) β-$BiNbO_4$.

http://dx.doi.org/10.5772/intechopen.68211

According to previous studies, the transition of low temperature β-$BiNbO_4$ to α-$BiNbO_4$ occurs at temperatures between 600 and 750°C, with the opposite, and supposed irreversible, transition occurring at 1020/1050°C [10, 11]. In 2007, Zhou et al. [8] reported the transformation from high temperature β-$BiNbO_4$ to α-$BiNbO_4$ in bulk samples; however, this transition is still considered irreversible in powder samples.

$BiNbO_4$ ceramics can have low sintering temperature; however, they are difficult to densify without sintering aids [5]. The first attempt to improve the bismuth niobate properties was developed by Kagata et al. [12] in 1992. They used V_2O_5 and CuO as additives, with the purpose of densifying the ceramic materials without compromising their dielectric properties. Since then, in order to obtain higher densification, better dielectric properties, and reduce the sintering temperature, various attempts have been undertaken to improve the bismuth niobate properties as a microwave dielectric ceramic [4]. Besides the addition of different oxides, such as CuO [6], ZnO [13], V_2O_5 [14, 15], PbO, Bi_2O_3 [16], and Fe_2O_3 [17], several researchers tried to improve bismuth niobate properties through the substitution of bismuth for metals or lanthanides. In order to accomplish this task, $(Bi_{1-x}R_x)NbO_4$ compositions were developed and studied, where R represents different metals, such as Fe [18] or Pb [19], and different lanthanides, such as Ce, Nd, Dy, Er, La, Sm, Ta, Gd, and La [20–25].

The solid-state reaction has been the most common method to process bismuth niobate [1, 7, 8], which means its benefits and drawbacks are already documented in the literature. $BiNbO_4$ powders prepared by solid-state reaction result in high crystallinity, but not always in pure and homogeneous phases, because some metastable phases can be obtained under the same conditions [25]. This method may cause other problems such as large grain growth, segregation of components, and loss of stoichiometry due to the volatilization of bismuth at a high temperature [10].

The wet-chemical techniques, such as co-precipitation, alkoxide and citrate sol-gel methods, and flux method, were found to present several advantages: nanosize particles, high reactivity, very good homogeneity in composition, and lower energy consumption, because they operate in mild temperatures. However, in these methods, metastable phases can also be formed [1, 7, 10].

Co-precipitation is considered one of the most successful techniques for the synthesis of ultrafine ceramic powders, having narrow particle size distribution. It is a less time-consuming method, since that can avoid complex steps such as refluxing of alkoxides. Nonetheless, an important drawback of co-precipitation is that all the cations should have similar solubility product [7, 10, 26].

The flux method is an alternative process that uses low-temperature liquid-phase reaction. According to previous studies, the $BiNbO_4$ prepared by the flux method exhibits higher crystallinity than that obtained by solid-state reaction; however, the successful application of this method depends on a large number of factors, like eutectic temperatures and solubility of solid components in the eutectic melts. Besides that, the decrease of the sintering temperature is not always achieved, since a proper combination of material/dopant is needed. In some cases, the mechanical separation between product and flux can be difficult [27, 28].

The sol-gel method, through citrate route, is considered a simple way to obtain stable precursors and stoichiometric fine powders which have been employed in the production of various oxides. This method is considered promising for achieving high homogeneity, since the formation of citrate complexes allows mixing the component cations in an atomic scale, thus they can immediately react with each other at relatively low temperatures [10, 13, 25].

In this work, the microwave dielectric properties of $(Bi_{1-x}Fe_x)NbO_4$ ($0.00 \leq x \leq 1.00$) samples, prepared by sol-gel method, through the citrate route, and heat-treated at specific temperatures, are studied. Since it is recognized that the behavior of functional materials is intrinsically linked to their structures [29], a structural characterization of the obtained samples was performed, using X-ray diffraction (XRD). Furthermore, to verify the structure consistency between low- and high-temperature β-$BiNbO_4$, and to better understand the phase transition process, the Raman spectroscopy was performed [10].

Since it is recognized that the dielectric properties can be related with density, porosity, and grain boundaries [20], the density of the samples was measured, using Archimedes method, and the morphology data was obtained by scanning electron microscopy (SEM).

2. Case study

2.1. Samples preparation

$(Bi_{1-x}Fe_x)NbO_4$ powders were prepared by the sol-gel method. Bismuth nitrate ($Bi(NO_3)_3{\cdot}5H_2O$), niobium chloride ($NbCl_5$), iron nitrate ($Fe(NO_3).9H_2O$), were used as starting materials and citric acid and ethylene glycol as chelating agent and reaction medium, respectively. All reagents used in the whole procedure were of analytical grade. The starting materials, with purities >99.9%, were supplied by Mateck and Merk and the chelating agent and reaction medium by Sigma-Aldrich. Bismuth nitrate, niobium chloride, and iron nitrate solutions were prepared in minor amounts of hydrogen peroxide (3%, V/V), combined in the corresponding molar ratio, and then added to citric acid and ethylene glycol. The obtained mixtures were stirred, to promote the solubility, until clear colloidal suspensions were obtained.

The obtained solutions were dried to evaporate the solvent: the solution of the reference sample, $x = 0.00$, was dried at 400°C, for 48 h; the solutions with $0.25 < x < 0.75$ were dried at 300°C, for 60 h; the solution with $x = 1.0$ was dried at 300°C, for 24 h. Subsequently, the obtained powders were thermally analyzed by differential thermal analysis (DTA), performed in a Linseis Apparatus type L92/095, in the temperature range of 20–1200°C, with a heating rate of 5 and 10°C/min, using Al_2O_3 as reference. Finally, the dry powders were pressed into cylinders and heat-treated, according to the DTA results, at temperatures between 500 and 1200°C, using a dwell time of 4 h, with a heating rate of 5°C/min.

2.2. Experimental methods

The structural characterization was performed using X-ray diffraction (XRD). The patterns data was obtained on a X'Pert MPD Philips diffractometer and on an Empyrean diffractometer (CuK α radiation, $\lambda = 1.54060$ Å) at 45 kV and 40 mA, in a Bragg–Brentano parafocusing optics configuration. The step counting method was used, with a step of 0.02° and a time per step of 1 s, with a 2θ angle range of 10–60°.

Raman spectroscopy measurements were performed at room temperature, in backscattering geometry, with a 441.6 nm laser line using a HR-800-UV Jobin Yvon Horiba spectrometer. A microscope objective focused the exciting light onto the sample.

The ceramics bulk density was measured using the Archimedes method in reference to water.

The morphology of the sintered samples was analyzed by scanning electron microscopy (SEM), using a TESCAN-Vega III. The samples were covered with carbon before microscopic observation.

The measurement of the complex permittivity was made using the small perturbation theory, with a cavity operating in TE_{105} mode, at resonant frequency of 2.7 GHz. These measurements were made using a HP 8753D coupled to the cavity resonator.

Microwave resonant cavities have been used for evaluating the dielectric properties of geometrically defined samples, when the cavity is calibrated with dimensionally identical sample of known permittivity.

For a given cavity and a sample of regular shape and well-defined dimensions, it is possible to determine the permittivity of the material. The permittivity values of a sample can be obtained through the changes in the resonant frequency, Δf, and in the inverse of the quality factor $\Delta(1/Q)$ of the resonant cavity, when introducing a sample in the cavity, where the electric field is maximal [30–33].

The shift in the resonant frequency of the cavity, Δf, can be related to the real part of the complex permittivity, ε', whereas the change in the inverse of the quality factor of the cavity, $\Delta(1/Q)$, can be related with the imaginary part, ε''.

Considering only the first-order perturbation in the electric field caused by the sample [30–33], we obtain

$$\varepsilon' = K\frac{\Delta f}{f_0}\frac{V}{v} + 1 \qquad (1)$$

$$\varepsilon'' = \frac{K}{2}\Delta\left(\frac{1}{Q}\right)\frac{V}{v} \qquad (2)$$

where K is a constant related to the depolarization factor, which depends upon the geometric parameters, v and V are the volumes of the sample and the cavity, respectively, and f_0 is the resonance frequency of the cavity. Using a sample of known permittivity, in our case polytetrafluorethylene (PTFE), we can determine the constant K.

2.3. Results and discussion

2.3.1. DTA

The DTA data of the powders with $x = 0.00$ ($BiNbO_4$) and $x = 1.00$ ($FeNbO_4$) are shown in **Figure 2**. In **Figure 2(a)**, one can see the occurrence of three exothermic peaks, at 649, 1043,

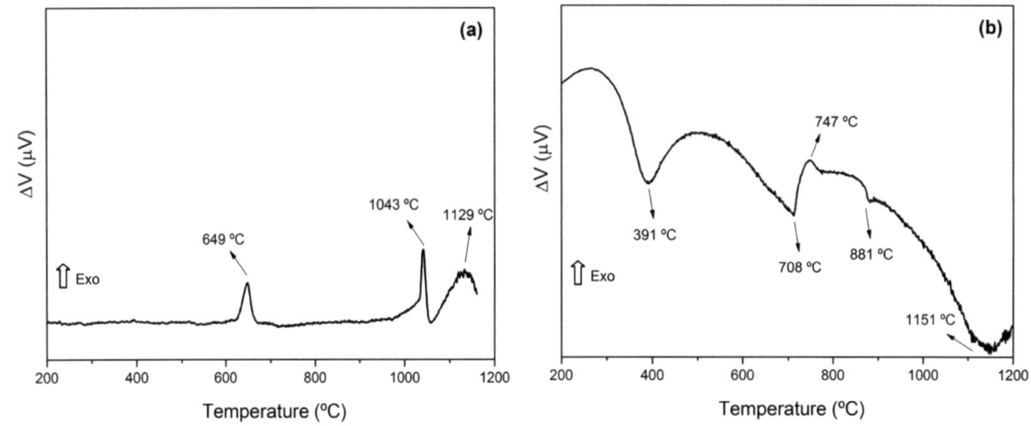

Figure 2. (a) DTA of $BiNbO_4$ powder, $x = 0.00$, heated at the rate of 10°C/min; (b) DTA of $FeNbO_4$ powder, $x = 1.00$, heated at the rate of 5°C/min.

and 1129°C. **Figure 2(b)** shows four endothermic phenomena at 391, 708, 881, and 1151°C. The peak that occurs at 391°C is due to the decomposition of residual organic compounds that were not released during the drying process.

The powders with $0.25 \leq x \leq 0.75$ were also thermally analyzed. When $x = 0.25$, three exothermic phenomena were identified, centered at 519, 637, and 1054°C. With $x = 0.50$, we can highlight once again three exothermic peaks, at 489, 556, and 637°C. The powder with $x = 0.75$ shows two exothermic peaks at 485 and 1157°C.

The heat treatments were made in agreement with these results, as shown in **Table 1**.

	Treatment temperatures (°C)								
x	500	650	800	850	1000	1050	1100	1150	1200
0.00	×	×	–	×	–	×	–	×	[a]
0.25	×	×	×	–	–	–	×	–	–
0.50	×	×	×	–	–	–	×	–	–
0.75	×	×	×	–	–	–	×	–	–
1.00	×	×	–	×	×	–	–	–	×

[a]The heat treatment performed at 1200°C promoted the fusion of the powders.

Table 1. Heat treatments performed in the cylindrical samples.

2.3.2. XRD pattern analysis

Figure 3 shows the X-ray diffraction patterns of $(Bi_{1-x}Fe_x)NbO_4$ powders, for $x = 0.00$, **Figure 3(a)**, and $x = 1.00$, **Figure 3(b)**.

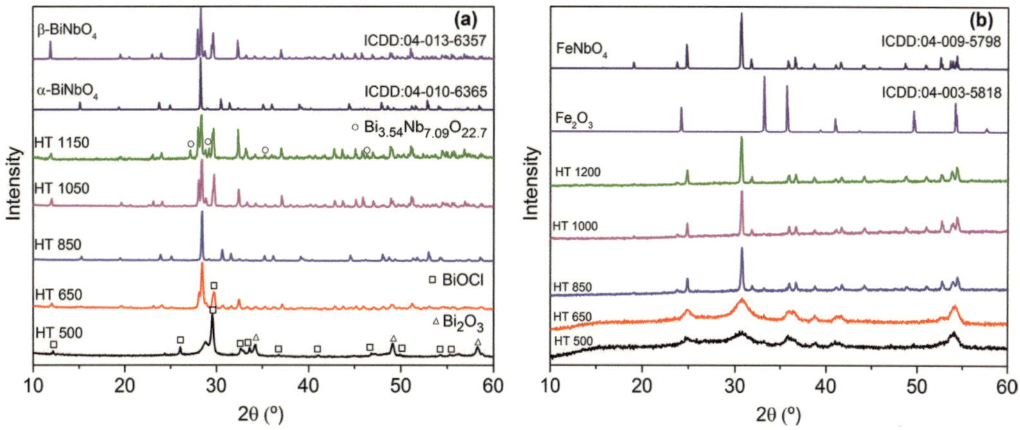

Figure 3. X-ray diffraction patterns of $(Bi_{1-x}Fe_x)NbO_4$ samples for (a) $x = 0.00$ and (b) $x = 1.00$.

Analyzing the diffraction patterns shown in **Figure 3(a)**, for the powders treated at 500°C, there are two secondary phases Bi_2O_3 and BiOCl, and the beginning of the β and α phases formation. The sample treated at 650°C is a combination of α- and β-$BiNbO_4$ phases, with the secondary phase BiOCl still present. After the treatment of the samples at 850°C, the X-ray diffraction pattern shows the formation of pure orthorhombic $BiNbO_4$ and, with a further temperature increase, this phase is converted to triclinic $BiNbO_4$. For the highest treatment temperature, the volatile nature of Bi^{3+} ions, at elevated temperatures, promotes the formation of a non-stoichiometric phase of bismuth niobate, $Bi_{3.54}Nb_{7.09}O_{22.7}$.

Therefore, the DTA exothermic band centered at 649°C can be due to the formation of α-$BiNbO_4$, and the exothermic reaction occurring at 1043°C can be attributed to the conversion of α-$BiNbO_4$ to β-$BiNbO_4$. The thermal treatment at 1200°C promoted the fusion of the samples; consequently, there is no exact information about the exothermic phenomenon at 1181°C. However, the formation of a non-stoichiometric phase, due to the volatile nature of bismuth [3], is predictable.

In **Figure 3(b)**, it is observed the formation of monoclinic $FeNbO_4$, even at the lowest heat-treatment temperature. For the treatment temperatures from 500 to 1000°C, one additional phase was detected, Fe_2O_3, with a content equal or inferior at 5%. For the 1200°C heat treatment, single-phase $FeNbO_4$ sample was obtained. Therefore, the DTA exothermic phenomena appearing at 747°C can be related with an improvement of the crystallization process of $FeNbO_4$ and the phenomena visible at 1151°C can be attributed to the decomposition of the Fe_2O_3, and consequent formation of $FeNbO_4$, with the niobium still present in the amorphous form. The increase of the amplitude and decrease of the half-width of the obtained XRD peaks, and of the Raman bands, shows that the increase of the heat treatment temperature promotes an increase of the crystallinity.

The samples with $0.25 \leq x \leq 0.75$ were also characterized using X-ray diffraction, with the diffraction patterns shown in **Figure 4** and the present phases identified in **Table 2**. Besides the

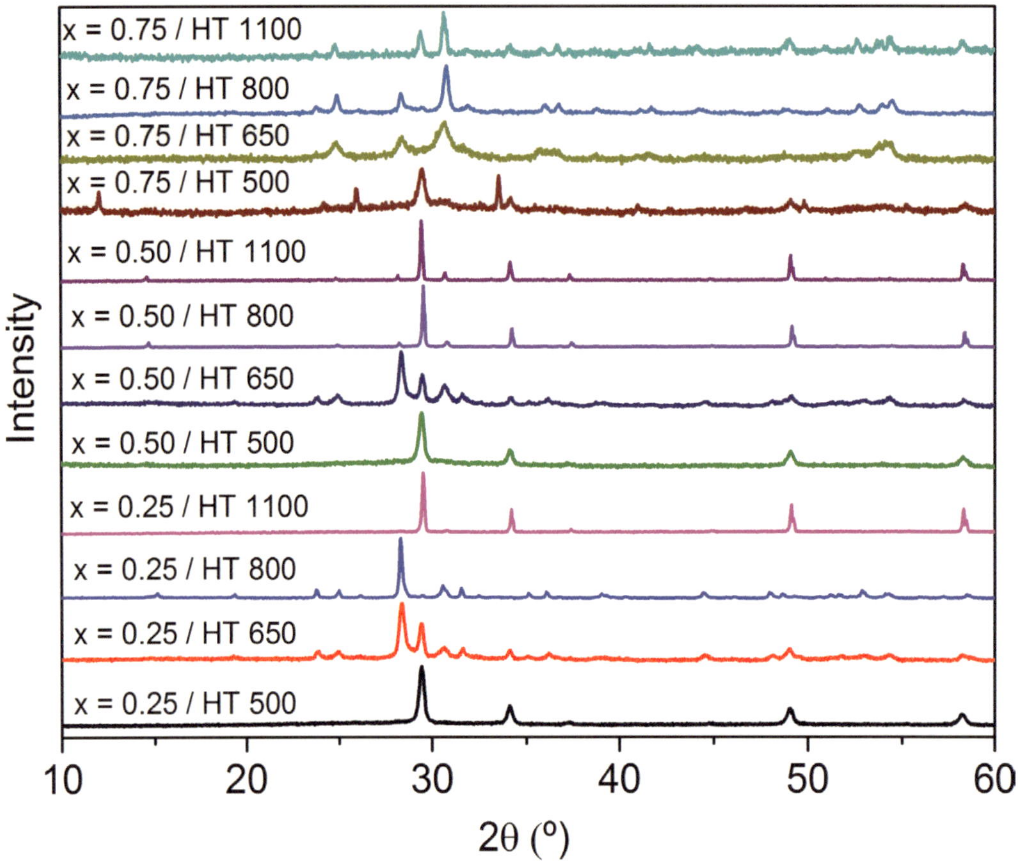

Figure 4. X-ray diffraction patterns of $(Bi_{1-x}Fe_x)NbO_4$ samples for $0.25 \leq x \leq 0.75$.

	HT 500	HT 650	HT 800	HT 1100
$x = 0.25$	$Bi_{1.34}Fe_{0.66}Nb_{1.34}O_{6.35}$	α-$BiNbO_4$ $Bi_{1.34}Fe_{0.66}Nb_{1.34}O_{6.35}$	α-$BiNbO_4$ $FeNbO_4$ Bi_2O_3	$Bi_{1.721}Fe_{1.056}Nb_{1.134}O_7$ $BiFeO_3$
$x = 0.50$	$Bi_{1.34}Fe_{0.66}Nb_{1.34}O_{6.35}$	α-$BiNbO_4$ $Bi_{1.34}Fe_{0.66}Nb_{1.34}O_{6.35}$	$Bi_{1.721}Fe_{1.056}Nb_{1.134}O_7$ $FeNbO_4$	$Bi_{1.721}Fe_{1.056}Nb_{1.134}O_7$ $FeNbO_4$
$x = 0.75$	$Bi_{15}NbO_{25}$ Fe_2O_3 $Fe_{0.33}Nb_{0.67}O_2$	$FeNbO_4$ Bi_2O_3	$FeNbO_4$ Bi_2O_3	$FeNbO_4$ $Bi_{15}NbO_{25}$

Table 2. Identification of the crystalline phases of $(Bi_{1-x}Fe_x)NbO_4$ samples for $0.25 \leq x \leq 0.75$.

expected phases like α-$BiNbO_4$ and $FeNbO_4$, there was the formation of two non-stoichiometric phases, $Bi_{1.34}Fe_{0.66}Nb_{1.34}O_{6.35}$ and $Bi_{1.721}Fe_{1.056}Nb_{1.134}O_7$, where the substitution of Bi^{3+} for Fe^{3+} ions was successful, and five more secondary phases. The absence of β-$BiNbO_4$ in the samples treated at 1100°C was unexpected, since the transition of α- to β-$BiNbO_4$ occurs, according to previous works, at 1020/1050°C [10, 11], and happened in the host sample at 1043°C.

Confronting the X-ray diffraction data with the DTA, it is possible to assume that for the samples with $x = 0.25$ and 0.50, the exothermic peaks can be related with the formation of the $Bi_{1.34}Fe_{0.66}Nb_{1.34}O_{6.35}$, α-$BiNbO_4$, and $Bi_{1.721}Fe_{1.056}Nb_{1.134}O_7$ phases, respectively. Since these three peaks are dislocated for lower temperatures in the sample with $x = 0.50$, it means that the energy required for these phase transformations decreases with the increase of the amount of iron. Considering the sample with $x = 0.75$, the peak occurring at 485°C can be assigned to the formation of $Bi_{15}NbO_{25}$ and the one centered at 1157°C should be related to a non-stoichiometric phase formation.

In the present study, one can conclude that the sol-gel method has the advantage of forming the orthorhombic $BiNbO_4$ phase at even lower temperatures, since the sample with $x = 0.00$, treated at 500°C, presents already the α phase, and that the inclusion of iron inhibits the formation of low- and high-temperature triclinic bismuth niobate.

2.3.3. *Raman spectroscopy analysis*

The Raman spectroscopy data of $(Bi_{1-x}Fe_x)NbO_4$ powders, for $x = 0.00$ and $x = 1.00$, are presented in **Figure 5**.

In **Figure 5(a)**, we can see that the Raman spectrum of the sample treated at 850°C, pure orthorhombic $BiNbO_4$, is less complex than the spectra were the triclinic $BiNbO_4$ is present. This result was expected since the lowered symmetry of the triclinic phase produces more Raman active modes [10]. The samples treated at 500, 650, 1050, and 1150°C possess peaks at 102, 450, and 687 cm^{-1} attributed to the vibration modes of β-$BiNbO_4$. This means that both low and high temperature β-$BiNbO_4$ have identical vibration modes. In the sample treated at 850°C, where only α-$BiNbO_4$ was detected, these peaks are absent. Furthermore, the peaks at 139 and 199 cm^{-1} are distinctive of the orthorhombic $BiNbO_4$, which explains their absence in the spectra of the samples treated at 1050 and 1150°C. All the peaks present in these two samples are attributed to triclinic $BiNbO_4$ [10, 34]. No vibration that can be assigned to non-stoichiometric phase $Bi_{3.54}Nb_{7.09}O_{22.7}$ was detected.

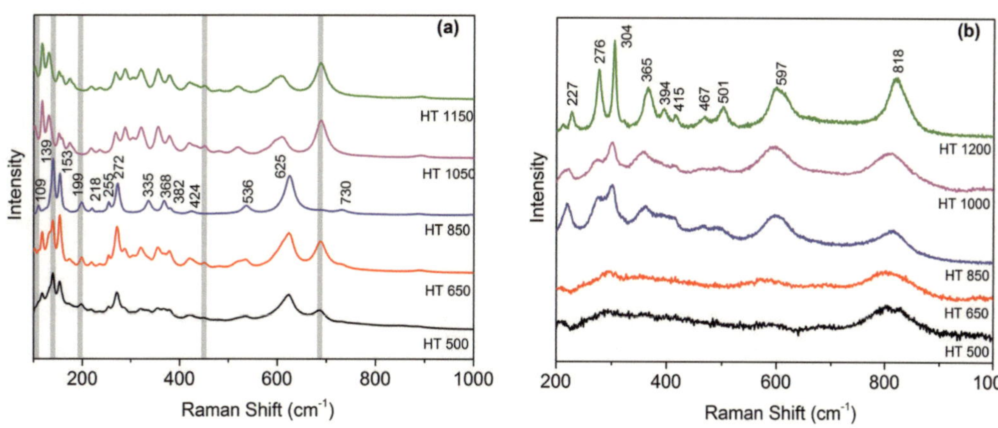

Figure 5. Raman spectra of $(Bi_{1-x}Fe_x)NbO_4$ samples for (a) $x = 0.00$ and (b) $x = 1.00$.

In the spectra presented in **Figure 5(b)**, all the detected bands are attributed vibrations of the base units of the monoclinic $FeNbO_4$ phase. The Nb—O stretching bands were observed at 276, 394, 597, and 818 cm^{-1}, and the bands visible at 365 and 467 cm^{-1} are exclusively for the monoclinic $FeNbO_4$ [35–38]. With the increase of the temperature, it can be seen a better definition of bands, which is consistent with the increase of the crystallinity of the samples.

The Raman spectroscopy data of $(Bi_{1-x}Fe_x)NbO_4$ powders, with $0.25 \leq x \leq 0.75$, are presented in **Figure 6**. Regardless of the x value, the obtained data shows that the crystallinity increases with the temperature.

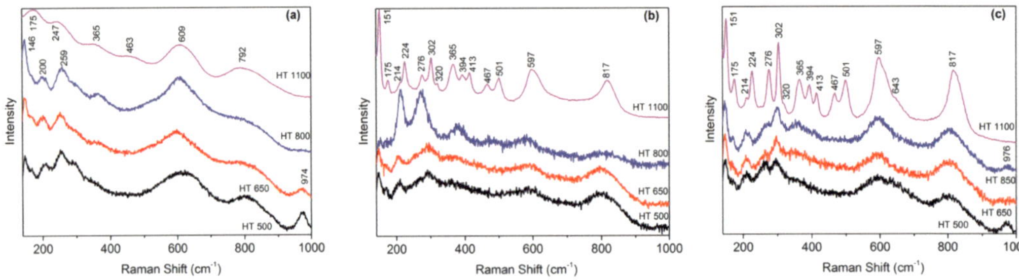

Figure 6. Raman spectra of $(Bi_{1-x}Fe_x)NbO_4$ samples for (a) $x = 0.25$, (b) $x = 0.50$, and (c) $x = 0.75$.

In the samples with $x = 0.25$ treated at 650 and 850°C and $x = 0.50$ treated at 650°C, the peak occurring at 200 cm^{-1} is distinctive of the orthorhombic $BiNbO_4$. On the other hand, the peaks visible at 224, 276, 304, 365, 394, 467, 501, 597, and 817 can be assigned to $FeNbO_4$.

2.3.4. *Density*

The bulk density values of $(Bi_{1-x}Fe_x)NbO_4$ ceramics, for $x = 0.00$ and $x = 1.00$, in reference to water versus treatment temperature are shown in **Figure 7**.

The bulk density increase with the treatment temperature increase is clearly visible for both the series of samples, and it allows to predict that porosity decreases at higher treatment temperatures.

The theoretical density of β-$BiNbO_4$ ceramic (7.5 g/cm^3) is higher than that of α-$BiNbO_4$ ceramic (7.345 g/cm^3) [21, 22]. Thus, and considering the samples with $x = 0.00$, the increase of bulk density of $BiNbO_4$ ceramics from 850 to 1050°C can be due to the orthorhombic-triclinic phase transformation of $BiNbO_4$. According to the literature, it is very difficult to densify the ceramics above 85% of theoretical density without sintering additives [24]. However, in the sample treated at 1150°C, the density is 85.3% of the theoretical density and that was slightly above the pointed level.

Considering now the samples with $x = 1.00$, and knowing that the theoretical density of $FeNbO_4$ ceramic is 5.40 g/cm^3 [39], we can determine that for the highest heat treatment, the only one where pure $FeNbO_4$ was formed, the density is 92.6% of the theoretical density.

For the remaining samples, the evolution of the bulk density values with the treatment temperature is shown in **Figure 8**. In this case, we can see the same trend for this set of samples.

http://dx.doi.org/10.5772/intechopen.68211

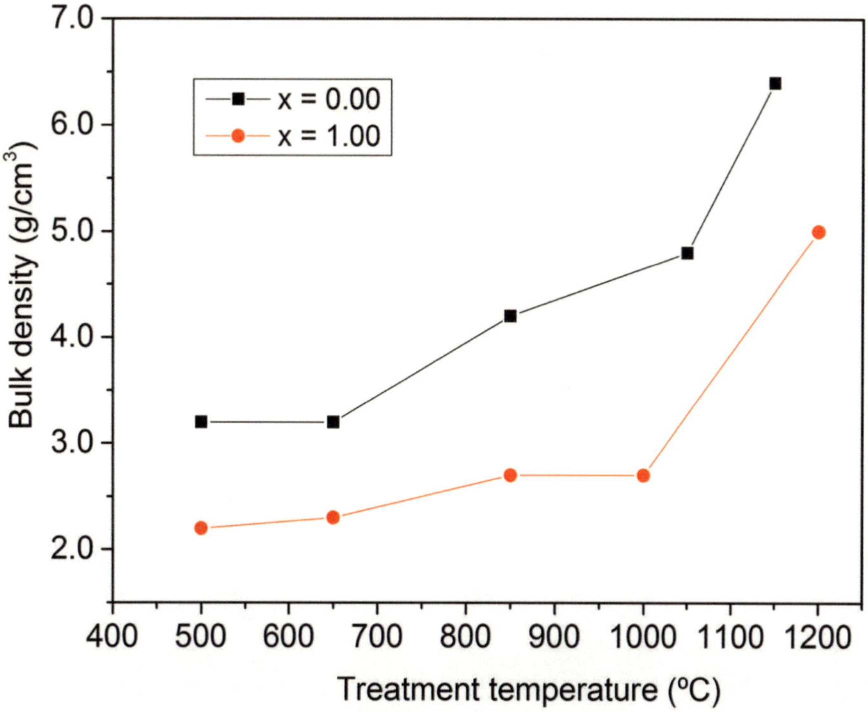

Figure 7. Bulk density of $(Bi_{1-x}Fe_x)NbO_4$ samples for $x = 0.00$ and $x = 1.00$.

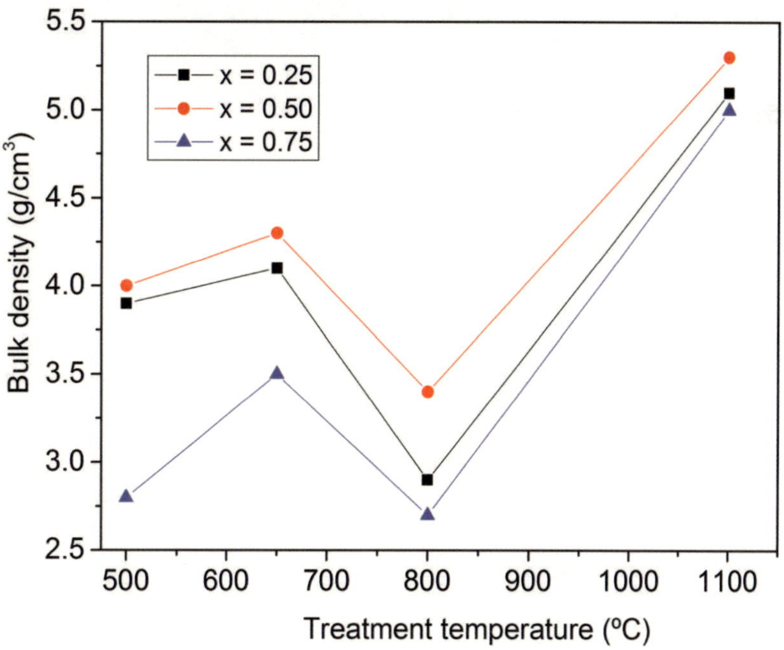

Figure 8. Bulk density of $(Bi_{1-x}Fe_x)NbO_4$ samples for $x = 0.25$, $x = 0.50$, and $x = 0.75$.

The bulk density increases when the treatment temperature increases for 650°C, decreases for the treatment at 800°C, reaching a minimum value, and increases again with the higher treatment temperature. Regardless of the heat treatment, the samples with higher bulk density are the ones with $x = 0.50$.

2.3.5. *Morphological characterization*

In **Figure 9**, SEM micrographs of the samples with $x = 0.25$ and 0.50, treated at 650°C, are shown. In these samples, where the same phases were identified, we can see particles with spherical shape and diameter of 0.2 μm. Increasing the amount of iron, a decrease in porosity is detectable, a result that is coherent with the increase in density, previously analyzed.

Figure 9. SEM micrographs of the samples with $x = 0.25$ and $x = 0.50$, treated at 650°C.

2.3.6. *Microwave dielectric properties*

In **Figure 10** is presented the transmission results obtained with the 2.7 GHz cavity, for the cases of the cavity only with the sample holder, with the sample holder filled with PTFE and filled with the samples with $0.25 \leq x \leq 0.75$, treated at 650°C. The expected shift in the resonant frequency of the cavity, Δf, is observed.

Since this shift, Δf, can be related to the real part of the complex permittivity, ε', it is expected that the sample with $x = 0.75$ presents the lowest ε', followed by the sample with $x = 0.25$ and finally the sample with $x = 0.50$, with the highest dielectric constant of this series of samples. This prediction will be confirmed later.

Figure 11 shows the real and the imaginary parts of the complex permittivity at 2.7 GHz, as function of the treatment temperatures, for the samples with $x = 0.00$ and $x = 1.00$.

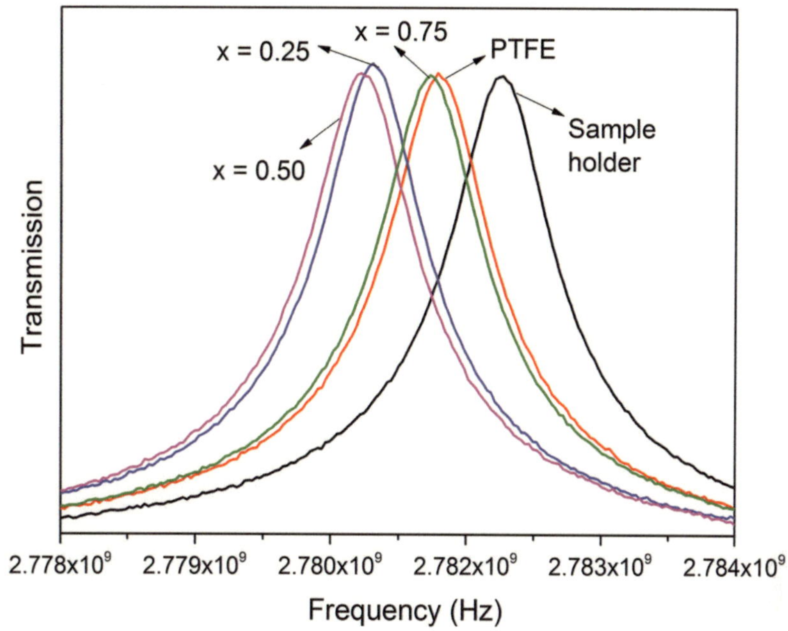

Figure 10. Transmission of the unperturbed and perturbed 2.7 GHz cavity, for $0.25 \leq x \leq 0.75$ $(Bi_{1-x}Fe_x)NbO_4$ samples, treated at 650°C.

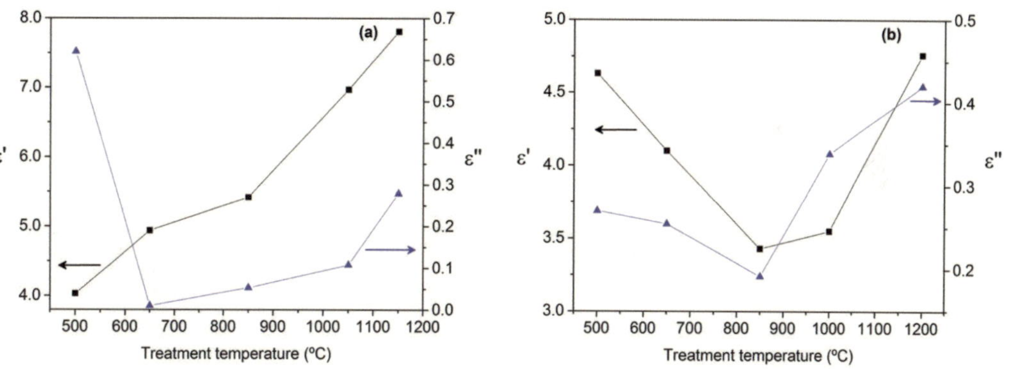

Figure 11. Calculated ε′ and ε″ of $(Bi_{1-x}Fe_x)NbO_4$ samples for (a) $x = 0.00$ and(b) $x = 1.00$, at room temperature.

For the samples with $x = 0.00$, **Figure 11(a)** shows that ε′ values increase with the treatment temperature. The same trend was previously seen for the bulk density, suggesting that the dielectric constant increase with temperature is due to the density increase. With the exception of the sample treated at 500°C, where $BiNbO_4$ presence is still reduced, the dielectric losses also show the same evolution, increasing with the treatment temperature.

Figure 11(b) shows that for the samples with $x = 1.00$, the dielectric constant and the dielectric losses follow the same trend.

In **Figure 12**, it is presented the dielectric constant for the samples with $0.25 \leq x \leq 0.75$, as function of the treatment temperatures. Once more, the experimental data suggests that the increase of the bulk density is determinant for the increase of the dielectric constant. Nevertheless, the samples treated at 1100°C, besides the higher ε' values, also show the higher losses.

Table 3 resumes the calculated ε' and ε'' for the different samples.

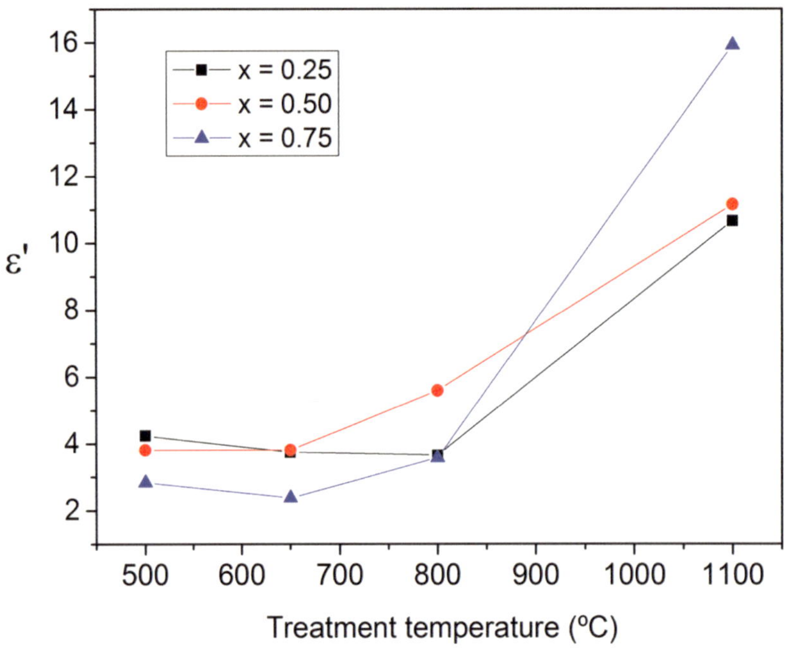

Figure 12. Calculated ε' for $0.25 \leq x \leq 0.75$ $(Bi_{1-x}Fe_x)NbO_4$ samples, at room temperature.

x		HT 500	HT 650	HT 800	HT 850	HT 1000	HT 1050	HT 1100	HT 1150	HT 1200
0.00	ε'	4.03	4.94	–	5.42	–	6.97	–	7.81	–
	ε''	0.62	0.01	–	0.05	–	0.11	–	0.28	–
0.25	ε'	4.24	3.74	3.65	–	–	–	10.66	–	–
	ε''	0.08	0.06	0.07	–	–	–	0.22	–	–
0.50	ε'	3.81	3.81	5.59	–	–	–	11.15	–	–
	ε''	0.09	0.09	0.15	–	–	–	0.35	–	–
0.75	ε'	2.83	2.37	3.57	–	–	–	15.91	–	–
	ε''	0.08	0.03	0.15	–	–	–	2.59	–	–
1.00	ε'	4.63	4.10	–	3.43	3.55	–	–	–	4.76
	ε''	0.27	0.25	–	0.19	0.34	–	–	–	0.42

Table 3. Calculated ε' and ε'' for different samples.

2.4. Conclusion

We can conclude that the $(Bi_{1-x}Fe_x)NbO_4$ ceramic powders were prepared successfully by the sol-gel method.

For $x = 0.00$, this method allows the formation of orthorhombic $BiNbO_4$ at lower temperatures than other preparation techniques. Both α and β phases can be stable in different temperature regions, below 850 and above 1050°C, respectively.

Furthermore, the density of the obtained samples was very promising, since no additives were used in the host samples.

The substitution of bismuth by iron was successful, since the two non-stoichiometric phases, $Bi_{1.34}Fe_{0.66}Nb_{1.34}O_{6.35}$ and $Bi_{1.721}Fe_{1.056}Nb_{1.134}O_7$, were obtained. Moreover, one can conclude that the inclusion of iron inhibits the formation of low- and high-temperature triclinic bismuth niobate.

Concerning the dielectric characterization, the small perturbation technique proved to be very practical for dielectric measurements on ceramic materials, featuring important advantages, as the easy sample preparation, the absence of measurement contacts, and the simplicity of the calculations.

The dielectric study revealed that, for $x = 0.00$, the dielectric constant increases with the increase of the sintering temperature, with the dielectric losses showing the same trend.

With the inclusion of iron, the highest ε' values were obtained for the highest treatment temperature; however, only the samples with $x = 0.25$ and 0.50 showed ε'' acceptable values.

Author details

Susana Devesa*, Manuel P.F. Graça and Luís C. Costa

*Address all correspondence to: susana_devesa@hotmail.com

I3N and Physics Department, University of Aveiro, Aveiro, Portugal

References

[1] Filho R.C., Araújo J.H., Ginani M.F., d'Assunção A.G., Martins R.A., d'Assunção A.G., Mendonça L.M. Simulation and measurement of inset-fed microstrip patch antennas on $BiNbO_4$ substrates. Microwave and Optical Technology Letters. 2010;**52**(5):1034–1036. DOI: 10.1002/mop.25098

[2] Sebastian M.T., Jantunen H. Low loss dielectric materials for LTCC applications: a review. International Materials Reviews. 2008;**53**(2):57–90. DOI: 10.1179/174328008X277524

[3] Płońska M., Czekaj D. Studies of temperature and fabrication methods influence on structure and microstructure of $BiNbO_4$ microwave electroceramics. Archives of Metallurgy and Materials. 2011;**56**(4):1169–1175. DOI: 10.2478/v10172-011-0131-8

[4] Kim E.S., Choi W. Effect of phase transition on the microwave dielectric properties of $BiNbO_4$. Journal of the European Ceramic Society. 2006;**26**(10–11):1761–1766. DOI: 10.1016/j.jeurceramsoc.2005.09.003

[5] Liou Y.C., Tsai W.C., Chen H.M. Low-temperature synthesis of $BiNbO_4$ ceramics using reaction-sintering process. Ceramics International. 2009;**35**(6):2119–2122. DOI: 10.1016/j.ceramint.2008.11.030

[6] Sales A.J.M., Oliveira P.W.S., Almeida J.S., Costa M.M., Rodrigues H.O., Sombra A.S.B. Copper concentration effect in the dielectric properties of $BiNbO_4$ for RF applications. Journal of Alloys and Compounds. 2012;**522**:264–270. DOI: 10.1016/j.jallcom.2012.07.025

[7] Radha R., Gupta U.N., Samuel V., Muthurajan H., Kumar H.H., Ravi V. A co-precipitation technique to prepare $BiNbO_4$ powders. Ceramics International. 2008;**34**(6):1565–1567. DOI: 10.1016/j.ceramint.2007.03.022

[8] Zhou D., Wang H., Yao X., Wei X., Xiang F., Pang L. Phase transformation in $BiNbO_4$ ceramics. Applied Physics Letters. 2007;**90**(17):2910. DOI: 10.1063/1.2732833

[9] Zhai H.F., Qian X., Kong J.Z., Li A.D., Gong Y.P., Li H., et al. Abnormal phase transition in $BiNbO_4$ powders prepared by a citrate method. Journal of Alloys and Compounds. 2011;**509**(42):10230–10233. DOI: 10.1016/j.jallcom.2011.08.077

[10] Radha R., Muthurajan H., Rao N.K., Pradhan S., Gupta U.N., Jha R.K., Mirji S.A., et al. Low temperature synthesis and characterization of $BiNbO_4$ powders. Materials Characterization. 2008;**59**(8):1083–1087. DOI: 10.1016/j.matchar.2007.08.026

[11] Xu C., He D., Liu C., Wang H., Zhang L., Wang P., et al. High pressure and high temperature study the phase transitions of $BiNbO_4$. Solid State Communications. 2013;**156**:21–24. DOI: 10.1016/j.ssc.2012.11.007

[12] Kagata H., Inoue T., Kato J., Kameyama I. Low-fire bismuth-based dielectric ceramics for microwave use. Japanese Journal of Applied Physic. 1992;**31**:3152–3155. DOI:10.1143/JJAP.31.315

[13] Wang N., Zhao M.Y., Yin Z.W., Li W. Low-temperature synthesis of β-$BiNbO_4$ powder by citrate sol–gel method. Materials Letters. 2003;**57**(24–25):4009–4013. DOI: 10.1016/S0167-577X(03)00256-8

[14] Tzou W.C., Yang C.F., Chen Y.C., Cheng P.S. Improvements in the sintering and microwave properties of $BiNbO_4$ microwave ceramics by V_2O_5 addition. Journal of the European Ceramic Society. 2000;**20**(7):991–996. DOI: 10.1016/S0955-2219(99)00228-9

[15] Zhou D., Wang H., Yao X., Liu Y. Microwave dielectric properties of low-firing $BiNbO_4$ ceramics with V_2O_5 substitution. Journal of Electroceramics. 2008;**21**(1–4):469–472. DOI: 10.1007/s10832-007-9223-2

[16] Almeida J.S., Fernandes T.S.M., Sales A.J.M., Silva M.A.S., Júnior G.P., Rodrigues H.O., et al. Study of the structural and dielectric properties of Bi_2O_3 and PbO addition on $BiNbO_4$ ceramic matrix for RF applications. Journal of Materials Science: Materials in Electronics. 2011;**22**(8):978–987. DOI: 10.1007/s10854-010-0247-z

[17] Yang Y., Ding S., Yao X. Influences of Fe_2O_3 additives on the dielectric properties of $BiNbO_4$ ceramics under different sintering atmosphere. Ceramics International. 2004;**30**(7):1341–1345. DOI: 10.1016/j.ceramint.2003.12.110

[18] Devesa S., Graça M.P., Henry F., Costa L.C. Microwave dielectric properties of $(Bi_{1-x}Fe_x)NbO_4$ ceramics prepared by the sol–gel method. Ceramics International. 2015;**41**(6):8186–8190. DOI: 10.1016/j.ceramint.2015.03.038

[19] Butee S.P., Kambale K.R., Upadhyay S., Bashaiah S., Raju K.J., Panda H. Synthesis and microwave dielectric behavior of $(Bi_{1-x}Pb_x)NbO_4$ ceramics. Journal of Advanced Dielectrics. 2016;**6**(01):1650006. DOI: 10.1142/S2010135X16500065

[20] Shihua D., Xi Y., Yu M., Puling L. Microwave dielectric properties of $(Bi_{1-x}R_x)NbO_4$ ceramics (R= Ce, Nd, Dy, Er). Journal of the European Ceramic Society. 2006;**26**(10):2003–2005. DOI: 10.1016/j.jeurceramsoc.2005.09.049

[21] Wang N., Zhao M.Y., Yin Z.W., Li W. Effects of complex substitution of La and Nd for Bi on the microwave dielectric properties of $BiNbO_4$ ceramics. Materials Research Bulletin. 2004;**39**(3):439–448. DOI: 10.1016/j.materresbull.2003.10.016

[22] Tzou W.C., Yang C.F., Chen Y.C., Cheng P.S. Microwave dielectric characteristics of $(Bi_{1-x}Sm_x)NbO_4$ ceramics. Ceramics international. 2002;**28**(1):105–110. DOI: 10.1016/S0272-8842(01)00064-5

[23] Wang N., Zhao M.Y., Yin Z.W. Effects of Ta_2O_5 on microwave dielectric properties of $BiNbO_4$ ceramics. Materials Science and Engineering: B. 2003;**99**(1):238–242. DOI: 10.1016/S0921-5107(02)00464-6

[24] Butee S., Kulkarni A.R., Prakash O., Aiyar R.P.R.C., Sudheendran K., Raju K.J. Effect of lanthanide ion substitution on RF and microwave dielectric properties of $BiNbO_4$ ceramics. Journal of Alloys and Compounds. 2010;**492**(1):351–357. DOI: 10.1016/j.jallcom.2009.11.093

[25] Almeida C.G., Andrade H.M.C., Mascarenhas A.J.S., Silva L.A. Synthesis of nanosized β-$BiTaO_4$ by the polymeric precursor method. Materials Letters. 2010;**64**(9):1088–1090. DOI: 10.1016/j.matlet.2010.02.020

[26] Gaikwad A.B., Navale S.C., Samuel V., Murugan A.V., Ravi V. A co-precipitation technique to prepare $BiNbO_4$, $MgTiO_3$ and $Mg_4Ta_2O_9$ powder. Materials Research Bulletin. 2006;**41**(2):347–353. DOI:10.1016/j.materresbull.2005.08.01

[27] Maruyama Y., Izawa C., Watanabe T. Synthesis of by the flux method. ISRN Materials Science. 2012;**2012**:170362. DOI: 10.5402/2012/170362

[28] Shlyakhtin O.A., Orlov A.V., Oh Y.J. Liquid phase low temperature sintering of niobate and cerate fine powders. Journal of Electroceramics. 2006;**17**(2–4):405–413. DOI: 10.1007/s10832-006-9711-9

[29] Siqueira K.P., Moreira R.L., Dias A. Synthesis and crystal structure of lanthanide orthoniobates studied by vibrational spectroscopy. Chemistry of Materials. Chemistry of Materials. 2010;**22**(8):2668–2674. DOI: 10.1021/cm100173p

[30] Henry F., Berteaud A.J. New measurement technique for the dielectric study of solutions and suspensions. Journal of Microwave Power. 1980;**15**(4):233–242.

[31] Henry F. Contribution a l'etude des processus d'hydratation [thesis]. France: Universite Pierre et Marie Curie; 1982.

[32] Costa L.C., Devesa S., André P., Henry F. Microwave dielectric properties of polybutylene terephtalate (PBT) with carbon black particles. Microwave and Optical Technology Letters. 2005;**46**(1):61–63. DOI: 10.1002/mop.20901

[33] Costa L.C., Henry F. The impact of blue inorganic pigments on the microwave electrical properties of polymer composites. International Journal of Microwave Science and Technology. 2012;**2012**:628237. DOI: 10.1155/2012/628237

[34] Devesa S., Graça M.P., Costa L.C. Structural, morphological and dielectric properties of $BiNbO_4$ ceramics prepared by the sol–gel method. Materials Research Bulletin. 2016;**78**:128–133. DOI: 10.1016/j.materresbull.2016.02.035

[35] Shim S.H., Duffy T.S. Raman spectroscopy of Fe_2O_3 to 62 GPa. American Mineralogist. 2002;**87**(2–3):318–326. DOI: 10.2138/am-2002-2-314

[36] Kumari T.S.D., Gandhi R.V., Rahul G., Kamalanathan G., Kumar T.P., Jeyakumar D., et al. Electrochemical lithium insertion behavior of $FeNbO_4$: Structural relations and in situ conversion into $FeNb_2O_6$ during carbon coating. Materials Chemistry and Physics. 2014;**145**(3):425–433. DOI: 10.1016/j.matchemphys.2014.02.031

[37] Cho I.S., Lee S., Noh J.H., Choi G.K., Jung H.S., Kim D.W., et al. Visible-light-induced photocatalytic activity in $FeNbO_4$ nanoparticles. The Journal of Physical Chemistry C. 2008;**112**(47):8393–18398. DOI: 10.1021/jp807006g

[38] Devesa S., Graça M.P., Henry F., Costa L.C. Dielectric properties of $FeNbO_4$ ceramics prepared by the sol–gel method. Solid State Sciences. 2016;**61**:44–50. DOI: 10.1016/j.solidstatesciences.2016.09.005

[39] Roth R.S., Waring J.L. Ixiolite + other polymorphic types of $FeNbO_4$. American Mineralogist. 1964;**49**(3–4):0003-004X.

Chapter 5

Fluorescence Properties of Rare-Earth-Doped Sol-Gel Glasses

Helena Cristina Vasconcelos and Afonso Silva Pinto

Additional information is available at the end of the chapter

http://dx.doi.org/10.5772/intechopen.68534

Abstract

Glasses may be prepared by sol-gel processing over a wide range of compositions and thick multilayer deposits may be used as waveguides for integrated optics. Doping these layers with rare-earth (RE) ions enables the fabrication of active devices for optical amplifiers; the incorporation of these ions into nanocrystallites offers possibilities for increased dopant concentration without fluorescence quenching, improved spectroscopic performance and high quantum yields. Rare-earth (RE) ions such as erbium (Er^{3+}), ytterbium (Yb^{3+}), neodymium (Nd^{3+}), thulium (Tm^{3+}), holmium (Ho^{3+}) and praseodymium (Pr^{3+}) have been widely used in optical applications and cover a range of wavelengths ranging from UV-visible to the near infrared. This chapter includes basic principles of fluorescence in RE doped glasses, fluorescence lifetimes, quantum yields and Judd-Ofelt analysis. A few information is given about the preparation and characterization of glasses, thin films and glass-ceramics (nanocrystallites embedded in glass matrix) prepared by sol-gel processing. The growth of nanocrystals in glassy sol-gel films through suitable heat treatments can avoid the influence of high phonon energy of silica glasses. The characterization of such materials can be evaluated by optical techniques, namely UV-Visible, FTIR, among other additional techniques that include Scanning Electron Microscopy (SEM), X-ray Diffraction (XRD) and Atomic Force Microscopy (AFM).

Keywords: sol-gel, rare earths, fluorescence, glass, nanocrystallites

1. Light, glasses, and rare earths

The need for high-speed data associated with the advance of telecommunication systems by optical transmission and fiber-optic connections has contributed to extend the optical regime to integrated circuits [1]; currently, waveguides and other devices necessary for a suitable operation of integrated optical circuits in transmission systems are increasingly being investigated. In the past, coaxial (electrical) cables were used for analog and digital signal transmission over long-distance communications. But the strong attenuation of the signal (about 30 dB/km at 400-MHz cable) [2] imposed the use of regenerative repeaters along the entire route. Fiber optic is now an easy way to transmit information, and the next generation of long-range communications will rely on optical amplifiers to the detriment of regenerative electronic repeaters. The discovery of new glasses of exceptional optical transparency and reduced attenuation to values lower than 1 dB/km boosted the great evolution of optical communication over the last years. Inorganic glasses have been used as optical materials for a long time due mainly to its high transparency in the visible and adjacent, ultraviolet (UV) and near-infrared (NIR) ranges. However, they do not exhibit electronic transitions in this region. For these transitions to take place, controlled introduction of optically active ions is used; therefore, the optical properties of rare-earth (RE) ion-doped inorganic glasses emerged in the field of materials physics. New optical materials suitable for the development of photonic devices based on RE-doped crystal or glassy hosts have thus attracted significant scientific and technological interest. Such wide technological applications are based on the interaction of light with matter where the fluorescent behavior is essential [3]. The most widely used RE ions in glass are erbium (Er^{3+}) [3], ytterbium (Yb^{3+}), and neodymium (Nd^{3+}). For instance, Er^{3+}-doped silica fiber is extensively used in optical communication; Yb^{3+}-doped silica fiber is used in engineering materials processing, and Nd^{3+} doped is applied in glass lasers used for inertial confinement fusion (ICF). The invention of erbium (Er^{3+})-doped fiber optic amplifiers in 1987, the so-called EDFA's, "*Erbium-Doped Fiber Amplifiers*" [3, 4], allowed the optical amplification of the signal around 1550 nm, the region where the propagation of light in silica (SiO_2) optical fibers is maximum, by using stimulated emission of optically excited Er^{3+} ions through transitions from the $^4I_{13/2}$ metastable energy level to the $^4I_{15/2}$ ground state. Furthermore, in integrated optical circuits, which are the equivalent optical of the electronic-integrated circuits ("*chips*"), the light is confined and routed to different optical components (lasers, electro-optical modulators, directional couplers, filters and multiplexers, etc.) through passive devices as *optical planar* (or slab) *waveguides* [3, 4]. A *slab* waveguide is similar to an optical fiber, except that it is a planar, rather than cylindrical. The transmission of light over slab waveguide is possible when a low refractive index glassy contains a slab (or channel) of higher index material, along which light is guided by total internal reflection. These devices can be easily prepared, for example, by sol-gel [5], in different shapes and sizes with uniform distribution of RE ion concentrations, without inducing crystallization, and exhibit large optical transparency window covering UV, visible, and IR regions. In particular, planar optical waveguides, as well as waveguides of different geometry channels, are currently manufactured in materials such as SiO_2 (silica-on-silicon) [5], Si (silicon-on-insulator) [6], GaAs, and $LiNbO_3$ [7]. In the field of photonics, silica-on-silicon technology stands out, where silica glass waveguides are manufactured on silicon substrates, allowing the combination of some discrete optical components with integrated ones [3]. Silica-on-silicon-integrated optical devices can be widely prepared by several methods, such as sputtering, silicon thermal oxidation, flame hydrolysis deposition (FHD), chemical vapor deposition (CVD), and sol-gel spin-coating and/or dip-coating [5]. The future of integrated optics therefore depends on the development of active

http://dx.doi.org/10.5772/intechopen.68534

(doped) materials signal amplification applications (e.g., optical amplifiers at 1550 nm, wavelength of great interest in fiber optic communications) [4], as well as materials with *passive* performance for the functions of detection, conduction, multiplexing, and de-multiplexing of optical signals, including ultra-fast switches, with response times on the order of picoseconds [8]. Silica-on-silicon passive devices have been highly developed, for instance by NTT (Japan) [9] using FHD. Presently, an alternative technique is sol-gel deposition, in which a liquid precursor is deposited and annealed to form the glass [10]. This process does not require high cost with respect to equipment, is environmentally tolerable, and is highly versatile in the control of final composition and microstructure. This advantage allows diverse modifications to the host glass and the addition of a wide variety of dopants, thus greatly increasing the range of applications of the resulting components. In fact, doping is an important technique used for the development of optical materials, by combining properties of a matrix (the host) with the dopant ones to give an "optical material" of unique properties. In both the cases of 1550-nm amplification and ultra-fast switching, the matrix used is generally a dielectric and the dopant introduced depends on the nature of the application. Thus, in the field of optical amplification, thulium (Tm^{3+}), holium (Ho^{3+}), and praseodymium (Pr^{3+}) have been accommodated by many simple and multicomponent glass matrices [3], displaying several spectroscopic properties, namely radiative transitions between energy levels that cover the UV, visible, and NIR spectral range. Rather, in the case of ultra-fast switching, it is worth noting the doping with metallic nanoparticles (Au, Ag, Cu, etc.), because they have excellent properties in nonlinear optics [11]. Over the past few decades, glasses have provided technological support for the spectrum of needs in optical and photonic applications, especially as host matrices. Glass-integrated optics is dominated by oxide glasses mainly due to its superior production techniques. SiO_2 glasses are of superior significance due to their application in actual optical fibers. These applications have also seriously increased the development of appropriate sources of pumping to meet the demands of the expanding field of communications. Diode lasers with 800- and 980-nm emissions [12] were developed because they coincide with electron transitions of Nd^{3+} and Er^{3+} ions, respectively. Since the discovery of the laser action in a glass-hosted $Nd^{3\,+}$ ions [13], the study of RE ion-doped glasses has developed enormously, especially in order to obtain (1) new transitions at different wavelengths, (2) new host glasses suitable for the miscibility of RE, and (3) new pump conditions from compact lasers capable of integration for applications in photoluminescence and amplification. The low solubility of RE in silica glass [3] leads to a possible formation of RE clusters, mostly if the doping concentration is higher than the equilibrium solubility. This phenomenon is particularly complex in integrated optical amplifiers since they need a much higher Er^{3+} concentration with respect to EDFAs in order to compensate the smaller interaction length. This high dopant concentration is also responsible for parasite effects caused by interactions between excited ions, notably cooperative up-conversion (UC) and quenching by energy transfer [3, 14], which is detrimental to the optical properties of laser glasses. Special actions are needed to suppress RE clustering. A promising approach is to add suitable network modifiers to SiO_2, such as Al_2O_3 and P_2O_5, in order to improve the RE solubility [15, 16]. The performances of the amplifier are governed by the electronic and optical characteristics of the RE ion but are also strongly influenced by matrix properties. Moreover, the matrix phonon energy is a crucial aspect, because it affects the amplification efficiency by non-radiative relaxation [17]. Despite the high chemical durability and superior chemical resistance of SiO_2 glasses, fluorescence studies are limited to only a few transitions owing to their higher phonon energies. Rather, non-oxides hosts (e.g., fluoride and chalcogenide glasses) exhibit more metastable states and undergo transparencies beyond 2000 nm wavelength [18]. New glass compositions with heavier glass ions and weaker bond strengths are presently suitable glass hosts for RE-doped optical devices, particularly for RE such as Pr^{3+}, which displays fast non-radiative relaxation paths in oxides [19]. Glasses with lower phonon energy, such as fluoride, chalcogenide, or tellurite glasses, are prone to crystallize and thus have proved hard to make in fibers with low optical loss.

Among the different glass, oxide and non-oxide systems, transparent glass-ceramics (GC) offer remarkable features to the field of photonics. Glass-ceramics are a class of hybrid materials consisting of nanocrystallites embedded in a glass matrix. Transparency is a key property, in particular for dielectric optical waveguides and optical fibers, and the effect of the nanocrystals activated by RE ions on the spectroscopic properties overcomes largely those of RE ions in a glass [20, 21]. Moreover, the transparent glass-ceramics still retains the properties of a glass and can be processed and shaped by techniques used for glasses. Among the various techniques used to fabricate optical planar waveguides and photonic devices, sol-gel process with top-down and bottom-up approaches demonstrated to be a suitable route to do it [5, 10, 22]. A high doping level of RE ions can be achieved by sol-gel process. Despite the presence of hydroxyl groups (OH) of inherent character in sol-gel silicate glasses, which are extremely effective at quenching excited RE ions [3, 23], even for a few hundred ppm of OH, sol-gel is a powerful technology in the development of glasses containing RE for various types of applications.

2. Sol-gel glass: the solid with a liquid background

The rapid cooling of some molten materials (e.g., SiO_2, BeF_2, B_2O_3, As_2O_3, P_2O_5, and GeO_2) [24] has become the traditional way to make glass, an inorganic material of fusion that has cooled to an amorphous solid without crystallization. So, unlike crystals, a glass lacks long-range order. The crystallization can be avoided when the cooling is fast, and so the viscosity of the molten liquid increases so much that the atomic *rearrangement* becomes difficult and the material becomes amorphous, which is a solid without long-range periodic order, which means that the spatial arrangement of atoms and ions does not exhibit a three-dimensional periodicity and the long-range order of the crystalline state. Nevertheless, glasses do not necessarily have to be formed from rapid melt cooling. In the mid-nineteenth century, the sol-gel process emerged as an innovative choice for glassmaking, using as an alternative to the traditional method of melting, a technique of chemical synthesis in aqueous media, very effective in the preparation of glasses at low temperatures [10]. Sol-gel glasses are prepared usually using alkoxides and soluble salts (precursors) by heat treatments at low temperatures [10]. The precursors are selected molecules that contain the elements of the future glass network and allow its formation by chemical reactions in liquid medium, silicates, aluminates, phosphates, and so on. Although the term alkoxide has been firstly applicable to products obtained from alcohols, by replacing the hydrogens of the hydroxide groups with a metal (Ti, Zr, etc.) or non-metal (Si, Ge, etc.), their use was generalized to compounds of the type $M(OR)_n$ wherein M represents the metal or nonmetallic atom and R the organic radical [10]. Because many alkoxides are soluble in a wide range of organic solvents, especially alcohols, they are extremely versatile and so are the most commonly used in the preparation of inorganic solids. However, there are many other precursors with interest in the sol-gel process, namely carboxylates ($M(OCOR)_n$), acetylacetonates ($M(CH_3COCHCOCH_3)_n$), and inorganic salts such as nitrates and chlorides. These two last kinds of precursors are a very common to introduce RE ions in the glass network. As an example, Er^{3+} ions can be easily added as an ethanol solution of $Er(NO_3)_3{\cdot}9(H_2O)$ or $ErCl_3{\cdot}6(H_2O)$. The traditional sol-gel process may be exemplified in the case of the preparation of silica glass [10]. So the process one needs an alkoxide precursor of M(OR)n type, for example, TEOS ($M = Si$; $R = C_2H_5$), water, an alcoholic solvent and a catalyst, acid or base [10]. The process begins with the hydrolysis of the alkoxide, which allows the formation of monomers (hydroxide *species*, M-OH), which are small solid fragments with a great potential to bind to each other during the condensation reactions (aggregates of several monomers) [10]. These fragments, initially very small ($\ll$100 nm), progressively increase in size until they occupy all the available volume of the solution, forming a three-dimensional (3D) network corresponding to the final structure of the glass. The process is called sol-gel because at a given moment a transition from a "*sol*" phase to a "*gel*" phase occurs [10]. Let's see some basic details: (1) the hydrolysis reaction consumes the initially added water, allowing the condensation reaction to occur, producing water as well. This water will be consumed again by the hydrolysis until all the alkoxides are hydrolyzed, which means that alkoxide ions are gradually replaced by OH^- ions. During this process, modifications of the polarity and viscosity of the medium occur, which give rise to a "*sol*"; (2) the condensation reactions continue until all monomers capable of binding are exhausted, which naturally lead to the formation of species of the —M—O—M— (or $(\text{—Si—O—Si-})_n$) type which remain "immobilized" at one stage, highly viscous, called "*gel*"; and finally (3) after drying the "gel" (loss of solvent by evaporation) [10], a high porosity solid is obtained which must be further densified (at a temperature close to glass transition temperature, T_g) to become a solid, dense glass. A brief look on the process is shown in **Figure 1**. We can define a sol-gel glass saying that it is solid with a liquid background! In fact, the liquid undergoes an evolution over time and reaches the equilibrium when it becomes structurally a solid. This approach enables to make of an interested analogy between the sol-gel process and *Darwin's Theory*. Charles Darwin was an English naturalist who published in 1859 the well-known book *Origin of Species*, explaining how living species beings adapt continually to the environment, assuring the survival of their descendants by reaching equilibrium. Through the sol-gel processing, a true evolution of species occurs and so the "*sol*" could well be called *Darwin's liquid*, since it reaches the "perfection" when a 3D cross-linked network is attained and becomes a glass. As in *Darwin's Theory*, the sol-gel evolution does not happen by chance, rather it is stimulated by several imbalances in the *alkoxide-water-alcohol-acid* system, such as changes in pH, viscosity, temperature, and so on, which force precursors to adapt continuously in the middle: *dimer* → *chain* → *ring* is the progression in the polymerization process. Each monomer bond that is established is a tiny evidence of *Species Evolution* in this process!

Sol-gel synthesis is nowadays used to produce glasses with a variety of compositions that are technically difficult or even impossible to produce by melting. Some binary and

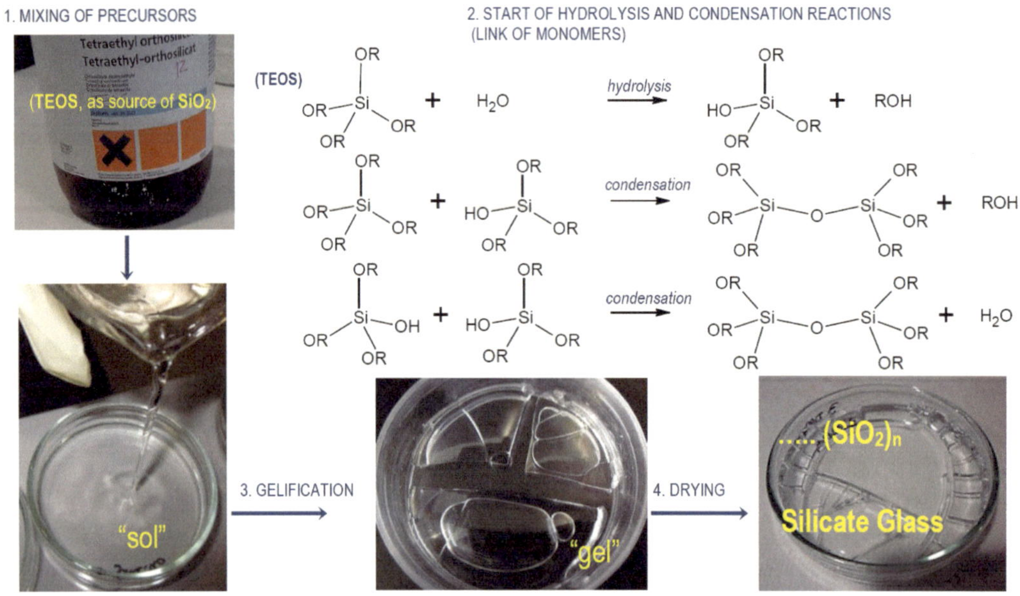

Figure 1. Scheme of the evolution sol → gel → glass.

multicomponent oxide systems exhibit liquid-liquid immiscibility and tend to phase separate in a short range of compositions and temperatures and thus can be successfully obtained by sol-gel. This process does indeed offer great possibilities to tailor the preparation of highly homogeneous glasses, such as RE-doped glasses. When RE ions are hosted by glasses, they exchange the network cation or they act as network modifiers [25]. It is therefore expected that the RE ions will favor non-bridging oxygen sites in the gel as the solvent is expelled. In addition, sol-gel method also offers the possibility of obtaining glassy coatings on silicon substrates to achieve planar waveguides [5]. A method of depositing thick sol-gel coatings is illustrated in **Figure 2a**.

Planar waveguides have a central rectangular region of higher refractive index n_1 (core) that is surrounded by other region, which has a lower refractive index n_2, as shown in **Figure 2b**. The deposition of multi-nano-silica-based glass layers on silicon substrates to obtain thick (micron) planar waveguides (which can later be coupled to external guide fibers) can be done by simple and multilayer deposition techniques (spin-coating and/or dip-coating) [26] of sol-gel solutions, allowing coatings with thickness and controlled refractive index. Indeed, by varying the precursor ratios in the sol, glasses of a wide range of compositions, and thus a desired refractive index, can be achieved; for example, SiO_2/TiO_2 of different molar ratios allows a wide index range from 1.46 to over 1.6. For waveguides, a bilayer is formed: a lower index buffer, n_2 (e.g., SiO_2) of sufficiently thick to prevent leakage of guided light into the substrate, and a higher index guiding layer, n_1 (e.g., SiO_2/TiO_2).

The refractive index in thin films containing SiO_2 (n ~1.46) increases linearly with the relative TiO_2 content, generally ca. 0.07 per 10 mol% of TiO_2 [27]. The guiding layer may also be doped with a functional material; for example, RE can be added to the sol through a variety of

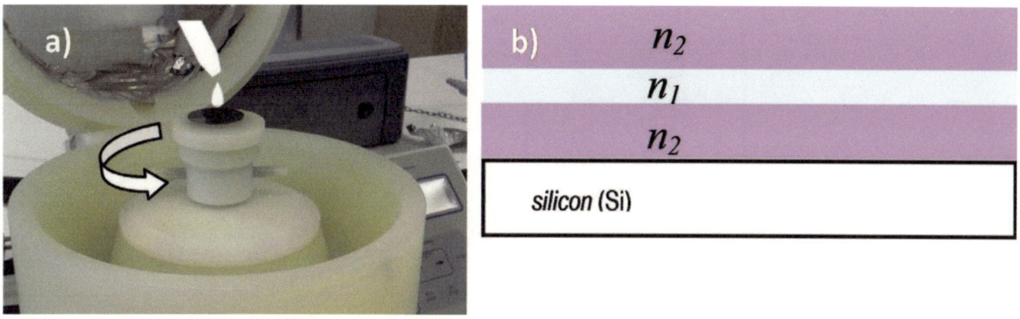

Figure 2. (a) Spin-coating process and (b) planar waveguide structure (on silicon substrate). n, refractive index; $n_1 > n_2$ (for total internal reflection).

precursors. A few drops of "*sol*" are spun to a thin layer on the Si-substrate, where upon the solvent evaporates and the condensation is speeded, giving a partially dense "*gel*" coating. This coating is subsequently heated to burn away remaining organics groups and achieves complete densification; repetition of this process can produce a multi-micron coating. Fully dense and homogeneous layers can be formed at relatively low temperatures, and the molecular structure formed chemically in the sol is maintained in the final glass. Thus, RE ions could be incorporated in the sol by condensation reactions, at the required high doping levels, and that this structure could be maintained in the glass. At low concentrations, the RE ions can be distributed homogeneously in an SiO_2 glass network with the formation of Si—O—RE structure. However, with an increasing concentration, due to close proximity between the RE ions and oxygen ions (~1.5–2 nm), RE—O—RE clusters may eventually form [28]. Sol-gel offers some advantages over FHD and CVD, namely greater flexibility in host glass composition and the possibility of homogeneously incorporating RE by chemical reaction in the sol. However, its major drawbacks are the excessive shrinkage of a wet gel upon drying which leads often to the crack, in particular of large monolithic pieces, inhomogeneous linkage during sol formation (in multicomponent glasses) due to differential reactivity of alkoxide precursors, and the presence of residual porosity and OH groups. Measurements of wavelength shape, intensity, and width of optical spectra and of excited state lifetimes can reveal the existence of RE clustering and the presence of OH in the neighborhood of the RE ions. Indeed, lifetime may be shortened if the hydroxyl content is not reduced to very low concentrations. The annealing process removes any excess water and OH groups as well as creating a dense, amorphous product, the sol-gel glass. However, the removal to a few ppm levels requires special chemical treatments with reactive gases like $CC1_4$, $C1_2$, and so on [29]. Residual porosity must also be removed by densification to prevent internal fluorescence quenching.

3. Basic principles of fluorescence: specificities of sol-gel glasses

Among usual forms of light interaction with matter (absorption, reflexion, etc.), fluorescence is a very special one that can be regarded as a source of light, or, in other words, a phenomenon in which a material radiates light (emission) at longer wavelength after a brief interval (termed

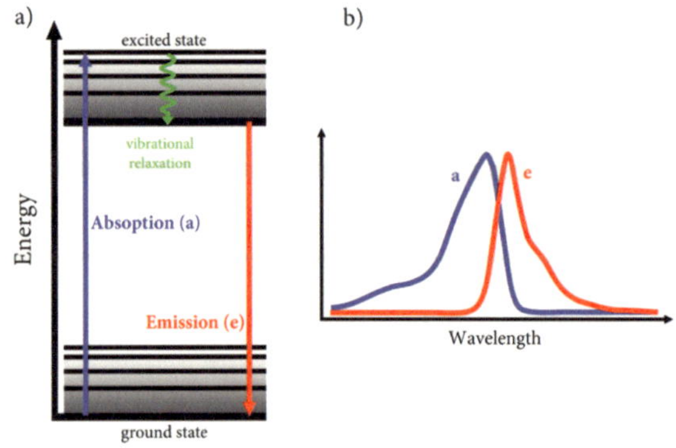

Figure 3. (a) Fluorescence scheme; (b) absorption spectrum (a) plus emission spectrum (e).

fluorescence lifetime), as a result of absorption of shorter wavelength. Fluorescence in optical glass is generated by the presence of RE ions. When these active ions are directly excited by incoming energy, the electron on it absorbs energy and is raised to an excited state (**Figure 3a**). The excited state returns to the ground state by emission of light of longer wavelength (**Figure 3b**). This behavior (absorption by short wave light, emission of longer wave light) is named "*fluorescence*" [30].

The processes that occur between the absorption and fluorescence emission of light are generally illustrated by the Jablonski [31] diagram. Indeed, **Figure 3a** shows a very simplified Jablonski diagram where the transitions between states are represented as vertical lines.

Following light absorption, emission usually occurs. Prior to emission, a rapid relax to the lowest vibrational level takes place, called vibrational relaxation, which is a non-radiative transition [32]. This process yields a relaxed excited state from which fluorescence emission originates. The fluorescence results from combined rates of radiative and non-radiative processes from a metastable excited state to the ground state (e.g., Er^{3+} transition $^4I_{13/2} \rightarrow {}^4I_{15/2}$ at 1550 nm). Radiative decay (W_R) deals with photon emission during de-excitation processes, while in a non-radiative decay (W_{NR}) the excited electronic states relax by energy dissipation via thermal processes such as vibrational relaxation and collisional quenching. Because RE ions may interact with vibrations of the matrix, either intrinsic or due to impurities (defect vibrations), non-radiative decay rates (W_{NR}) can occur. The total decay rate of an excited state (W_T) is then given by

$$W_T = W_R + W_{NR} \tag{1}$$

where

$$W_{NR} = W_{MP} + W_{ET} + W_{CR} + W_{OH} \tag{2}$$

W_{MP}, W_{ET}, W_{CR}, and W_{OH} are the rates of multiphonon decay, energy transfer, cross relaxation, and decay due to water (OH groups) in glasses, respectively. The W_{OH} component for the non-radiative decay cannot be neglected for sol-gel glasses [23] since being a very effective fluorescent-quenching mechanism of transferring energy to OH groups, by non-radiative multiphonon relaxation. Moreover, W_{MP} can be relevant because it deals with the phonon density of states, which is high in solids, so the high-energy losses come from the highest-energy phonons of the matrix [14, 30, 32]. Oxide glasses have superior W_{MP} due to their higher network vibrational frequencies, when compared to halides and chalcogenides. For glasses, in particular for SiO_2-based glasses, the vibrations causing multiphonon relaxation are the high frequency, localized stretching modes of their oxygen-based Si—O—Si structural bonds. W_{ET} strongly depends on the average distance and thus the concentration of active ions (due to ion-ion interactions). This process is particularly important when the RE ion concentration increases (up to concentration quenching). Moreover, cross-relaxation or cooperative up-conversion processes [32] can occur due to concentration quenching. The fluorescence efficiency is degraded if the non-radiative decay is similar to the radiative one. This means, in practice, there is a significant reduction in the quantum yield (η) of RE ions and a shortening of the measured metastable level lifetime (τ_{meas}). Quantum yield is the number of emitted photons relative to the number of absorbed (pump) photons. W_R and W_{NR} both depopulate the excited state through emission, and hence the η is given by the ratio between the radiative decay rate and the total (radiative + non-radiative) decay rate, according to Eq. (3):

$$\eta = \frac{W_R}{W_R + W_{NR}} \tag{3}$$

Glasses with the largest quantum yields, approaching unity, display emissions with major fluorescence intensity; the fluorescence lifetime (τ_{meas}), instead, refers to the average time a population of N-active ions stays in its excited state before emission and is usually determined by using fluorescence decay measurements [30]. The temporal evolution of the population of excited states follows Eqs. (**4**) and (**5**) [30]:

$$\frac{dN(t)}{dt} = -W_T N(t) \tag{4}$$

where

$$N(t) = N_0 e^{-W_T t} \tag{5}$$

N_0 is the density of excited ions at $t = 0$, just after the pulse light is absorbed. The de-excitation process can be experimentally observed by analyzing the temporal decay of the emitted light, $I(t)$, as shown in **Figure 4a**. Therefore, the fluorescence decay is exponential given by Eq. (6) [30]:

$$I(t) = C \times W_R N(t) = I_0 e^{-W_T t} \tag{6}$$

where C is a constant of proportionality and $I_0 = C \times W_R \times N_0$ is the intensity at $t = 0$. As shown in **Figure 4a**, τ_{meas} is the measured lifetime that also corresponds to the time in which the emitted intensity decays I_0/e.

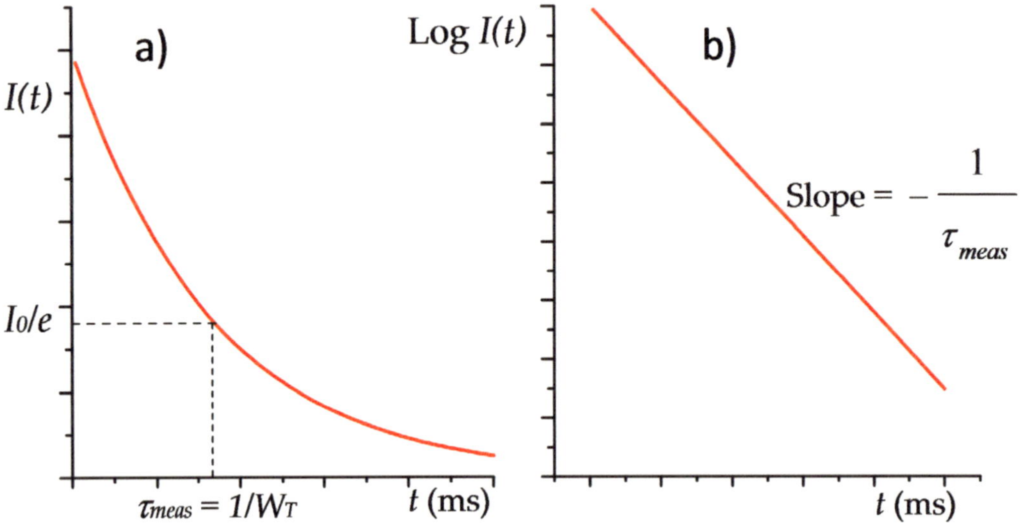

Figure 4. (a) Fluorescence decay; (b) linear plot of $I(t) = I_0 e^{-W_T t}$.

The lifetime, τ_{meas}, also corresponds to the slope of the linear plot of Eq. (6), exhibited in **Figure 4b**. Therefore, Eq. (1) can be rewritten by Eq. (7):

$$\frac{1}{\tau_{meas}} = W_R + W_{NR} = \frac{1}{\tau_R} + W_{NR} \tag{7}$$

where τ_R is the value of the lifetime before the non-radiative processes has been reached (also named radiative lifetime or excited state lifetime). Hence, η calculated from Eq. (3) can also be given by Eq. (8):

$$\eta = \frac{W_R}{W_R + W_{NR}} = \frac{\tau_{meas}}{\tau_R} \tag{8}$$

Because the fluorescence intensity is directly proportional to the number of molecules in the excited state, lifetime measurements can be done by measuring fluorescence decay after a brief pulse of excitation [30]. **Figure 5** shows the fluorescence decay of Ho^{3+} ($^5I_7$) at 2000 nm, fitted with a single exponential curve, measured for Ho^{3+}-doped sol-gel glass. However, in this case, visual inspection indicates a poor fit to the experimental data, confirmed by an R^2 of 0.97506. Therefore, in such cases, the fluorescence decay profile can be better fitted with the double exponential function, composed by the sum of two exponentials, as shown in Eq. (9):

$$I(t) = A_1 e^{(-t/\tau_1)} + A_2 e^{(-t/\tau_2)} \tag{9}$$

where τ_1 and τ_2 are often defined as the fast and slow components of a fluorescence decay profile. In RE-doped glass-ceramics (GCs), the decay curves of the luminescence are usually adjusted to a double exponential. As previously mentioned, sol-gel GCs are obtained from a glass matrix when an accurate thermal treatment is applied [20, 21]. During this thermal process,

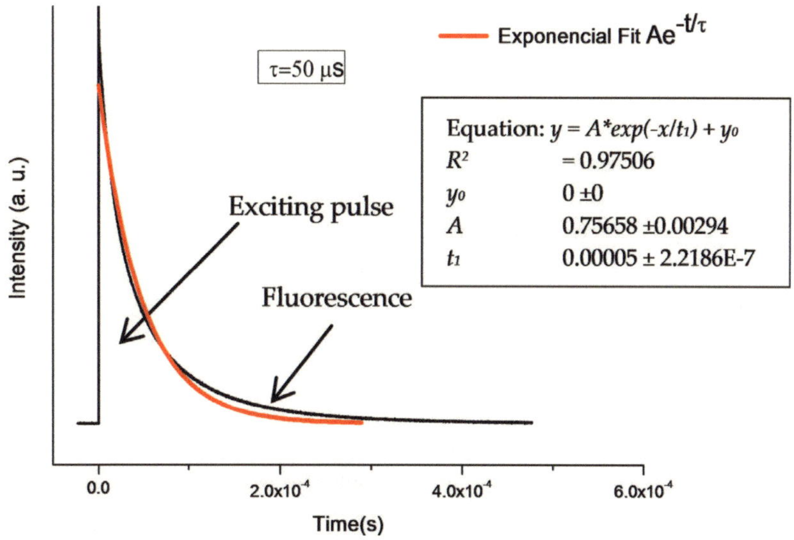

Figure 5. Decay curve of the holmium fluorescence ($^5I_7$) in a sol-gel glass; fit by single exponential decay.

nanocrystals precipitate into the host matrix and the RE dopants divide into the nanocrystalline and glassy phases.

Some of the ions predominantly doping the nanocrystals and another part remain in the vitreous state, so the average distance between them increases reducing concentration quenching. This approach has been widely reported in the literature [33], assigning the lifetime of the remaining RE ions in the vitreous phase to the decay of the fast component, while the slow component will correspond to the ions within the crystalline environment. For the multi-exponential decay is usually defined a mean life, which is given by Eq. (10) [34]:

$$\tau_{\mathrm{av}} = \frac{A_1\tau_1^2 + A_2\tau_2^2 + A_3\tau_3^2}{A_1\tau_1 + A_2\tau_2 + A_3\tau_3} \tag{10}$$

where A_3 and τ_3 come from Eq. (9), duly adapted to three components. In the absence of a host (the nanocrystals) in the vicinity of the RE ions to optimize spectroscopic properties (e.g., phonon energies, etc.), RE ion-doped sol-gel glasses have normally ions concentration, which is relatively high and so the mean distance between two of them may be small enough to allow a non-negligible probability of energy interaction. In fact, if one of the ions is excited it can transfer its energy to a nearby one. Frequently, the excitation is also transferred from a sensitizer ion (e.g., Yb^{3+}) to another active ion (Er^{3+}). The influence of the concentration quenching on the reduction of RE ion quantum yield was first evidenced by Förster and Dexter [35, 36] who showed that the electronic energy transfer probability strongly depends on the ion-ion distance through an empirical criterion applied to a microscopic system of two ions interacting among them. The empirical formula, which relates the measured lifetime to the ion concentration, c, is

$$\tau_{\text{meas}} = \frac{\tau_{\text{co}}}{1 + \left(\frac{c}{Q}\right)^{p}} \tag{11}$$

where τ_{co} is the lifetime in the limit of zero concentration (c~0) and Q is the quenching concentration, at which $\tau_{meas} = \tau_{co}/2$. Energy transfer significantly affects the luminescence properties of a material. However, in some particular cases, energy transfer processes occurring between RE ions could be advantageous, as those between Yb^{3+} and Er^{3+} ions, which are widely used for optical amplification at 1550 nm. In such systems, the pump radiation (at 980 nm) is strongly absorbed by Yb^{3+} ions, which acts as the sensitizer. Subsequently, the Yb^{3+} ions efficiently transfer their energy to the Er^{3+} ions at ground state by emitting light at 1550 nm. This process is known as down-conversion [34, 37] because high-energy excitation photons are transformed to lower-energy photons. Rather, up-conversion (UC) is the most common sense of this phenomenon whereby photons of lower energy are absorbed by a material, to be re-emitted as a higher-energy photon. An important feature is that materials can be tuned to respond to NIR (low energy) to emit at high energies at visible wavelengths range. Up-conversion laser emission can be easily achieved, for instance, in pairs Er^{3+}-Yb^{3+} or Ho^{3+}-Yb^{3+} co-doped glasses, where high-energy photons (green or blue light) can be obtained from red or infrared light (low-energy photons) [38]. A typical experimental setup to record UC visible emission spectra is shown in **Figure 6**. The UC visible spectrum of Er^{3+}-Yb^{3+} doped sol-gel glass excited at 980 nm, relax emitting light in the visible; the $^{2}H_{11/2}/^{4}S_{3/2} \rightarrow {}^{4}I_{15/2}$ green emission lines are clearly dominant, as shown in **Figure 6**. Various combinations with different RE ions and appropriate thermal treatments offer the possibility to optimize the fluorescence intensity, lifetime, and the energy transfer between doping ions. From the various fluorescence-quenching processes, some of them have a special significance in glasses made by sol-gel which are (1) concentration quenching via cross-relaxation, (2) non-radiative vibrational excitation of hydroxyl groups (OH), and (3) multiphonon relaxation due to impurities (defects vibrations). These mechanisms decrease fluorescence efficiency [34]. Moreover, it was shown that RE-host interactions result from defects (porosity and/or passive crystallites dispersed in matrix), expressed by an intensity decrease in the fluorescence emission; symmetry factors that arise from host nature (amorphous or crystalline), expressed by the character of the lifetime decay curve, and cross-relaxation, expressed by the broadening/narrowing (shape) of a specific transition [15]. All of these processes depend on the type and composition of the glass used. The high concentration of OH groups that remain in sol-gel glasses decreases the fluorescence quantum yield and shortens lifetimes of RE ions in glasses [3, 34] affecting negatively the optical device performance.

OH quenching arises from the residual water, solvents, and silanol groups (Si—OH) of the early stages of sol-gel glasses. This leads to an enhancement of non-radiative decays due to the coupling between the RE states and the high vibrational energy of OH (3200 cm^{-1}) [3, 4].

Annealing the doped gels changes the local environment of the RE ions, which results in changes in the fluorescence emission shape [15].

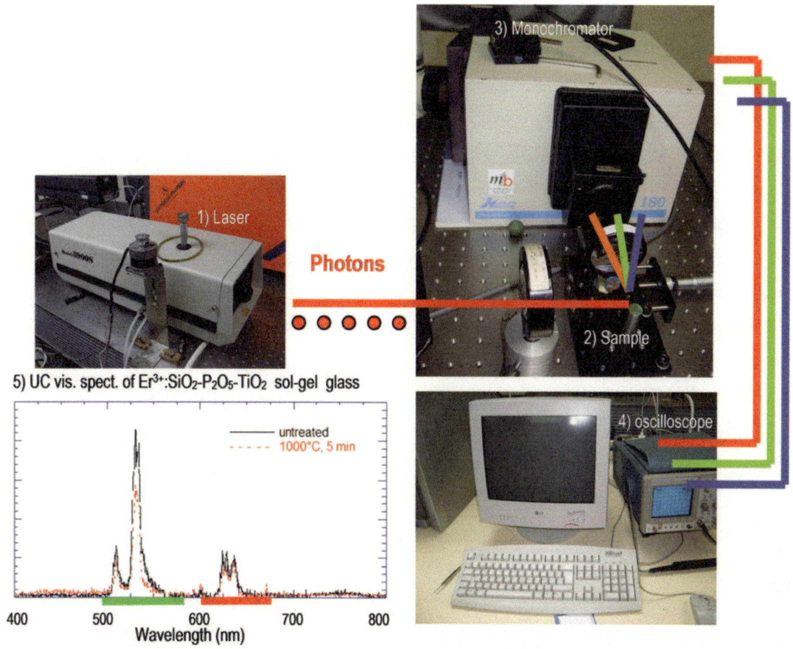

Figure 6. Experimental setup to record UC emission spectra: (1) excitation source (laser); (2) sample (Er^{3+}: SiO_2-P_2O_5-TiO_2 sol-gel glass); (3) monochromator/PMT; (4) oscilloscope (to acquire the signal); and (5) sample UC spectrum.

4. The 4f-4f transitions of rare earths and intensity probabilities

The REs are a group of 15 chemical elements from the Lanthanide series (atomic number, Z, ranging from 57 to 71, at the sixth row of the periodic table), starting in lanthanum (La) and ending in lutetium (Lu), which exhibit similar chemical behavior and high stability in their triply ionized form (3+). When incorporated in crystalline or amorphous hosts, the RE maintains their most stable ionized form [39]. All REs have the general configuration $[Xe]4f^n5s^25p^6$ where [Xe] is the electronic configuration of Xenon and n represents the number of internal electrons (the optical active electrons) in level 4f [39]. Their most remarkable characteristic is the partially filled 4f shell that is shielded from external fields by the $5s^2$ and $5p^6$ electrons, allowing a 4f electrons behavior similar to that exhibited in a free RE ion [39]. The energy levels of RE are therefore largely insensitive to the environment in which they are placed [39–41]; the effect of the static crystal field of the environment becomes very weak, causing only small host-induced splitting [3, 41]. The crystal field is the electric field interaction between the RE ion and surrounding ions of the host matrix. An inherent characteristic of the RE ions is that their optical transitions occur within the 4f shell reducing the influence of the host lattice on the wavelengths, bandwidths, and cross sections of the relevant optical transitions, which correspond therefore to weak, sharp absorption and emission lines [3]. In a crystal, each ion in the lattice is affected by the same intensity of the crystalline field, whose influence on the electronic transitions leads to narrow bands, resembling those of the free ions. In glasses, however, the bands are considerably broadened [41]. Each RE ion is affected by a different crystalline field and therefore exhibits unlike Stark levels. Each ion has its own absorption and emission spectra and its own radiative decay rates. Therefore, the glass fluorescence properties can be viewed as a kind of average of all ion fluorescent behaviors. Thus, there is a site-to-site variation in the energy levels of these ions and hence in their radiative and non-radiative transition probabilities lead to the band broadening. Although the shielding of the 4f shell means that the RE ion energy levels are largely insensitive to the host, Stark splitting broadens the levels as a result of the crystal field. Stark splitting is observed for RE ions in glasses, as a result of a large quantity of sites that differ from each other in the number and position of the surrounding anions [41]. In order to characterize the energy levels associated with the electronic levels 4f, the Russell-Saunders notation (or LS-coupled states) is often used, represented as $^{2S+1}L_J$ [41]. This notation couples the angular momenta (*L*) and spins of the valence electrons (*S*) to a total angular momentum *J*. The energy levels are then identified according to the $^{2S+1}L_J$ notation, where the term $2S + 1$ is the spin multiplicity and $L = 0,1,2,3,4,5,6 \ldots$ correspond to S,P, D,F,G,H,I..., respectively. The free Er^{3+} ion, for instance, has an electron configuration $4f^{11}$ for which the lower term is $^4I_{15/2}$, and the first excited multiplet $^4I_{13/2}$, which is about 6500 cm^{-1} higher than $^4I_{15/2}$. So when $L = 6$ (state I) and $S = 3/2$ (the total spin quantum number for $4f^{11}$ electrons can be 3/2, with three unpaired, or 1/2, with only one unpaired) [39], *J* may assume several terms (*J* multiplets) as $^4I_{15/2}$, $^4I_{13/2}$, $^4I_{11/2}$, and $^4I_{9/2}$ which correspond to the lowest-energy states of the Er^{3+} ion. In 1964, Dieke published the energy diagram of all trivalent RE ions within $LaCl_3$, knowing as "*Dieke diagram*." Due to the negligible influence of the host lattice on the energy levels, the diagram is applicable to RE ions in any host matrix [42], including glass, crystals, and glass-ceramics, very useful to estimate the symmetry of the environment where the ion is placed. The absorption spectrum of Er^{3+}:$80SiO_2$ $20TiO_2$ (mol%) sol-gel glass is shown in **Figure 7**. From optical absorption, it is possible to determine energy levels of the RE ions and identify them through Dieke diagram, as shown in the inset of **Figure 7**. The absorption transitions between specific Stark components of different *J* multiplets can be detected at room temperature as separate peaks in RE-doped crystals, but not in glasses [3, 41], where a broad band overlaps them. In 1962, Judd and Ofelt proposed a theory [43, 44] to estimate transition probabilities, or radiative decays (W_R). This theory offers an alternative approach to obtain radiative transition probability of all excited states, radiative lifetime, and quantum yield by calculating phenomenological parameters (Ω_i, $i = 2, 4, 6$), known as the Judd-Ofelt (JO) parameters, from experimentally observed absorption data. In addition, this theory can also be used to predict oscillator strengths (*f*) of and branching ratios (β) [30]. This *f* can be expressed under specific conditions as a function of Ω_i whose values are directly influenced by the host matrix [45] and provide a quantitative measure of the asymmetry of the local crystal field of RE ion. The parameter Ω_2 is sensitive to the covalent bonding between the RE ions and the ligands anions, for example, RE-O bonding [45], which is an indication of the asymmetry of the local environment of the RE sites. This means that it is small for ionic host material, for example, fluoride glasses, while it is large for covalent host materials, such as silicate glasses. Moreover, for a glass co-doped with more than one type of RE ions, for example, Yb^{3+} and Er^{3+}, the possibility of a reduction on the number of non-bridged oxygens to coordinate with Er^{3+} ions is high, which can cause the decrease of the covalency between Er-O oxygen bonds, and, therefore, decrease Ω_2. On the other hand, Ω_4 and Ω_6 reveal

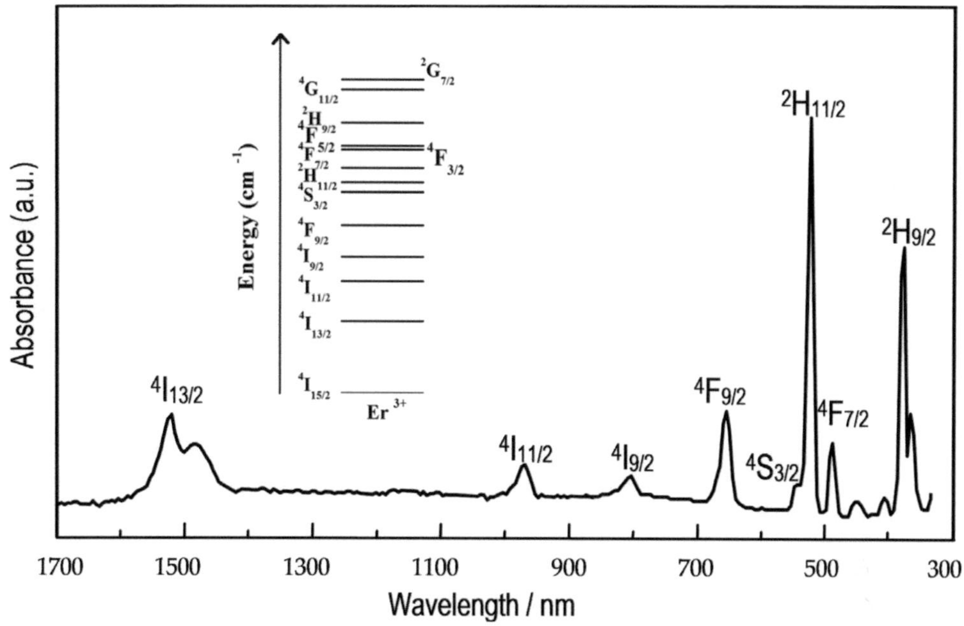

Figure 7. Absorption spectrum of Er^{3+}-doped $80SiO_2$-$20TiO_2$ glass. Inset: schematic energy level diagram of Er^{3+} ion (from Dieke diagram).

broad properties of the host glass such as rigidity and viscosity. Moreover, the ratio Ω_4/Ω_6 is a spectroscopic indicator of the quality of the parameter used to predict the strength of the radiative emission of the first transition (e.g., $^4I_{13/2} \rightarrow {}^4I_{15/2}$: 1550 nm in Er^{3+}: SiO_2 glass). For sol-gel glasses, the calculated Judd-Ofelt parameters are average values from a multiplicity of sites for RE ions in different environments. From the absorption spectrum (**Figure 7**), the experimental oscillator strengths (f_{exp}) of the electronic transitions can be calculated using the expression:

$$f_{exp} = \frac{mc}{\pi e^2 N} \int \frac{2.303OD(\upsilon)}{d} d\upsilon \tag{12}$$

where m and e are the electron mass and charge, respectively, c is the light speed, N is the concentration of absorbing ions, υ is the light frequency, $OD(\upsilon)$ is the optical density, which is proportional to the absorption coefficient (α), and d is the thickness of the sample. The f_{exp} values obtained using Eq. (11) and the experimental absorption spectrum can therefore be used to obtain the JO parameters [46]. The experimental oscillator strength, f, is fitted to the theoretical expressions in the frame of the JO theory using a statistical least-square algorithm or chi-squared methods. The radiative emission probabilities between two electronic states can be calculated, according to the JO theory, if they are known.

5. Tailoring sol-gel-derived SiO_2 glassy optical waveguides

Nowadays, great attention still is given to the development of efficient and compact planar optical waveguides, with active (doped) and/or passive (undoped) purposes, for integrated optical devices [4]. Light propagation in passive waveguides depends strongly on the layer structure and morphology, whereas propagation losses or attenuation must be kept as low as possible. Several factors are considered to disturb the light propagation [4, 47]: (1) absorption losses (AL_1) due to light absorption in the glass waveguide (due to electronic (UV-visible) absorption at the Urbach edge and vibrational (IR) absorption at the multiphonon edge); (2) scattering losses (SL_1) due to refractive index fluctuations; (3) absorption losses (AL_2) by impurities like OH; and (4) scattering losses (SL_2) due to the imperfection of the waveguide structure such as defects or surface roughness.

Porosity, dopants, and crystalline phases within the volume of the waveguide cause volume scattering while surface scattering loss can be significant for rough surfaces. Indeed, these kinds of imperfections can appear in sol-gel glasses-derived waveguides in the form of porosity, cracks, polycrystalline phases, dust contamination, and so on, as a result of deposition operation. However, particularly for sol-gel-derived waveguides, some of the imperfections such as pores and cracks are almost inevitable, as shown in **Figure 8**. Moreover, the porous removal requires high annealing temperatures to densify the waveguides, and thus cracks can occur from internal stress [10].

The principal types of defects that may cause scattering losses in sol-gel waveguides are shown in **Figure 9**.

They can be classified as textural defects (cracks, pores, dust, and surface roughness), compositional defects (nanocrystallites and discontinuous refractive index), or structural defects (thickness changes, internal stress and interaction between film and substrate, etc.). The size and the optical properties of these imperfections control their degree of contribution for the scattering loss with the operational wavelength.

Furthermore, light scattering tends to increase with λ^{-1}. According to Rayleigh's law, when the imperfection size is smaller than λ, the light diffusion is roughly proportional to λ^{-4}.

Therefore, the longer wavelength range produces a weaker propagation loss. Silica glass is transparent in NIR and exhibits a non-negligible attenuation. Alternatives to this material are heavy metal fluorides (ZBLAN) glasses, transparent in the mid-infrared (MIR) wavelength range. For SiO_2 glasses, here are two λ_{min} which correspond to the communication channels at 1310 and 1550 nm, respectively. The minimum absorption depends on the vibrational modes (intrinsic "phonons") of the atomic network and shifts to higher wavelengths with increasing atomic mass. However, although the theoretical predictions are notably more favorable to ZBLAN, in practice, these materials are difficult to obtain either in fiber or in waveguide configuration. SiO_2 gives high gain levels per unit length in silica-based waveguides and thus remains a very attractive optical glass even though their emissions from the energy levels of RE dopants are decidedly linked to be non-radiative than radiative ones. This is because the phonon energy of the sol-gel silica glass is higher than most of other glasses. One way to overcome this problem is to use a low vibrational energy matrix with heavy oxides such as GeO_2 or TeO_2. While sol-gel is well suited to oxides, it is still little studied for other systems. Despite this, germanium sulfide films, based on sol-gel preparation and dip or spin coating, have achieved waveguiding features [48]. Glass hosts typically exhibiting higher RE solubility

http://dx.doi.org/10.5772/intechopen.68534

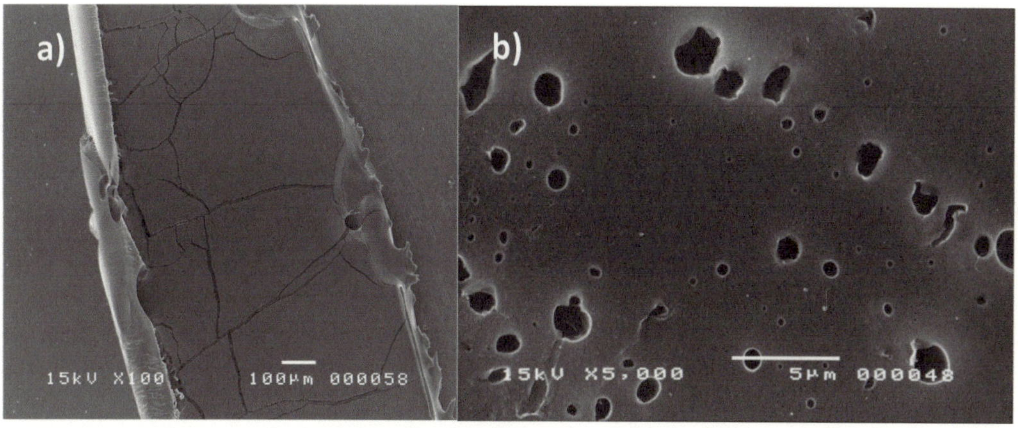

Figure 8. SEM images of Er^{3+}:SiO_2-TiO_2-Al_2O_3 sol-gel glasses: (a) at 500°C thermal treated and (b) as deposited, where defects as micro-cracks and pores can be observed, respectively.

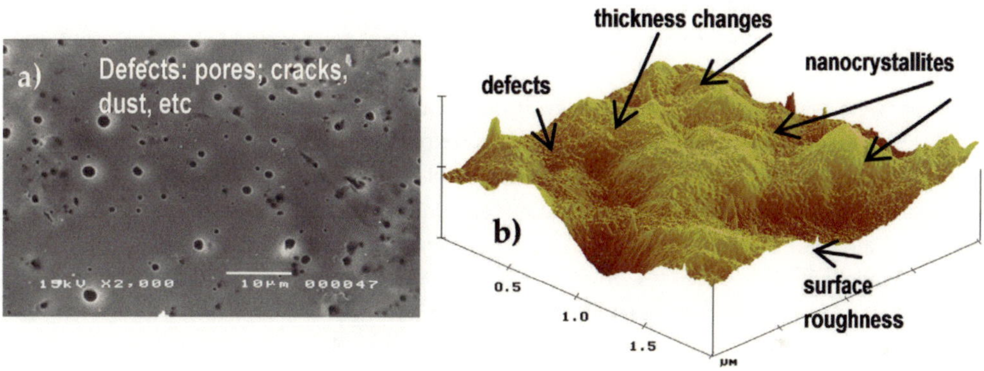

Figure 9. Different textural, compositional, and structural defects on sol-gel thin films of Er^{3+}:SiO_2-TiO_2-P_2O_5: (a) SEM images of the thin film annealed at 200°C, exhibiting a highly porous surface; (b) AFM of the thin film annealed at 700°C exhibiting a singular morphology and an average roughness.

and lower phonon energies are highly favorites to develop active waveguides as they result in longer excited state lifetimes of RE emissions. However, beyond the energy lost to host glass through lattice vibrations, there are two critical additional quenching mechanisms in sol-gel glasses: (1) energy transfer to hydroxyl (OH) groups owing to the some fundamental vibrations (between 2700 and 4000 nm) which are unavoidable due to the use of water and alcoholic solvents during the preparation process and (2) energy transfer between RE ions due to cluster formation even at lower concentration (concentration quenching). Indeed, a fluorescence intensity decrease is commonly attributed to cross-relaxation mechanisms of RE ions in clustering RE—O—RE bonds [27]. The Er^{3+} fluorescence emission of the $4I_{13/2} \rightarrow {}^4I_{15/2}$ transition in silicate glasses exhibits a characteristic shape, as depicted in **Figure 10** [21], which can, however, change its intensity due to non-radiative rates related to ion clustering [27] but also related to the presence of pores and passive or not optically active crystalline phases (e.g., anatase or rutile in the case of SiO_2-TiO_2 glasses) within the matrix [21].

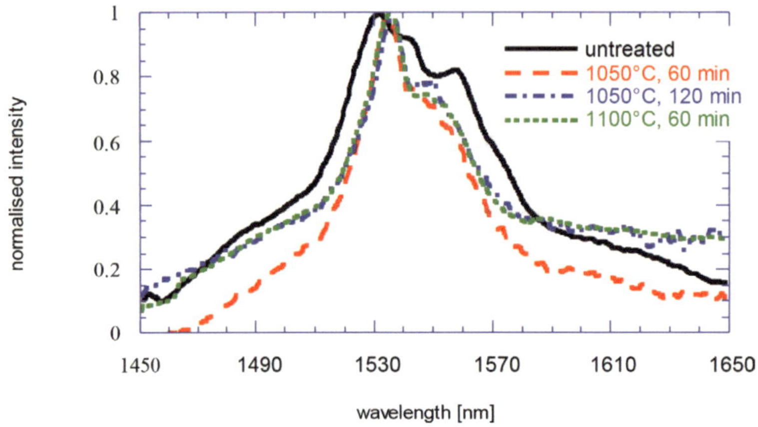

Figure 10. Erbium fluorescence in Er^{3+}:SiO_2-TiO_2 films for different heat treatments.

Fourier transform infrared spectroscopy (FTIR) is one of the most popular analytical techniques used to study the microstructural evolution of sol-gel glasses as a function of temperature and synthesis parameters. The FTIR spectrum of $80SiO_2$-$20TiO_2$ glass, as soon after deposition (wet), is shown in **Figure 11a**. The broad peak around 3300 cm^{-1} is the fundamental stretching vibration of OH group, which reveals, as expected, the presence of hydroxyl groups in the gel, the *fingerprint* of sol-gel glasses. Hydroxyl concentrations, in ppm, can be determined using the Beer-Lambert law [49]. Moreover, the peak around 1600 cm^{-1} is assigned to the bending mode of the H_2O molecule. The other bands relate to characteristic vibrations of the glass structure itself, namely the dominant band at 1080 cm^{-1} with a shoulder at 1185 cm^{-1} is assigned to asymmetric stretching vibrations of the tetrahedral SiO_4^- unit (Si—O—Si); the peak arising at 465 cm^{-1} results from rocking vibrations of Si—O—Si bonds and the band around 980 cm^{-1} is assigned as stretching vibration of silanol groups, Si—OH, owing to non-bridging oxygen which overlaps stretching vibrations of Si—O—Ti bonds at ~950 cm^{-1}. This latter is usually taken as an indicator of the degree of condensation revealing the homogeneity in the glass network, whereas the 1080 cm^{-1} band is indicative of SiO_2-rich regions. The effect of temperature on OH band, exhibiting a pronounced lowering at temperatures higher than 600°C, is clear in **Figure 11b**. In fact, the OH-stretching vibration in Er^{3+}-doped silicate glasses affects the fluorescence decay at 1550 nm because only two OH vibrations are enough to bridge the gap of ~6500 cm^{-1} between the $^4I_{15/2}$ and the $^4I_{13/2}$ states. The presence of OH species in the matrix also leads to reduction of quantum yield at the first excited state of Er^{3+} ions. However, there are chemical limitations to the solubility of RE in various glassy materials which become even more significant with an increase in the concentration of RE ions. This limitation is explained by a disparity in size and valence between the RE ions and the SiO_2 network components. Hence, an effective way to increase the amounts of RE ions in a SiO_2 glass and avoid clustering effects is the addition of co-doping oxides, such as P_2O_5 or $A1_2O_3$ [28, 50].

When a non-glass-forming oxide such as Al_2O_3 is added to an SiO_2 glass, the Si—O—Si bond is broken, which leads to the formation of two types of oxygen: the oxygen that is attached to

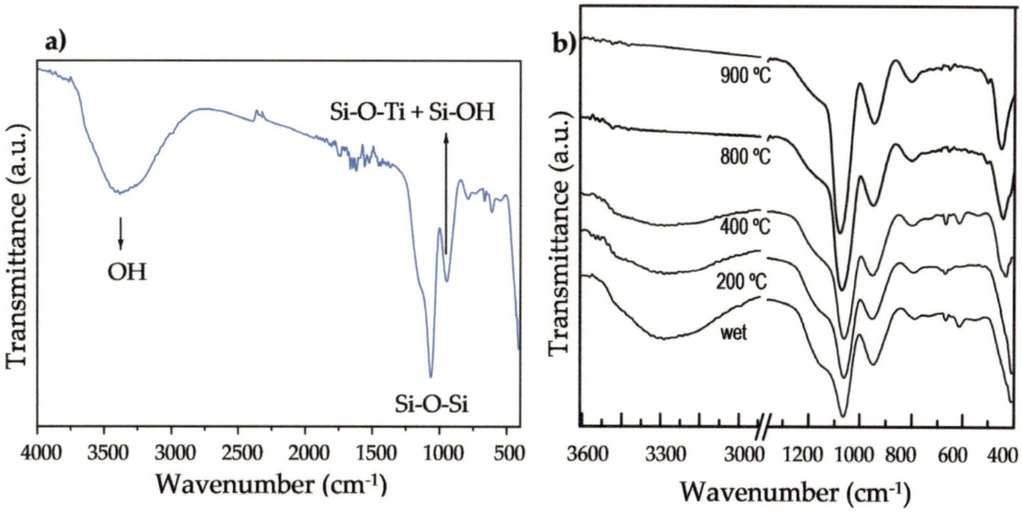

Figure 11. (a) Transmission IR spectrum for the composition $80SiO_2$-$20TiO_2$ immediately after deposition; (b) FTIR spectra of Er^{3+}:$80SiO_2$-$20TiO_2$ glasses heat treated up to 900°C.

two Si, called the bridge oxygen (BO) and the other, which is connected to one Si, called non-oxygen bridge (NBO). Each RE ion solubilized in the glass matrix needs three NBOs to compensate the 3+ charges. Since RE ions cannot induce the necessary coordination number of NBO, therefore the RE clustering becomes energetically more favorable to share the limited number of NBO. Rather, the Al^{3+} ions embedded in the silicate glass matrix can be incorporated in two local bonding configurations such as tetrahedrally coordinated ($AlO_{4/2}$), as a network former or octahedrally coordinated ($AlO_{6/2}$), as a network modifier. Therefore, Al—O—RE bonds are formed instead of RE—O—RE ones, thus increasing the RE solubility. Also, the phonon energy of the Al—O—Si bonds is smaller than that of Si—O—Si vibration. Besides, P_2O_5 is also used as co-doping agent to improve the fluorescent properties of RE ions in sol-gel glasses [16, 21]. In fact, the breaking up of the RE—O—RE regions and formation of RE—O—P or RE—O—Al bonds is highly likely and allows, for example, Nd^{3+} contents as high as 7 wt% before phase separation or clustering are observed [51]. FTIR and X-ray photoelectron spectroscopy (XPS) can provide valuable information about bonding configuration of glasses and a quantitative estimation of the Si—O—NBO species and thus contribute to improve the design of new active waveguides. **Figure 12** shows the FTIR spectra of $80SiO_2$-$20TiO_2$-xP_2O_5 (x = 0, 5, 10 mol%) sol-gel films. The growing incorporation of phosphorus in the SiO_2-TiO_2 matrix leads to a decrease of the band at 950 cm^{-1} which reveals an inhibition of the formation of Si—O—Ti^{4+} terminal species in favor of Si—O—P—O —Ti^{4+} species [28]. XPS measurements allow estimation of the concentrations of the different types of oxygen atoms present in the Si—O—Ti, Si—O—P and Si—O—Al bonding sequences. The O 1s XPS spectra for $80SiO_2$-$20TiO_2$-xP_2O_5 (x = 0–15 mol%) sol-gel films are shown in **Figure 13**, which displays the existence of different chemical environments for the oxygen atoms. Based on the electronegativities of the different elements involved (O, Ti, Si, and P), the individual bond types are assigned as follows: 530.7 eV, to Ti—O—Ti bonds; 531.5 eV, to Si—O—Ti bonds; 533.8 eV, to Si—O—Si bonds; and 534.4 eV, to Si—O—P bonds, since higher the ionic character of the

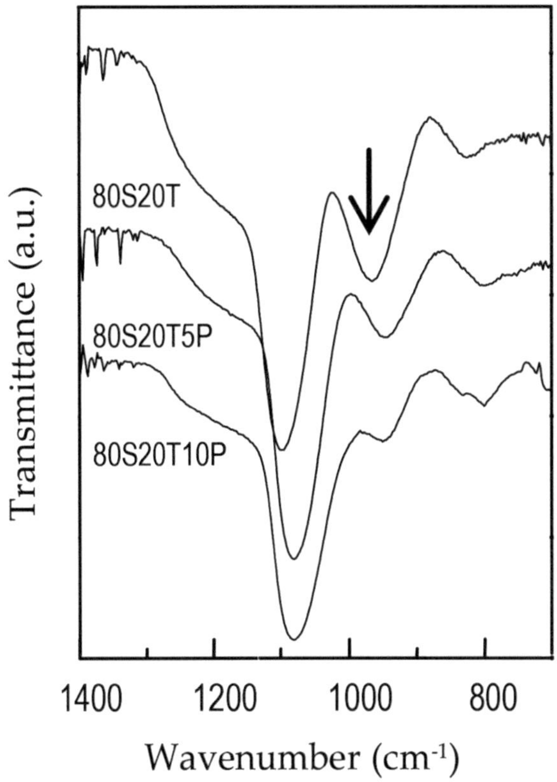

Figure 12. FTIR spectra of $80SiO_2$-$20TiO_2$-xP_2O_5 (x = 0–10 mol%) glasses heat treated at 900°C, 2 min.

bonds, lower is the O 1s binding energy. The fraction of oxygen atoms in Si—O—Ti and Si—O—P bonds can be calculated from the measured areas of each of the two components relative to the total O 1s peak area.

The influence of the high vibrational energy of SiO_2-based glasses on non-radiative relaxation can be minimized by the incorporation of RE ions into nanocrystals with low-energy phonons dispersed in the matrix. This ordered environment would avoid any RE ions clustering. The interest demonstrated in recent years by nanocomposite GC is due to the possibility of creating a low-vibrational energy neighborhood around RE ions leading to a higher luminescent efficiency. Nevertheless, a nanocrystal larger than 10–15 nm drastically increases the propagation losses. This alternative becomes especially interesting for the following reasons: (1) a host material (the nanocrystal) is used in the vicinity of the RE ions in order to optimize the spectroscopic properties; (2) the matrix is still a suitable fiber junction material, with superior processing and high stability. The optical properties of waveguides doped with nanocrystals (fluorescence spectrum, fluorescence life, and optical losses) were found to be basically dependent on the following parameters: (a) nature of the nanocrystalline phases obtained; (b) crystallite sizes; (c) volume fraction of crystalline phases dispersed in the amorphous matrix; and (d) residual concentration of RE ions retained in the amorphous matrix. The XRD patterns of $80SiO_2$-$20TiO_2$-$ErO_{1.5}$ sol-gel thin films in **Figure 14** show different crystalline phases. At 1000°C, only anatase

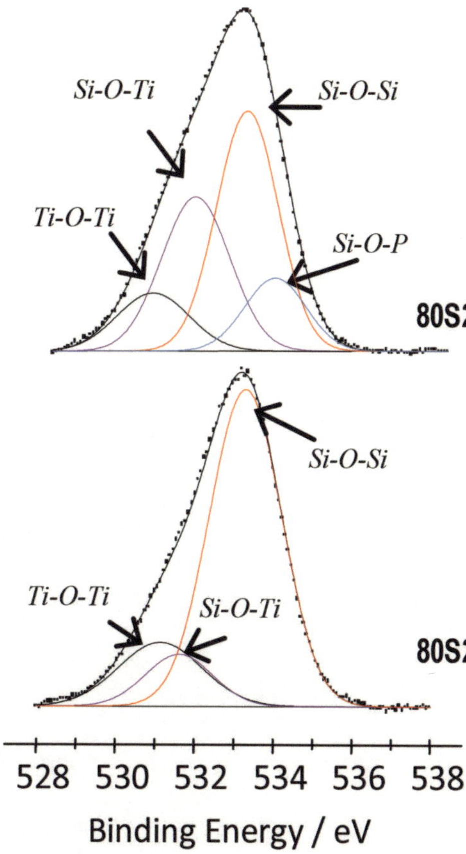

Figure 13. Deconvoluted spectra of O 1s $80SiO_2$-$20TiO_2$-xP_2O_5 (x = 0–10 mol%) glasses.

phase, one of the crystalline polymorphic forms of TiO_2, among rutile and brookite, is evidenced by its main peak at 25.28, which corresponds to (101) plane. By increasing the annealing temperature, the decomposition of anatase phase took place and a transition from anatase to rutile crystalline phase occurs.

The main peak of the film annealed at 1050°C is at 27.44, which corresponds to rutile (110) plane. The active phase $Er_2Ti_2O_7$ precipitates together with rutile at 1100°C. Both anatase and rutile are passive phases that can damage the fluorescence quantum yield by scattering losses. In fact, high treatment annealing lowers the OH content but can increase the losses scattering from non-optical crystals such as rutile.

The interest in $Er_2Ti_2O_7$ is obviously because Er^{3+} ions are inserted in a locally well-ordered phase thus giving relatively sharp photoluminescence emissions in a wide range of spectral bands from infrared to the blue region.

Results indicated clearly that the luminescence was significantly improved while erbium was present in the form of $Er_2Ti_2O_7$ crystallites dispersed in an amorphous matrix [21]. When the content of Er_2O_3 (the most usual erbium precursor) greatly exceeds the solubility limit of the

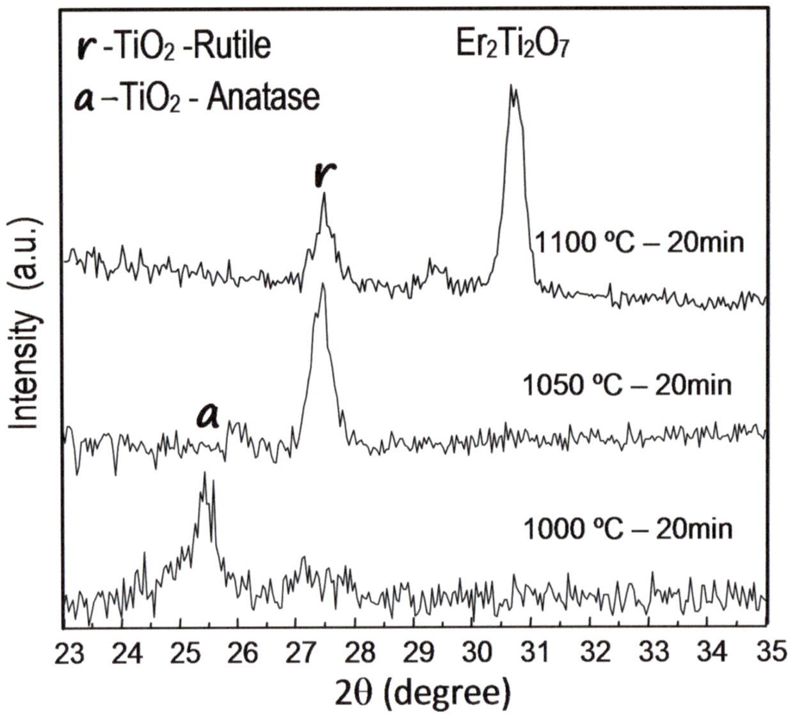

Figure 14. X-ray diffraction pattern of $80SiO_2$-$20TiO_2$-$ErO_{1.5}$ waveguides annealed at 1000, 1050, and 1100°C.

glass, it reacts almost entirely with P_2O_5 forming $ErPO_4$ (EPO) crystallites. In the SiO_2-TiO_2-P_2O_5 system, the $ErPO_4$ particles are crystallized during the glass-annealing process. It has been shown that $ErPO_4$ nanocrystals resulted in an increase in the fluorescence lifetime at 1550 nm greater than 200% with a maximum value of 9 ms [21]. **Figure 15** shows the room temperature fluorescence spectra relative to the $^4I_{13/2} \rightarrow {}^4I_{15/2}$ obtained upon 514.5-nm excitation, for $80SiO_2$-$20TiO_2$-$ErO_{1.5}$ waveguides. All the spectra exhibit a main emission peak at 1530 nm with a shoulder at

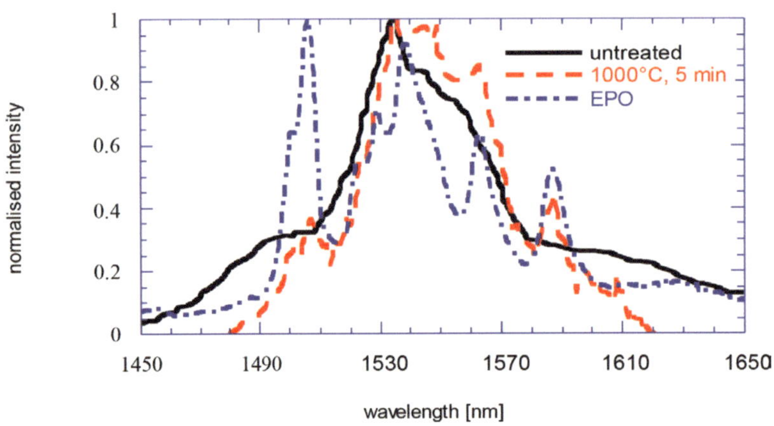

Figure 15. Fluorescence spectrums relative to the $^4I_{13/2} \rightarrow {}^4I_{15/2}$ transition of the Er^{3+} ions for $80SiO_2$-$20TiO_2$-P_2O_5-$ErO_{1.5}$ waveguides at different heat treatments, upon excitation at 514.5 nm.

about 1550 nm. $ErPO_4$, on the other hand, emits a well-detectable fluorescence band around 1535 nm, with four peaks due to the Stark splitting of the $^4I_{13/2}$ level in the crystalline compound.

6. Conclusions

Glass-integrated optics is controlled by oxide glasses, in part because of advantages of production techniques. At present, the Er^{3+}-doped optical silicate glass is a very important light-amplifying element. However, non-oxides are receiving increasing interest for optical applications. Their low phonon energies make them useful for RE doping for lasers and amplifiers, particularly for doping with Pr^{3+} which have rapid non-radiative relaxation rates in oxides. A suitable approach to improve fluorescence lifetime of RE ion-doped glasses is to shield ions from the glass matrix in a local environment, such as a nanocrystallite, allowing optical properties of the host glass to be maintained. For the future, sol-gel has high potential for optical devices. A deep knowledge about materials physics and quantum optics is therefore essential to achieve these challenges.

Acknowledgements

The authors gratefully acknowledge the financial support from (1) FEDER, through Programa Operacional Factores de Competitividade—COMPETE and Fundação para a Ciência e a Tecnologia—FCT, by the project UID/FIS/00068/2013 and (2) Government of the Azores (DRCT - Programme PRO-SCIENTIA, Ref. M3.3.c/Edições/009/2017). The authors also wish to thank Dr. F. Rivera-López (ULL) for providing images used in this manuscript.

Author details

Helena Cristina Vasconcelos[1,2,3]* and Afonso Silva Pinto[3]

*Address all correspondence to: helena.cs.vasconcelos@uac.pt

1 Biotechnology Centre of Azores (CBA), Azores University, Ponta Delgada, Açores, Portugal

2 Centre of Physics and Technological Research (CEFITEC), FCT/UNL—Department of Physics, Quinta da Torre, Caparica, Portugal

3 Faculty of Science and Technology, Azores University, Ponta Delgada, Açores, Portugal

References

[1] Righini JC. 25 years of integrated optics: where we are and where we will go. Proc. SPIE. 2212, Linear and Nonlinear Integrated Optics, Giancarlo C. Righini, David Yevick, Editors, 2. (1994) DOI: 10.1117/12.185098

[2] Rimoldi, B. (2016) Principles of Digital Communication: A Top-Down Approach, Published by Cambridge University Press, United Kingdom. https://doi.org/10.1017/CBO9781316337387

[3] Digonnet MJF. Rare Earth Doped Fiber Lasers and Amplifiers. 2nd ed. New York, NY: Marcel Dekker Inc.: 1993. DOI: 10.1201/9780203904657

[4] Righini GC. Passive and active glasses for integrated optics. In: Mazzoldi P, editor. From Galielo's Ochialino to Optoelectronics. Singapore: World Scientific; 1993. pp. 272-294. DOI: 10.1604/9789810213329

[5] Orignac X, Almeida RM. Silica-Based Sol-Gel Optical Waveguides on Silicon in IEE Proceedings—Optoelectronics. 1996;**143**(5):287–292. DOI: 10.1049/ip-opt:19960834

[6] Colinge J-P. Silicon-On-Insulator Technology: Materials to VLSI. Boston, MA, USA: Kluwer Academic Publishers; 2004. DOI: 10.1007/978-1-4757-2611-4

[7] Hu H, Ricken R, Sohler W, Wehrspohn RB. Lithium niobate ridge wave-guides fabricated by wet etching. IEEE Photonics Technology Letters. 2007;**19**:6. DOI: 10.1109/LPT.2007.892886

[8] Bindra KS, Bookey HT, Kar AK, Wherrett BS. Nonlinear optical properties of chalcogenide glasses: Observation of multiphoton absorption Applied Physics Letters. 2016;**79**:1939. DOI: http://dx.doi.org/10.1063/1.1402158

[9] Kawachi M. Silica waveguides on silicon and their application to integrated-optic components. M. Optical and Quantum Electronics. 1990;**22**:392. DOI: 10.1007/BF02113964

[10] Brinker CJ, Scherer GW. Sol-Gel Science-The Physics and Chemistry of Sol-Gel Processing. San Diego; Academic Press: 1990. DOI: 10.1002/adma.19910031025

[11] Kreibig U, Vollmer M. Optical properties of metal clusters. Spring Series in Materials Science. 1995;**25**:13–201. DOI: 10.1007/978-3-662-09109-8

[12] Diehl R. High-Power Diode Lasers: Fundamentals, Technology, Applications. Springer Berlin Heidelberg; 2000. ISBN: 978-3-540-66693-6, DOI: 10.1007/3-540-47852-3

[13] Snitzer E. Optical maser action of Nd^{+3} in a barium crown glass. Physics Review Letters. 1961;**7**:444. DOI: https://doi.org/10.1103/PhysRevLett.7.444

[14] Layne CB, Lowdermilk WH, Weber MJ. Multiphonon relaxation of rare-earth ions in oxide glasses. Journal of Material Science Letters. 1994;**13**:615. DOI: https://doi.org/10.1103/PhysRevB.16.10

[15] Vasconcelos HC, Meirelles MG, Rivera-López F. Erbium photoluminescence response related to nanoscale heterogeneities in sol-gel silicates. Journal of Rare-Earths. 2013;**31**:18-26. DOI: 10.1016/S1002-0721(12)60228-2

[16] Vasconcelos HC. The effect of PO2,5 and AlO1,5 additions on structural changes and crystallization behaviour of SiO_2-TiO_2 sol-gel derived glasses and thin films. Journal of Sol-Gel Science and Technology. 2010;**55**:126-136. DOI: 10.1007/s10971-010-2223-8

[17] Miniscalco WJ. Erbium-doped glasses for fiber amplifiers at 1500 nm. Journal of Lightwave Technologies. 1991;**9**:234-250. DOI: 10.1109/50.65882

[18] Santa Cruz P, Morin D, Dexpert-Ghys J, Sadoc A, Glas F, Auzel F. New lanthanide-doped fluoride-based vitreous materials for laser applications. Journal of Non-Crystalline Solids. 1995;**190**:238-243. DOI: 10.1016/0022-3093(95)00273-1

[19] Kenyon AJ. Recent developments in rare-earth doped materials for optoelectronics. Progress in Quantum Electronics. 2002;**26**:225-284. DOI: 10.1016/S0079-6727(02)00014-9

[20] Santana-Alonso A, Yanes AC, Méndez-Ramos J, del-Castillo J, Rodríguez VD. Sol–gel transparent nano-glass-ceramics comprising rare-earth-doped $NaYF_4$ nanocrystals. Physica Status Solidi (a). 2009;**206**:2249-2254. DOI:10.1002/pssa.200881717

[21] Strohhöfer C, Fick J, Vasconcelos HC, Almeida RM. Active optical properties of Er-containing crystallites in sol-gel derived glass films. Journal of Non-Crystalline Solids. 1998;**226**:182. DOI: 10.1016/S0022-3093(98)00365-2

[22] X. Orignac, D. Barbier. Potential for fabrication of sol-gel-derived integrated optical amplifiers. Proc. SPIE 2997, Integrated Optics Devices: Potential for Commercialization, Edited by S. Iraj Najafi and Mario Nicola Armenise, 271 (January 23, 1997); DOI:10.1117/12.264157

[23] Yan Y, Faber AJ, de Waal H. Luminescence quenching by OH groups in highly Er-doped phosphate glasses. Journal of Non-Crystalline Solids. 1995;**181**:283-290. DOI: 10.1016/S0022-3093(94)00528-1

[24] Rawson H. In: Glass Chemistry. Inorganic Glass-Forming Systems. London, UK: Academic Press; 1967. pp. 123-207. DOI: 10.1007/978-3-642-78723-2_7

[25] Weber MJ. Science and technology of laser glass. Journal of Non-Crystalline Solids. 1990;**123**(1–3):208-222. DOI: 10.1016/0022-3093(90)90786-L

[26] Scriven LE. Physics and applications of dip coating and spin coating. Materials Research Society Symposium Proceedings. 1988;**121**:717. DOI: 10.1557/PROC-121-717

[27] cbv Weisembach L, Zelinski BJ, O'Kelly J, Morreale J, Roncone RL, Burke JJ. The influence of processing variables on the optical properties of SiO_2-TiO_2 planar waveguides. SPIE. 1991;**1590**:50. DOI: 10.1117/12.50201

[28] McGahay V, Tomozawa M. Phase separation in rare-earth-doped SiO_2 glasses. Journal of Non-Crystalline Solids. 1993;**159**(3):246–252. DOI: 10.1016/0022-3093(93)90230-U

[29] Mukher Jee SP. In: Klein LC, editor. Sol-Gel Technology for Thin Films, Fibers, Preform, Electronics and Specialty Shapes. NJ: Noyes Publications; 1988. DOI: 10.1002/adma.19890010816

[30] García Solé J, Bausá LE, Jaque D. An introduction to the optical spectroscopy of inorganic solids. New Delhi, India: John Wiley & Sons, Ltd.; 2005. p. 283. DOI: 10.1002/0470016043.app3

[31] Jablonski A. Über den Mechanisms des Photolumineszenz von Farbstoffphosphoren. Journal of Physics. 1935;**94**:38-46. DOI: 10.1007/BF01330795

[32] Reisfeld R. Radiative and non-radiative transitions of rare-earth ions in glasses. Structure & Bonding. 1975;**22**:123-175. DOI: 10.1007/BFb0116557

[33] Haro-González P, Lahoz F, González-Platas J, Cáceres JM, González-Pérez S, Marrero-López D, Capuj N, Martín IR. Optical properties of Er^{3+}-doped strontium barium niobate nano-crystals obtained by thermal treatment in glass. Journal of Luminescence. 2008;**128**:908-910, 2008. DOI: 10.1016/j.jlumin.2007.12.014

[34] Lakowicz JR. Principles of Fluorescence Spectroscopy. 3rd ed. New York, NY: Springer; 2006. DOI: 10.1007/978-0-387-46312-4

[35] Förster T. Zwischenmolekulare Energiewanderung und Fluoreszenz, Annals in Physics. 1948;**2**:55. DOI: 10.1002/andp.19484370105

[36] Dexter DL. A Theory of Sensitized Luminescence in Solids, Journal of Chemistry and Physics. 1953;**21**:836. DOI: 10.1063/1.1699044

[37] Bo F, Céline P, Jean-Luc A, Xianghua Z, Xianping F, Hongli M. Near-infrared down-conversion in rare-earth-doped chloro-sulfide glass GeS_2-Ga_2S_3-CsCl: Er, Yb. Journal of Applied Physics. 2011;**110**(11):113107. DOI: 10.1063/1.3665638

[38] Orignac X, Barbier D, Du XM, Almeida RM, McCarthy O, Yeatman E. A Theory of Sensitized Luminescence in Solids, Optical Materials 1999;**12**(1):1–18. DOI: 10.1016/S0925-3467(98)00076-7

[39] Wybourne BG. Spectroscopic properties of rare earths. Physics Today. 1965;**18**(9):70. DOI: 10.1063/1.3047727

[40] Peacock RD, Nieboer E, Jørgensen CK, Reisfeld R. The intensity of lanthanide f↔f transitions. Structure and Bonding. 2007;**22**:83–122. DOI: 10.1007/BFb0116556

[41] Weber MJ. The Role of Lanthanides in Optical Materials. Report Number: LBL-37536. 1995. Available from: https://publications.lbl.gov/islandora/object/ir%3A102074

[42] Dieke GH. Spectra and Energy Levels of Rare-Earth Ions in Crystals. New York, NY: Wiley; 1968. DOI: 10.1119/1.1976350

[43] Judd BR. Optical Absorption Intensities of Rare-Earth Ions, Physics Review B. 1962;**127**:750. DOI: 10.1103/PhysRev.127.750

[44] Ofelt GS. Intensities of Crystal Spectra of Rare-Earth Ions, Journal of Chemistry and Physics. 1962;**37**:511. DOI: 10.1063/1.1701366

[45] Jorgensen CK, Reisfeld R. Judd-Ofelt parameters and chemical bonding. Journal of the Less Common Metals. 1983;**93**(1):107–112. DOI: 10.1016/0022-5088(83)90454-X

[46] Lahoz F, Capuj N, Haro-Gonzalez P, Martın IR, Perez-Rodrıguez C, Caceres JM. Stimulated emission in the red, green, and blue in a nanostructured glass ceramics, Journal of Applied Physics. 2011;**109**:043102. DOI: 10.1063/1.3549157

[47] Almeida RM, Morais PJ, Vasconcelos HC. Optical loss mechanisms in nanocomposite sol-gel planar waveguides. Proceedings of the SPIE. 1997;**3136**:296–303. DOI: 10.1117/12.284127

[48] R. M. Almeida , J. Xu, Sol–Gel Processing of Sulfide Materials, Handbook of Sol-Gel Science and Technology, Published by Springer, pp 1–26, 2016. DOI: 10.1007/978-3-319-19454-7_11-1

[49] Davis K, Agarwal A, Tomozawa M, Hirao K. Quantitative infrared spectroscopic measurements of hydroxyl concentrations in silica glass. Journal of Non-Crystalline Solids. 1996;**203**:27–36. DOI: 10.1016/0022-3093(96)00330-4

[50] Almeida RM, Vasconcelos HC, Gonçalves MC, Santos LF. XPS and NEXAFS studies of rare-earth doped amorphous sol-gel films. Journal of Non-Crystalline Solids. 1998;**232–234**:65–71. DOI: 10.1016/S0022-3093(98)00545-6

[51] Bowron DT, Newport RJ, Rigden JS, Tarbox EJ, Oversluizen M. An X-ray absorption study of doped silicate glass, fibre optic preforms, Journal of Materials Science. 1996;**31**:485. DOI:10.1007/BF01139168

Chapter 6

Recent Characterisation of Sol-Gel Synthesised TiO_2 Nanoparticles

Muhamad Zamri Yahaya, Mohd Asyadi Azam,
Mohd Asri Mat Teridi, Pramod Kumar Singh and
Ahmad Azmin Mohamad

Additional information is available at the end of the chapter

http://dx.doi.org/10.5772/67822

Abstract

High demand and current applications have led to continuous study and subsequent improvement of TiO_2 nanoparticles. The versatility of the sol-gel method allows employing different process parameters to influence the resultant properties of TiO_2 nanoparticles. The evaluation and characterisation process of the synthesised TiO_2 nanoparticles commonly involves a series of methods and techniques. Such characterisation methods include phase, structural, morphology and size analysis. A combination of data from these evaluations provides the relationship between the synthesis parameters and the end properties of TiO_2 nanoparticles. Apart from the research findings on TiO_2 nanoparticles, the characterisation used to obtain these findings is equally important. Thus, this chapter highlights the recent characterisation techniques and practices employed for TiO_2 nanoparticles synthesised by the sol-gel method.

Keywords: TiO_2 nanoparticles, sol-gel, phase analysis, Rietveld refinement, morphology, particle size

1. Introduction

Developments in the polymorphic TiO_2 nanoparticles have extensively drawn major interest of researchers and scholars. The wide exploitation of TiO_2 nanoparticles includes the study and application in photovoltaics [1], photocatalysis [2], batteries [3], filler material in composites [4] and biomedical products [5]. The key issues addressed focused on achieving better characteristics of TiO_2 nanoparticles as well as improved performance of end devices. This explains the large number of ongoing research studies considering the versatility of TiO_2 nanoparticles in a broad range of applications.

The significant interest on TiO_2 nanoparticles was driven by the unique characteristics of the material, such as low preparation cost, non-toxic, favourable band edge positions and diverse morphologies possibilities [6–8]. Factors determining these end properties are mainly dependent on the synthesis routes used to produce TiO_2 nanoparticles. Since 1971, the sol-gel has been the most significant synthesis method known and applied in producing multi-component oxides such as TiO_2 [9]. Parameters such as the type of precursors [10], the pH of solution [11], preliminary solution treatment [12] and calcination temperatures [13] yielded varying properties of the resultant TiO_2 nanoparticles.

The process of evaluating the performance/defects of TiO_2 nanoparticles involves a series of characterisation methods. To ensure sufficient data, the selection of characterisation methods is highly important. Highly practised characterisations attributed to the evaluation of TiO_2 nanoparticles include:

(i) Structural and phase analysis

(ii) Morphological observations

(iii) Particle size analysis

These analyses allow researchers to determine the effects of the sol-gel parameters for the synthesised TiO_2 nanoparticles. Such information is vital to continuously develop TiO_2 nanoparticles. This explains the reasons why such characterisation methods are highly preferred in the current research works related to TiO_2 nanoparticles.

Currently, there are several available review articles on TiO_2 nanoparticles, which discuss the photocatalytic performances in various applications [14–16]. The preparation and synthesis were critically reviewed as well [17–19]. Other available reviews include the phase and structural transformation of TiO_2 nanoparticles [20, 21]. Technical review on TiO_2 nanoparticles characterisation, however, has received only little attention although it is significantly relevant to the evaluation of TiO_2 nanoparticles.

Thus, this chapter focuses on recent characterisation of TiO_2 nanoparticles synthesised by the sol-gel method. Commonly practised configuration of characterisation corresponding to TiO_2 nanoparticles for over 7 years (2010–2016) is selectively reviewed in this chapter. Few earlier articles were also referred to strengthen the overall understanding on the subject matter.

2. Chemical reaction of the sol-gel method

The sol-gel method is the process of transforming sols (solid particles suspended in liquid) into gels (particulate networks of sols). This involves two main reactions: hydrolysis and condensation, prior to obtaining crystalline TiO_2 nanoparticles by calcination (**Figure 1**). For synthesising TiO_2 nanoparticles, commonly used precursors include $Ti(OBu)_4$ [22], $TiCl_3$ [10],

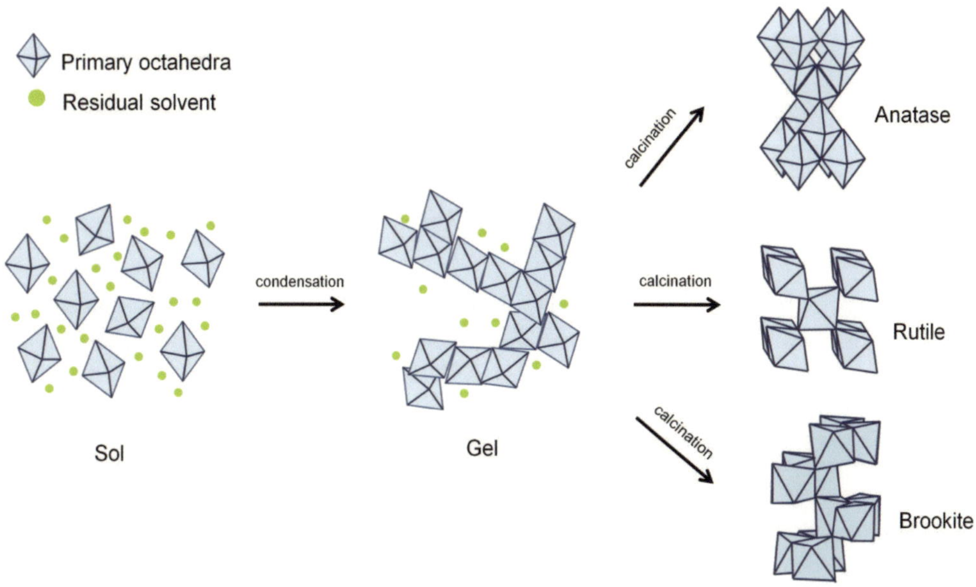

Figure 1. The hydrolysis, condensation and calcination process of the sol-gel method in synthesising the crystalline anatase, rutile and brookite TiO_2 nanoparticles.

$TiCl_4$ [23], $TiBr_4$ [24] and $Ti[OCH(CH_3)_2]_4$ (TTIP) [25]. These precursors were then hydrolysed by adding water (hydrolysis), resulting in the formation of complex three-dimensional network (condensation) as shown in the following equations:

Hydrolysis:

$$Ti(OR)_4 + 4H_2O \rightarrow 2Ti(OH)_4 + 4ROH \tag{1}$$

Condensation:

$$Ti(OH)_4 + Ti(OH)_4 \rightarrow 2TiO_2 + 4H_2O \text{ (oxolation)} \tag{2}$$

$$Ti(OH)_4 + Ti(OR)_4 \rightarrow 2TiO_2 + 4ROH \text{ (alcoxolation)} \tag{3}$$

where R in the equation represents ethyl, i-propyl, n-butyl and so on [26, 27]. The titanium precursor is often diluted before adding water. This reduces the rapid reaction rate of the hydrolysis process.

Size and morphology of the end TiO_2 nanoparticles are highly influenced by the precursor–water ratio [28]. Lower ratio of water–precursor resulted in monodisperse particle of 0.5–1 mm in diameter [11]. For higher ratio values, unstable colloidal and predicates would form and aggregate. Peptisation is commonly carried out for these aggregates to achieve the final size, which is usually less than 100 nm [29]. Higher pH of solution contributed to increased particle size of TiO_2 nanoparticles [30]. The calcination process should be carefully determined as the phase transformation of TiO_2 is highly influenced by the employed temperature [31]. The end structures of crystalline TiO_2 polymorphs (anatase, rutile or brookite) are thus formed from the colloidal suspension, depending on the above parameters.

3. Phase and structural characterisation

The phase and structural analyses are significant characterisation techniques that are usually associated with the main discussion in analysing the current sol-gel-synthesised TiO_2 nanoparticles. The means of X-ray diffraction (XRD) are utilised to qualitatively identify the phases obtained by referring to the XRD databases [32]. Additionally, the data then can be subjected to the Rietveld refinement to yield significant fitting parameters for quantitative evaluations.

Due to nanosized TiO_2 nanoparticles (<100 nm), deviations on the diffraction signal can be avoided to achieve a reliable XRD analysis. Physically, the fine powder form of the TiO_2 nanoparticles provides relatively easy sample handling and preparation, ensuring smooth and flat surface. This is important as sample displacement is the main factor contributing to errors in the determination of structural parameters. The dominant TiO_2 diffraction peaks were found to lie between the angle of $2\theta = 25–30°$. Thus, Bragg angle range of $2\theta = 20–80°$ was practically applied in analysing TiO_2 nanoparticles [33, 34]. Due to the increase in crystallinity, TiO_2 nanoparticles subjected to calcination were commonly associated to higher peak intensities compared with the untreated TiO_2 nanoparticles [35]. The lattice plane corresponding to the particular peak reflects the preferable growth orientation of TiO_2 nanoparticles crystal [32, 36, 37]. Additionally, by employing the Debye-Scherrer equation, the crystallite sizes of the synthesised TiO_2 nanoparticles can be calculated directly.

Examples of phase determination (by search and match) can be observed from the work by Banerjee et al. [38] in examining the addition of Pd and Ga on synthesised TiO_2 nanoparticles. In addition to the observation on the intensities reduction (**Figure 2a**), the observed peak broadening (**Figure 2b**) and deconvolution of peaks (**Figure 2c–e**) in the XRD pattern is also another indication that can be discussed in parallel to the phase identification. These phenomena, however, are often related to structural alteration due to the presence of non-uniform strain, posed by the substitutional and interstitial dopants [39, 40]. Similarly, the work by Chen et al. [33] also utilise the phase analysis in investigating the Fe^{3+}-doped TiO_2 nanoparticles. With reference matched only to the JCPDS File No. 21-1272, the formation of pure anatase phase was reported. Additionally, as secondary phases were absent in the observed XRD pattern, it can be concluded that the added Fe^{3+} from the ferric nitrate had been completely incorporated into the TiO_2 nanoparticles. Relating to the current sol-gel practices, such phase identification is essential as most sol-gel synthesis approaches involve a wide diversity in chemical variations. Apart from that, other effects of Fe^{3+} addition, such as reduction of the peak intensities, were also compared in explaining the crystallinity of TiO_2 nanoparticles.

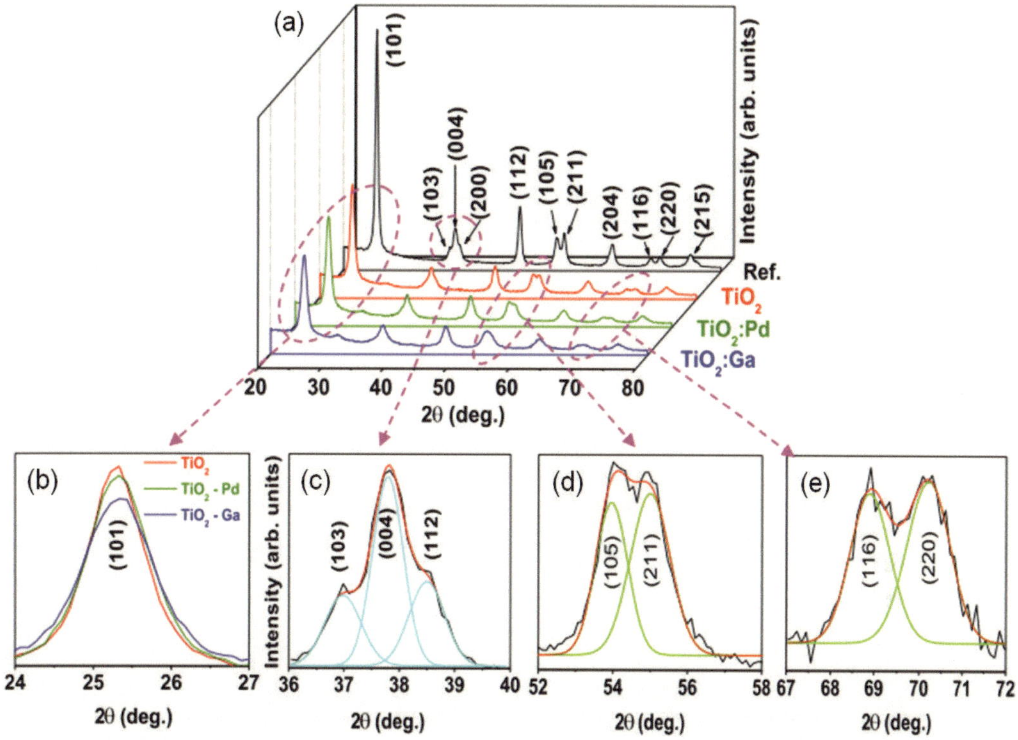

Figure 2. Comparison of XRD patterns for (a) pure TiO_2 nanoparticles and TiO_2 nanoparticles doped with Pd and Ga, (b–e) overlapped XRD curves representing peak broadening and deconvolution with respect to the crystallographic orientation. Reprinted with permission from Ref. [38]. Copyright 2016, Elsevier.

Furthermore, comparison by the phase analysis is highly preferable in investigating large variation of sol-gel techniques. As an example, from the spectra comparison, synthesis with different precipitation conditions clearly portrayed a significant influence on the resultant structural analysis of the TiO_2 nanoparticles (**Figure 3a–b**). It is shown that the storage condition favours earlier rutile transformation as compared to the centrifuge condition [41]. Other example includes the studies on the use of weak/strong acids in favouring different mixtures of TiO_2 polymorphs. Interestingly, pure anatase phase was reported to be more favourable with the usage of milder acid [42]. In another case, different mixtures of anatase and rutile phases were also obtained in the XRD spectra. However, these preferences were mainly associated to the different ranges of applied calcination temperature [43]. In most cases, higher calcination temperature tended to produce rutile, mainly due to its stability compared with anatase and brookite [43, 44]. Apart from that, Potlog et al. [45] reported that the reverse transformation of rutile to anatase is possible by utilising the H_2 environment. Such behaviour can be distinguished clearly from the XRD curves after annealing, which indicates the absence of the rutile (110) peak when compared to before annealing in the H_2 environment.

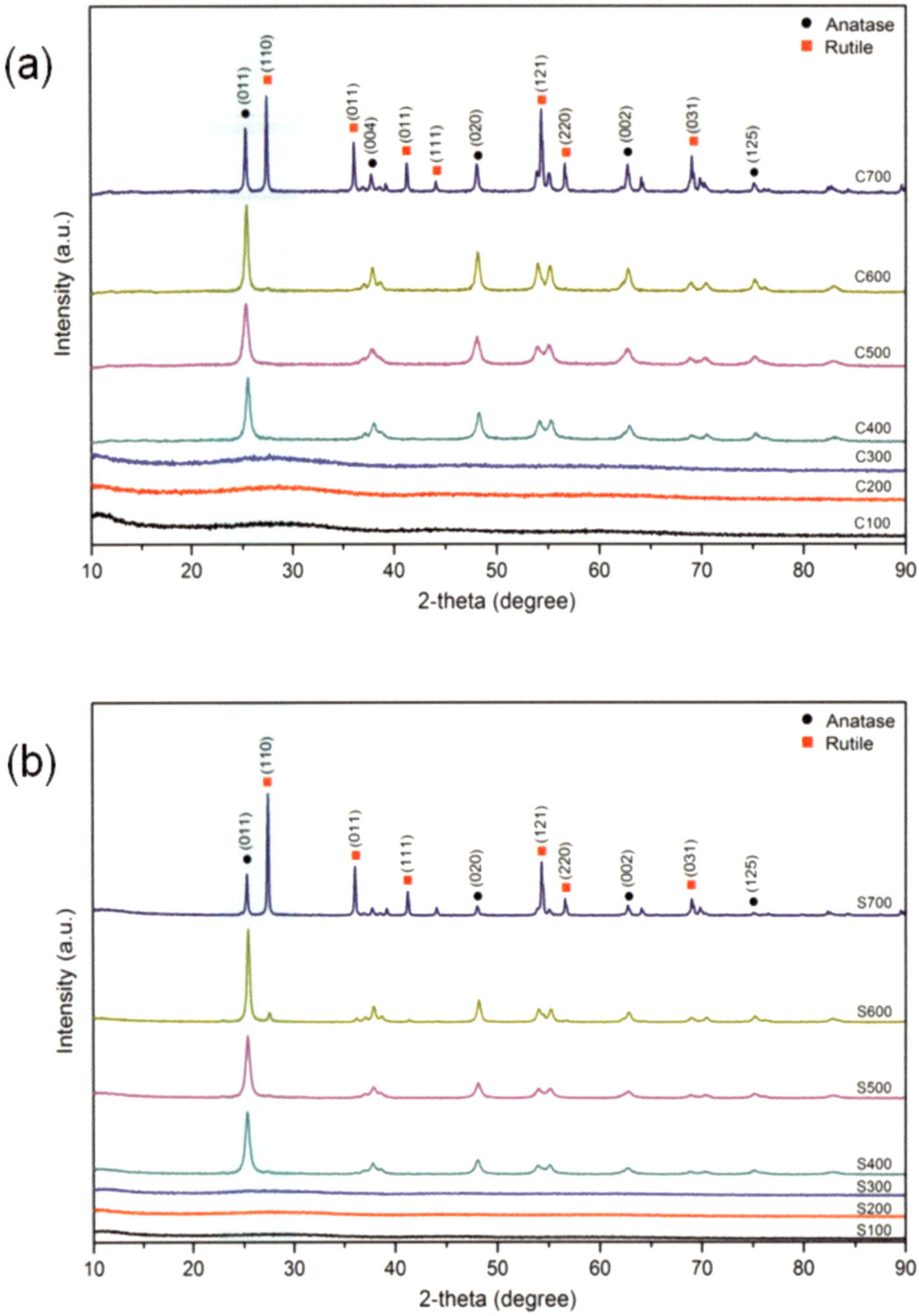

Figure 3. X-ray diffraction patterns of TiO_2 nanoparticles at various calcination temperatures synthesised through (a) centrifuge and (b) storage precipitation conditions (the shaded area represents changes on the anatase and rutile peak). Reprinted with permission from Ref. [41]. Copyright 2015, Elsevier.

4. Rietveld refinement

In Rietveld analysis, the experimental XRD pattern is fitted onto a reference data, yielding a resulted/simulated XRD model. Differences between the experimental and simulated XRD produce the fitting parameters through the Rietveld algorithm. Such parameters (goodness of fit (GOF), R_{exp} and R_{wp}) reflect the reliability of the corresponding XRD pattern. To relate, the analysis was significantly practised to further characterise TiO_2 nanoparticles, indicating the reliability of the XRD analysis and the synthesis process as a whole.

The refinement process utilises the pattern and data obtained from XRD. Thus, factors that affect the XRD analysis are also responsible in determining the accuracy of the Reitveld refinement. In the phase quantification analysis, the evaluations are attributed to the intensities of the diffraction peaks, yielding the phase percentages of TiO_2 anatase, rutile or brookite. Additionally, the increment on the full width at half maximum (FWHM) of TiO_2 polymorph peaks reduces the resultant crystallite sizes of these phases [46].

During refinement, various parameters (peak shape, lattice parameter and atomic position) were adjusted to achieve the best-fit model, commonly discussed as the visual fit process. These optimised models were generated from the differences between the observed and computed intensities to yield fitting parameters such as the weighted R profile (R_{wp}), R expected (R_{exp}) and the goodness of fit (GOF) [47]. The values vary accordingly to the different sol-gel approach employed in synthesising TiO_2 nanoparticles. For instance, typical GOF values obtained for the anatase TiO_2 synthesised by various heat treatments were in the range of 1.0–2.0 [41, 48].

As an example, the fitted model obtained by Yahaya et al. [41] was used in determining the structural parameters of TiO_2 nanoparticles synthesised by centrifuge and storage precipitation (**Figure 4a**). Most Reitveld discussions focused on this crucial plot (as it is the sole figure directly representing the visual fit) by observing the displacement of the peaks between the calculated and observed pattern. Then, from such displacement, the fitting parameters were generated (**Figure 4b–c**). From these quality indicators ($R_{wp,}$ R_{exp} and GOF), the centrifuge condition was quantitatively determined as the synthesis route to ensure better reliability in the XRD analysis. In most refinement works, these parameters are the commonly reported results, which, however, are significant towards the phase and structural analysis/discussion of the synthesised TiO_2 nanoparticles [49].

In different cases, the usefulness of the Rietveld analysis was capitalised in investigating chemical reactions of TiO_2 composite coatings [50]. The approach was to compare the refined pattern of pure rutile and anatase TiO_2 with the refined pattern of the hydroxyapatite (HAP)/ TiO_2 composite coatings. Two main points were used to prove that the chemical reactions did not occur. The first was by obtaining similar structural parameters of TiO_2 and HAP in the composites compared with the parameters of the corresponding phases individually. The second was by obtaining comparable occupancy factor value (acquired from the refinement) of the HAP for all HAP/TiO_2 composite samples (similar values indicate retaining of its structure by HAP, which reflect the absence of any reactions) [51]. In complex TiO_2 sol-gel synthesis,

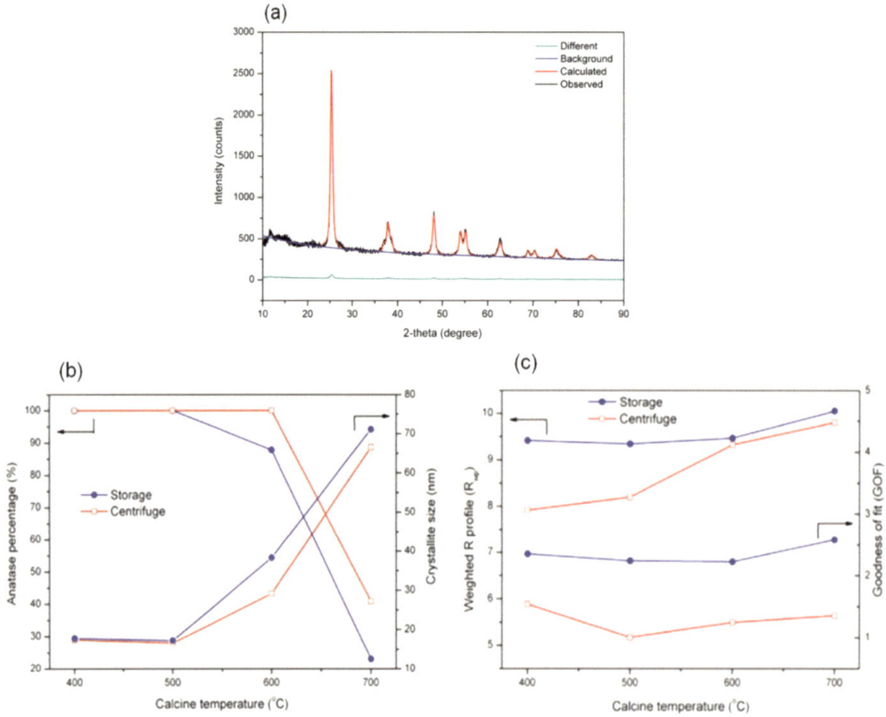

Figure 4. Example of the fitted profile for (a) the TiO_2 nanoparticles synthesised by centrifuging and 500°C calcination temperature and the Rietveld analysis for the synthesised TiO_2 nanoparticles, (b) Anatase percentage and crystallite size and (c) Weighted R profile. Reprinted with permission from Ref. [41]. Copyright 2015, Elsevier.

this approach (occupancy value determination) is highly essential in phase investigation/studies, as it provides quantitative comparison—an added bonus apart from the conventional phase analysis from the XRD.

5. Microstructural and elemental characterisation

Morphologies observation is an important qualitative characterisation in analysing TiO_2 nanoparticles. Observation on the microstructural behaviour of nanoparticles was done based on the micrograph obtained from electron microscopy, commonly the FESEM [52] and HRTEM [53]. Selection of the characterisation method usually depends on the samples' suitability and resolution of the techniques. Methods such as the FESEM and HRTEM offer higher versatility, as both can be further exploited to compensate additional analysis such as the elemental analysis.

5.1. Field emission scanning electron microscope

In FESEM imaging, the secondary electron (SE) mode is more preferable compared with the backscattered electron (BSE) mode. SE mode allows more detailed surface images of TiO_2 nanoparticles, which is more significant compared with the BSE mode [52]. This is significant in allowing better interpretation on micrographs as TiO_2 nanoparticles are commonly subjected to agglomerations.

The observation of the TiO_2 nanoparticles' morphology commonly requires magnification up to 30,000 times [54]. However, higher magnification up to 200,000 times has also been reported in characterising the surface morphology of TiO_2 nanoparticles [52, 55]. These mainly depend on the condition of TiO_2 nanoparticles prior to characterisation. In addition, dry TiO_2 nanoparticles resulting from calcination allow higher magnification observation compared with uncalcined nanoparticles. Coatings with conductive metals such as gold are required prior to the imaging process. In terms of image acquisition, less crystalline TiO_2 nanoparticles were commonly subjected to higher charging effects during imaging, causing whitish disturbance, which can be wrongly interpreted to the actual TiO_2 nanoparticles.

Good examples of how SE images are much preferable for TiO_2 nanoparticles can be demonstrated by the significant differences on the morphology of the rutile and anatase phases which was attained by different route of sol-gel synthesis [41]. With magnification of up to 30,000 times, the shapes and sizes of both centrifuge and storage TiO_2 nanoparticles were clearly distinguishable from the micrographs (**Figure 5a–f**). Additionally, the capability to go as high as such magnification was also contributed by high calcination temperature (450–700°C) [52]. Apart from providing easier observation and interpretation on TiO_2 nanoparticles, the ability to go for high magnification allows accurate size determination of the particles (size determination from SEM micrograph currently gained significant interest in TiO_2 nanoparticles characterisation). A similar approach was also demonstrated by Khatun et al. [55] in observing agglomerated TiO_2 nanoparticles. Images obtained with magnification up to 200,000 times clearly distinguished the morphological behaviour of the TiO_2 nanoparticles. Due to that, the fusing of individual TiO_2 nanoparticles into larger agglomerates was proven in the work. The initial decarburisation process at 450°C for 6 h allowed such SEM observation. Furthermore, most relatively high heat-treating processes (mainly above 400°C) allowed good observation on the sol-gel synthesised TiO_2 nanoparticles during SEM imaging [35, 56].

5.2. High-resolution transmission electron microscope

The main difference of the HRTEM technique from that of the FESEM is how images are generated by the transmission of electron through TiO_2 nanoparticles. Thus, a very thin sample is required for observation through HRTEM.

In the initial preparation, TiO_2 nanoparticles were suspended in a solution and then deposited onto a silicon oxide film supported by a Cu mesh. This allowed observing individual and non-agglomerate particles of TiO_2. Apart from morphological and surface behaviour characterisations, higher resolution of HRTEM provided observation and measurement on the lattice fringes of TiO_2 nanoparticles as well. This was used for the validation of the structural analysis from XRD and to observe the possible growth plane of TiO_2 nanoparticles [34]. The common magnification range used to observe TiO_2 nanoparticles ranged from 195,000 times for shape and distribution observation to 610,000 times for lattice fringes analysis [57].

Observation on the shape of TiO_2 nanoparticles synthesised by the centrifuge precipitation sol-gel method was obtained by Yahaya et al. [41] by utilising a magnification of 610,000 times.

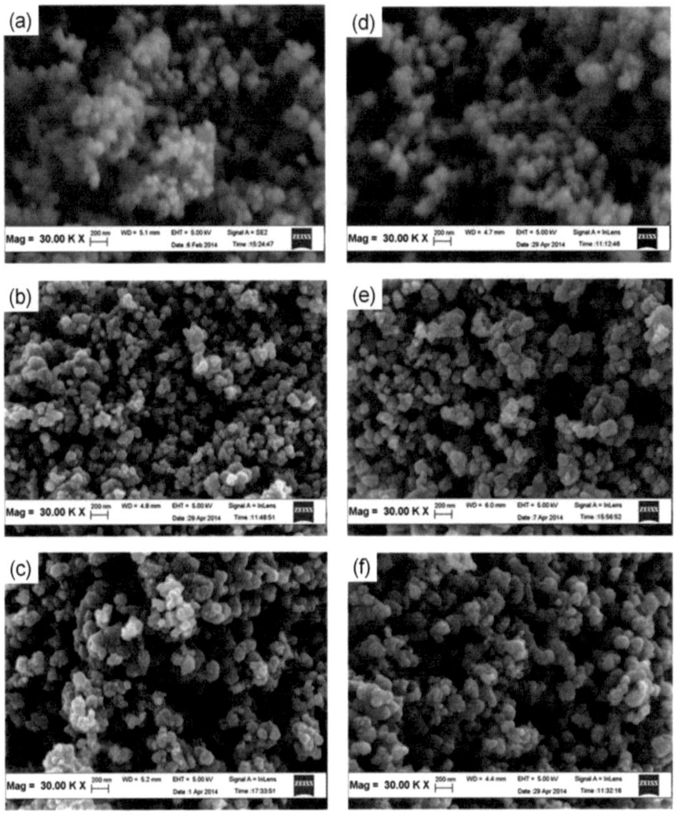

Figure 5. SEM images of the TiO_2 nanoparticles synthesised by the sol-gel method for (a) as synthesised centrifuge, (b) centrifuge calcined at 500°C, (c) centrifuge calcined at 600°C, (d) as synthesised storage (e) storage calcined at 500°C, and (f) storage calcined at 500°C. Reprinted with permission from Ref. [41]. Copyright 2015, Elsevier.

The preferable circular shapes of the TiO_2 nanoparticles favour by such synthesis method were obviously revealed with such high magnification (**Figure 6a, b**). Similar magnification was also employed by Shen et al. [34] in observing the rectangular shape of the microwave-assisted sol-gel-synthesised TiO_2 nanoparticles. Making such observation is the main benefit of employing HRTEM when compared with SEM. However, HRTEM is mainly employed for sol-gel-synthesised TiO_2 nanoparticles, which also depend on the lattice fringes measurement and observation [43]. Additionally, high-resolution images (**Figure 6c**) of the TiO_2 nanoparticles indicated favourable crystal growth along the (011), evaluated by the lattice fringes (**Figure 6d**).

Apart from morphological analysis, common practices involving HRTEM micrograph include the validation of structural analysis from XRD [32]. For instance, in this work, the measured lattice distance was comparable with the d-spacing value, thus supporting the initial XRD evaluation by the author. As an example, lattice fringes of anatase (0.347 nm) were evaluated from the HRTEM images (**Figure 6d**), indicating a good correlation to the values attained from the structural analysis [41]. Other than single-phase studies, such approaches were also widely applied in investigating mixtures of TiO_2 polymorphs [58]. This kind of evaluation was indeed to further prove the co-existence of the phases in the TiO_2 mixtures [59]. It

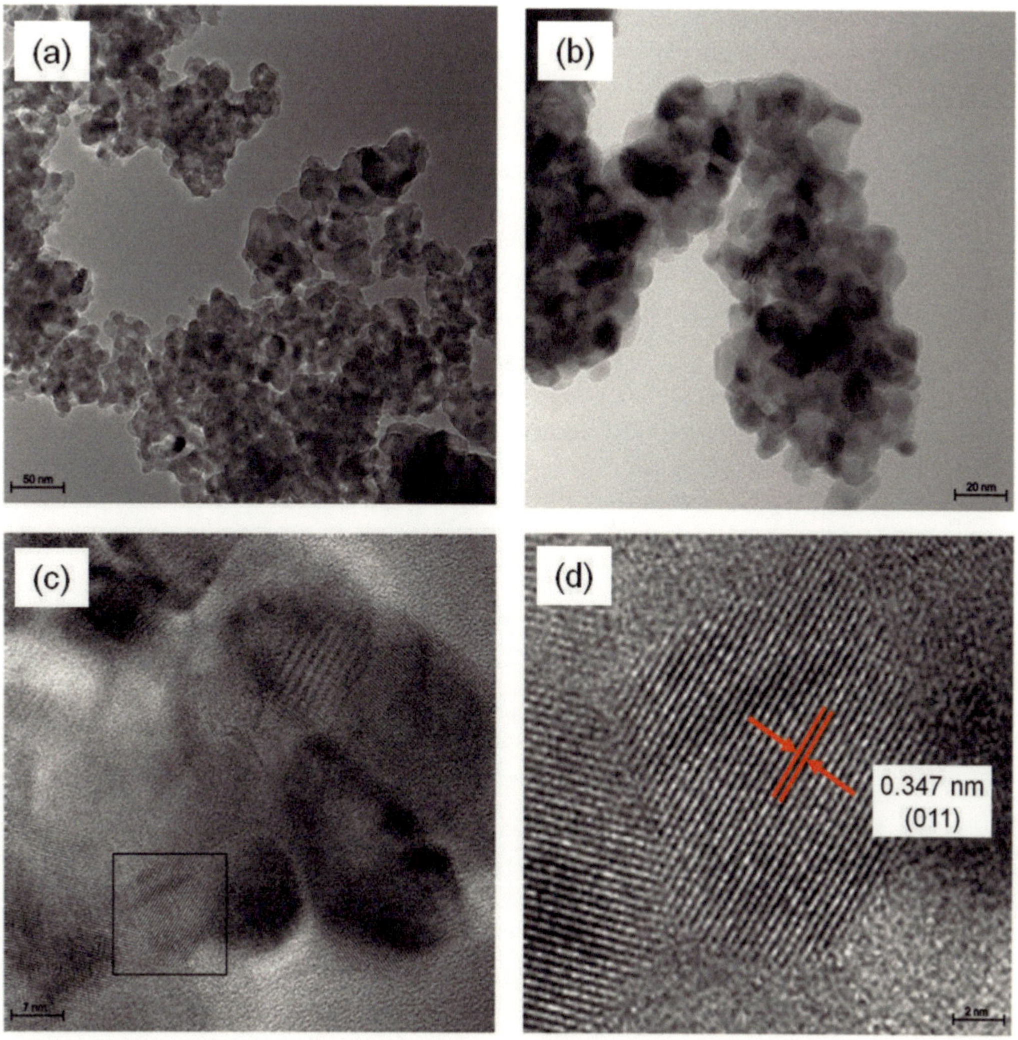

Figure 6. TEM and HRTEM images of TiO_2 nanoparticles synthesised by centrifuge sol-gel method at magnification of (a) 97 kx, (b) 195 kx, (c) 610 kx and (d) magnified image of selected square in (c). Reprinted with permission from Ref. [41]. Copyright 2015, Elsevier.

was clear that HRTEM provides a direct relation between the qualitative analysis (from the images) and the quantitative analysis (structural parameters) from the XRD. In the context of sol-gel synthesis, this clearly opened up more plausible explanation when investigating and characterising TiO_2 nanoparticles.

5.3. Elemental analysis

Parallel to FESEM and HRTEM, the electron dispersive X-ray (EDX) paired to the equipment can simultaneously provide the elemental analysis based on the emitted characteristic X-rays. Area scan EDX is commonly practised in providing the overall elemental analysis of TiO_2

nanoparticles due to the involvement of only single phase [41]. The elemental analysis generally follows subsequently after FESEM or HRTEM characterisation by employing the micrograph selected by the two techniques.

In sample preparation, suitable coating elements should be carefully selected to avoid coatings being detected by EDX to avoid misleading analysis. The EDX analysis provides elemental data strictly on atomic behaviour as opposed to molecular information [60]. The information from area under the EDX spectrum, the accelerating voltage and sensitivity factor were mathematically translated into the elemental information consisting of the atomic percent (at. %) and the weight percent (wt. %) values.

For instance, the elemental analyses conducted on the SEM micrograph were used to confirm the existence of titanium and oxygen in the synthesised TiO_2 nanoparticles (**Figure 7a–f**) [41]. In general, such presence of the particular Ti and O element from the EDX spectrum also highly support other characterisation especially the phase and structural analysis [61]. Additionally, the EDX analysis is useful in explaining the effects of certain parameters on

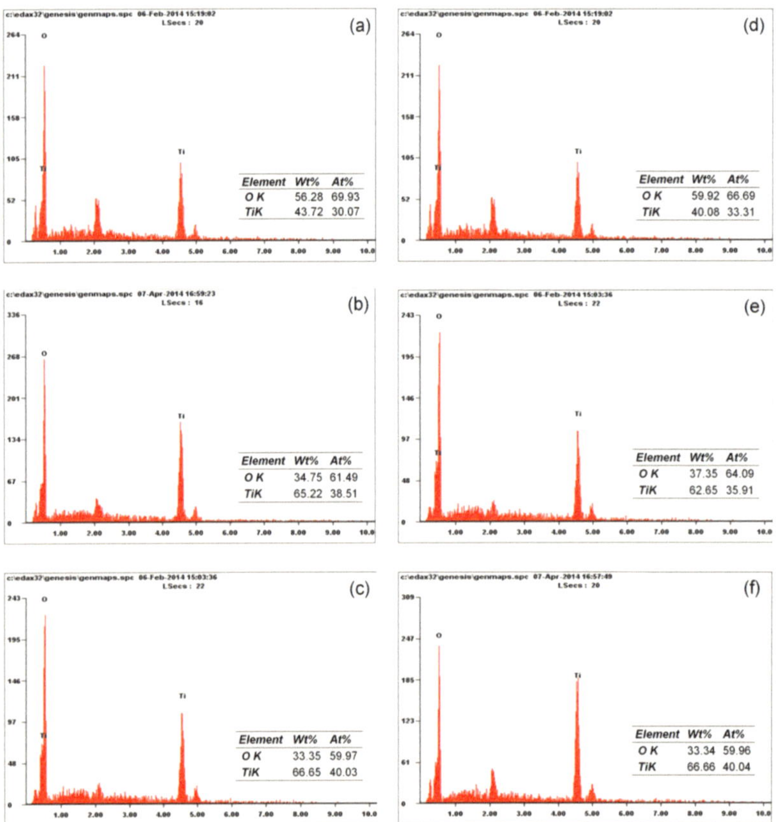

Figure 7. EDX spectrum of the TiO_2 nanoparticles, obtained by centrifuge and calcination at (a) 0°C, (b) 500°C and (c) 600°C, by storage and calcination at (d) 0°C, (e) 500°C and (f) 600°C temperature under ambient atmosphere. Reprinted with permission from Ref. [41]. Copyright 2015, Elsevier.

the corresponding elements such as heat treatment [41]. As an example, the variations on the elemental composition from the EDX spectra obtained in this work explained the influence of calcination temperatures on the resultant TiO_2 nanoparticles. It was clearly observed that higher calcination temperature depleted the oxygen content in the EDX spectra. As current TiO_2 nanoparticles synthesis by the sol-gel involves various changes in chemical parameters, the EDX acts as an important tool to validate the presence/absence of any additional elements from such a complexity. This can be viewed from the work by Martins et al. [62], where the EDX analysis was used to investigate TiO_2 nanoparticles synthesised with the addition of activated carbon. The presence of oxygen, titanium and carbon on the EDX patterns in such cases confirmed the elements presence on the material surface, which in turn indicates the occurrence of carbon reaction during the synthesis process.

6. Particle size analysis

Size determination of TiO_2 nanoparticles was highly practised in conjunction to other characterisation methods. This analysis is significantly crucial as the size factor greatly influences the end properties of TiO_2 nanoparticles. There are few available techniques that allow such analysis to be conducted on the synthesised TiO_2 nanoparticles. However, each selected application is mainly dependent on the research requirement and suitability.

Size determination through FESEM and HRTEM is commonly practised and utilised. The drawback is the need to carry out the initial morphology observation before the actual size determination. A more straightforward approach can be espoused through the use of the particle size analyzer. For example, the use of laser diffraction provides a fast and reproducible particle size analysis [63]. To aid the analysis, current practices are commonly incorporated in the preliminary dispersion process, which helps eliminate agglomeration issues of TiO_2 nanoparticles [64]. Other methods include the electrospray scanning mobility particle sizer (ES-SMPS) and the dynamic light scattering (DLS). As the DLS is suitable for evaluating hydrodynamic radius of particles, ES-SMPS was proven to yield reliable size distribution of nanoparticles larger than 5 nm [65].

A good example of the reliability of the technique can be observed from the investigation on the particle size by Abbas et al. [66], which employed the ES-SMPS method. In the report, high reproducibility of the particle size measurement had been obtained with negligible variations for all TiO_2 nanoparticles. The importance of the consistency is mainly to provide an accurate interpretation on the research parameters. For instance, in this work, the size distribution plotted from the reliable data of ES-SMPS evaluation reflects solely on the influence of the storage condition towards the resultant particle size (**Figure 8a, b**). However, recent work in size determination of TiO_2 nanoparticles mostly presents the measurement obtained from SEM/HRTEM micrographs [57, 67]. Although the processes are more convenient due to the readily available images from SEM/HRTEM, the average size obtained was limited to the particles' presence in the micrograph. Unlike from the particle size analyzer, larger samples quantity can be analysed, thus practically representing better average size value of TiO_2 nanoparticles.

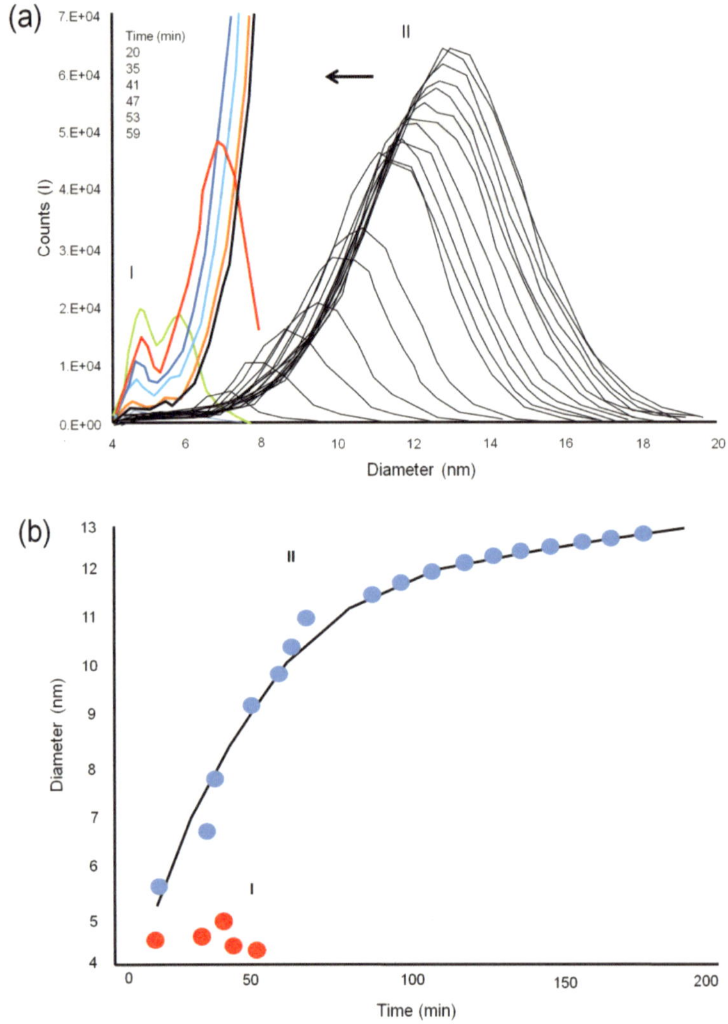

Figure 8. Particle size distribution obtained by the ES-SMPS method for (a) growth over the period of 200 min (II). The data for short times are enlarged by a factor of 10 (I), (b) the changes in the average diameter of the particles during growth. Reprinted with permission from Ref. [66]. Copyright 2011, Elsevier.

Considering the one-step approach rather than measurement from the image (two steps: imaging from SEM/HRTEM then measurement), the particle size analyzer is most suitable for research work focusing on particle size, in addition to higher reliability and accuracy.

7. Conclusion and perspective

Driven by the potential of TiO_2 nanoparticles, continuous development involving numerous research works is expected to rapidly increase in the future. The versatility provided by the

http://dx.doi.org/10.5772/67822

sol-gel method makes these efforts to synthesise TiO_2 nanoparticles better and more relevant, considering the demands in the current applications. Parallel to the research and development process, characterisation of the synthesised TiO_2 nanoparticles is unavoidable. Despite the large availability of experimental alternatives, characterisation methods such as phase, structural, morphologies and size analysis were highly popular among researchers in different TiO_2-related fields. It is apparent that the characterisation techniques will be constantly used in research studies of TiO_2 nanoparticles.

For future work, it is suggested that the studies on the synthesised TiO_2 nanoparticles should be augmented towards structural modelling and simulation. With recent literatures focusing on the use of dopants, such theoretical works are essential in structurally visualising the doping mechanism. This is practically more significant in terms of discussing the performance and properties of the synthesised TiO_2 nanoparticles. Due to advances in computers and software technologies, the properties of the modelled TiO_2 nanoparticles can be completely simulated. Thus, another interesting approach would be to provide a comparison between the generated properties with the results from the experimental work. Additionally, simulation works can eventually show the potential of a particular formulation, thus providing an insight for researchers before venturing into the actual synthesis of TiO_2 nanoparticles.

Acknowledgements

M.Z.Y. would like to acknowledge the MyBrain15 Fellowship scheme. The authors appreciate the financial support provided by the USM-RUI grant (1001/PBahan/814262).

Author details

Muhamad Zamri Yahaya[1], Mohd Asyadi Azam[2], Mohd Asri Mat Teridi[3], Pramod Kumar Singh[4] and Ahmad Azmin Mohamad[1]*

*Address all correspondence to: aam@usm.my

1 School of Materials and Mineral Resources Engineering, Universiti Sains Malaysia, Nibong Tebal, Pulau Pinang, Malaysia

2 Carbon Research Technology Research Group, Advanced Manufacturing Centre, Faculty of Manufacturing Engineering, Universiti Teknikal Malaysia Melaka, Durian Tunggal, Melaka, Malaysia

3 Solar Energy Research Institute, National University of Malaysia (UKM), Bangi, Selangor Darul Ehsan, Malaysia

4 Material Research Laboratory, School of Basic Sciences & Research, Sharda University, Greater Noida, India

References

[1] Lee Y, Chae J, Kang M. Comparison of the photovoltaic efficiency on DSSC for nanometer sized TiO_2 using a conventional sol–gel and solvothermal methods. Journal of Industrial and Engineering Chemistry. 2010;**16**:609-614. DOI: http://dx.doi.org/10.1016/j.jiec.2010.03.008

[2] Metzler DM, Li M, Erdem A, Huang C. Responses of algae to photocatalytic nano-TiO_2 particles with an emphasis on the effect of particle size. Chemical Engineering Journal. 2011;**170**:538-546. DOI: http://dx.doi.org/10.1016/j.cej.2011.02.002

[3] Panda SK, Yoon Y, Jung HS, Yoon W-S, Shin H. Nanoscale size effect of titania (anatase) nanotubes with uniform wall thickness as high performance anode for lithium-ion secondary battery. Journal of Power Sources. 2012;**204**:162-167. DOI: http://dx.doi.org/10.1016/j.jpowsour.2011.12.048

[4] Tang Y, Li G, Pan Y. Effects of TiO_2 nanoparticles addition on microstructure, microhardness and tensile properties of Sn–3.0 Ag–0.5 Cu–xTiO_2 composite solder. Materials & Design. 2014;**55**:574-582. DOI: http://dx.doi.org/10.1016/j.matdes.2013.10.033

[5] Wu S, Weng Z, Liu X, Yeung K, Chu P. Functionalized TiO_2 based nanomaterials for biomedical applications. Advanced Functional Materials. 2014;**24**:5464-5481. DOI: 10.1002/adfm.201400706

[6] Wu L, Buchholz D, Bresser D, Chagas LG, Passerini S. Anatase TiO_2 nanoparticles for high power sodium-ion anodes. Journal of Power Sources. 2014;**251**:379-385. DOI: http://dx.doi.org/10.1016/j.jpowsour.2013.11.083

[7] Sadovnikov S, Gusev A. Chemical deposition of nanocrystalline lead sulfide powders with controllable particle size. Journal of Alloys and Compounds. 2014;**586**:105-112. DOI: http://dx.doi.org/10.1016/j.jallcom.2013.10.008

[8] Kubiak P, Geserick J, Hüsing N, Wohlfahrt-Mehrens M. Electrochemical performance of mesoporous TiO_2 anatase. Journal of Power Sources. 2008;**175**:510-516. DOI: http://dx.doi.org/10.1016/j.jpowsour.2007.09.044

[9] Dislich H. New routes to multicomponent oxide glasses. Angewandte Chemie International Edition in English. 1971;**10**:363-370. DOI: 10.1002/anie.197103631

[10] Kumar SG, Rao KK. Polymorphic phase transition among the titania crystal structures using a solution-based approach: From precursor chemistry to nucleation process. Nanoscale. 2014;**6**:11574-11632. DOI: 10.1039/C4NR01657B

[11] Mahshid S, Ghamsari MS, Askari M, Afshar N, Lahuti S. Synthesis of TiO_2 nanoparticles by hydrolysis and peptization of titanium isopropoxide solution. Journal of Materials Processing Technology. 2007;**189**:296-300. DOI: http://dx.doi.org/10.1016/j.jmatprotec.2007.01.040

[12] Latt KK, Kobayashi T. TiO_2 nanosized powders controlling by ultrasound sol–gel reaction. Ultrasonics Sonochemistry. 2008;**15**:484-491. DOI: http://dx.doi.org/10.1016/j.ultsonch.2007.08.001

[13] Loryuenyong V, Angamnuaysiri K, Sukcharoenpong J, Suwannasri A. Sol–gel derived mesoporous titania nanoparticles: Effects of calcination temperature and alcoholic solvent on the photocatalytic behavior. Ceramics International. 2012;**38**:2233-2237. DOI: http://dx.doi.org/10.1016/j.ceramint.2011.10.072

[14] Kumar SG, Devi LG. Review on modified TiO_2 photocatalysis under UV/visible light: Selected results and related mechanisms on interfacial charge carrier transfer dynamics. Journal of Physical Chemistry A. 2011;**115**:13211-13241. DOI: 10.1021/jp204364a

[15] Daghrir R, Drogui P, Robert D. Modified TiO_2 for environmental photocatalytic applications: A review. Industrial & Engineering Chemistry Research. 2013;**52**:3581-3599. DOI: 10.1021/ie303468t

[16] Ni M, Leung MKH, Leung DYC, Sumathy K. A review and recent developments in photocatalytic water-splitting using TiO_2 for hydrogen production. Renewable and Sustainable Energy Reviews. 2007;**11**:401-425. DOI: http://dx.doi.org/10.1016/j.rser.2005.01.009

[17] Diebold U. Structure and properties of TiO_2 surfaces: A brief review. Applied Physics A: Materials Science & Processing. 2003;**76**:681-687. DOI: 10.1007/s00339-002-2004-5

[18] Gupta SM, Tripathi M. A review of TiO_2 nanoparticles. Chinese Science Bulletin. 2011;**56**:1639-1657. DOI: 10.1007/s11434-011-4476-1

[19] Macwan D, Dave PN, Chaturvedi S. A review on nano-TiO_2 sol–gel type syntheses and its applications. Journal of Materials Science. 2011;**46**:3669-3686. DOI: 10.1007/s10853-011-5378-y

[20] Gupta SM, Tripathi M. A review on the synthesis of TiO_2 nanoparticles by solution route. Central European Journal of Chemistry. 2012;**10**:279-294. DOI: 10.2478/s11532-011-0155-y

[21] Reyes-Coronado D, Rodríguez-Gattorno G, Espinosa-Pesqueira ME, Cab C, De Coss R, Oskam G. Phase-pure TiO_2 nanoparticles: Anatase, brookite and rutile. Nanotechnology. 2008;**19**:145605. DOI: https://doi.org/10.1088/0957-4484/19/14/145605

[22] Liao D, Wu G, Liao B. Zeta potential of shape-controlled TiO_2 nanoparticles with surfactants. Colloids and Surfaces A. 2009;**348**:270-275. DOI: http://dx.doi.org/10.1016/j.colsurfa.2009.07.036

[23] Fang C-S, Chen YW. Preparation of titania particles by thermal hydrolysis of $TiCl_4$ in n-propanol solution. Materials Chemistry and Physics. 2003;**78**:739-745. DOI: http://dx.doi.org/10.1016/S0254-0584(02)00416-9

[24] Németh Z, Horváth E, Magrez A, Réti B, Berki P, Forró L, Hernádi K. Preparation of titania covered multi-walled carbon nanotube thin films. Materials & Design. 2015;**86**:198-203. DOI: http://dx.doi.org/10.1016/j.matdes.2015.07.048

[25] Liu AR, Wang SM, Zhao YR, Zhen Z. Low-temperature preparation of nanocrystalline TiO_2 photocatalyst with a very large specific surface area. Materials Chemistry and Physics. 2006;**99**:131-134. DOI: http://dx.doi.org/10.1016/j.matchemphys.2005.10.003

[26] Vorkapic D, Matsoukas T. Effect of temperature and alcohols in the preparation of titania nanoparticles from alkoxides. Journal of the American Ceramic Society. 1998;**81**:2815-2820. DOI: 10.1111/j.1151-2916.1998.tb02701.x

[27] Brinker CJ, Scherer GW. Sol-gel Science: The Physics and Chemistry of Sol-gel Processing. 1st ed. USA: Academic Press; 2013. p. 912 DOI: http://dx.doi.org/10.1016/B978-0-08-057103-4.50007-6

[28] Watson SS, Beydoun D, Scott JA, Amal R. The effect of preparation method on the photoactivity of crystalline titanium dioxide particles. Chemical Engineering Journal. 2003;**95**:213-220. DOI: http://dx.doi.org/10.1016/S1385-8947(03)00107-4

[29] Look JL, Zukoski C. Colloidal stability and titania precipitate morphology: Influence of short-range repulsions. Journal of the American Ceramic Society. 1995;**78**:21-32. DOI: 10.1111/j.1151-2916.1995.tb08356.x

[30] Li Z, Zhu Y, Wang J, Guo Q, Li J. Size-controlled synthesis of dispersed equiaxed amorphous TiO_2 nanoparticles. Ceramics International. 2015;**41**:9057-9062. DOI: http://dx.doi.org/10.1016/j.ceramint.2015.03.278

[31] Naghibi S, Sani MAF, Hosseini HRM. Application of the statistical Taguchi method to optimize TiO_2 nanoparticles synthesis by the hydrothermal assisted sol–gel technique. Ceramics International. 2014;**40**:4193-4201. DOI: http://dx.doi.org/10.1016/j.ceramint.2013.08.077

[32] Liang D, Cui C, Hu H, Wang Y, Xu S, Ying B, Li P, Lu B, Shen H. One-step hydrothermal synthesis of anatase TiO_2/reduced graphene oxide nanocomposites with enhanced photocatalytic activity. Journal of Alloys and Compounds. 2014;**582**:236-240. DOI: http://dx.doi.org/10.1016/j.jallcom.2013.08.062

[33] Chen C-C, Hu S-H, Fu YP. Effects of surface hydroxyl group density on the photocatalytic activity of Fe^{3+}-doped TiO_2. Journal of Alloys and Compounds. 2015;**632**:326-334. DOI: http://dx.doi.org/10.1016/j.jallcom.2015.01.206

[34] Shen P-S, Tai Y-C, Chen P, Wu Y-C. Clean and time-effective synthesis of anatase TiO_2 nanocrystalline by microwave-assisted solvothermal method for dye-sensitized solar cells. Journal of Power Sources. 2014;**247**:444-451. DOI: http://dx.doi.org/10.1016/j.jpowsour.2013.08.104

[35] He F, Ma F, Li J, Li T, Li G. Effect of calcination temperature on the structural properties and photocatalytic activities of solvothermal synthesized TiO_2 hollow nanoparticles. Ceramics International. 2014;**40**:6441-6446. DOI: http://dx.doi.org/10.1016/j.ceramint.2013.11.094

[36] Maurya A, Chauhan P, Mishra SK, Srivastava RK. Structural, optical and charge transport study of rutile TiO_2 nanocrystals at two calcination temperatures. Journal of Alloys and Compounds. 2011;**509**:8433-8440. DOI: http://dx.doi.org/10.1016/j.jallcom.2011.05.108

[37] Gao Y, Wang H, Wu J, Zhao R, Lu Y, Xin B. Controlled facile synthesis and photocatalytic activity of ultrafine high crystallinity TiO_2 nanocrystals with tunable anatase/rutile ratios. Applied Surface Science. 2014;**294**:36-41. DOI: http://dx.doi.org/10.1016/j.apsusc.2013.12.107

[38] Banerjee AN, Hamnabard N, Joo SW. A comparative study of the effect of Pd-doping on the structural, optical, and photocatalytic properties of sol–gel derived anatase TiO_2 nanoparticles. Ceramics International. 2016. DOI: http://dx.doi.org/10.1016/j.ceramint.2016.04.128

[39] Husain S, Alkhtaby LA, Giorgetti E, Zoppi A, Miranda MM. Investigation of the role of iron doping on the structural, optical and photoluminescence properties of sol–gel derived TiO_2 nanoparticles. Journal of Luminescence. 2016;**172**:258-263. DOI: http://dx.doi.org/10.1016/j.jlumin.2015.12.002

[40] Ivanov S, Barylyak A, Besaha K, Dimitrova A, Krischok S, Bund A, Bobitski J. Enhanced lithium ion storage in TiO_2 nanoparticles, induced by sulphur and carbon co-doping. Journal of Power Sources. 2016;**326**:270-278. DOI: http://dx.doi.org/10.1016/j.jpowsour.2016.06.116

[41] Yahaya MZ, Abdullah MZ, Mohamad AA. Centrifuge and storage precipitation of TiO_2 nanoparticles by the sol–gel method. Journal of Alloys and Compounds. 2015;**651**:557-564. DOI: http://dx.doi.org/10.1016/j.jallcom.2015.08.110

[42] Vinogradov AV, Vinogradov VV. Low-temperature sol–gel synthesis of crystalline materials. Royal Society of Chemistry Advances. 2014;**4**:45903-45919. DOI: 10.1039/C4RA04454A

[43] Wang C-L, Hwang W-S, Chu H-L, Lin H-J, Ko H-H, Wang M-C. Kinetics of anatase transition to rutile TiO_2 from titanium dioxide precursor powders synthesized by a sol-gel process. Ceramics International. 2016;**42**:13136-13143. DOI: http://dx.doi.org/10.1016/j.ceramint.2016.05.101

[44] Sarma BK, Pal AR, Bailung H, Chutia J. Effect of post-deposition annealing on the growth of nanocrystalline TiO_2 thin films and elastic anisotropy of rutile phase at different temperatures. Journal of Alloys and Compounds. 2013;**577**:261-268. DOI: http://dx.doi.org/10.1016/j.jallcom.2013.05.137

[45] Potlog T, Dobromir M, Luca D, Onufrijevs P, Medvids A, Shamardin A. Rutile to anatase phase transition in TiO_2: Nb thin films by annealing in H_2 atmosphere. Current Applied Physics. 2016;**16**:826-829. DOI: http://dx.doi.org/10.1016/j.cap.2016.04.021

[46] Delekar S, Yadav H, Achary S, Meena S, Pawar S. Structural refinement and photocatalytic activity of Fe-doped anatase TiO_2 nanoparticles. Applied Surface Science. 2012;**263**:536-545. DOI: http://dx.doi.org/10.1016/j.apsusc.2012.09.102

[47] Toby BH. R factors in Rietveld analysis: How good is good enough? Powder Diffraction. 2006;**21**:67-70. DOI: https://doi.org/10.1154/1.2179804

[48] Djerdj I, Tonejc A. Structural investigations of nanocrystalline TiO_2 samples. Journal of Alloys and Compounds. 2006;**413**:159-174. DOI: http://dx.doi.org/10.1016/j.jallcom.2005.02.105

[49] Masson O, Rieux V, Guinebretière R, Dauger A. Size and shape characterization of TiO_2 aerogel nanocrystals. Nanostructured Materials. 1996;**7**:725-731. DOI: http://dx.doi.org/10.1016/S0965-9773(96)00049-9

[50] Sarkar A, Kannan S. In situ synthesis, fabrication and Rietveld refinement of the hydroxyapatite/titania composite coatings on 316L SS. Ceramics International. 2014;**40**:6453-6463. DOI: http://dx.doi.org/10.1016/j.ceramint.2013.11.096

[51] Tripathi AK, Mathpal MC, Kumar P, Singh MK, Mishra SK, Srivastava RK, Chung JS, Verma G, Ahmad M, Agarwal A. Synthesis based structural and optical behavior of anatase TiO_2 nanoparticles. Materials Science in Semiconductor Processing. 2014;**23**:136-143. DOI: http://dx.doi.org/10.1016/j.mssp.2014.02.041

[52] Ning F, He Y-B, Li B, Du H, Zhai D, Kang F. Effects of TiO_2 crystal structure on the performance of $Li_4Ti_5O_{12}$ anode material. Journal of Alloys and Compounds. 2012;**513**:524-529. DOI: http://dx.doi.org/10.1016/j.jallcom.2011.10.109

[53] Ghows N, Entezari MH. Ultrasound with low intensity assisted the synthesis of nanocrystalline TiO_2 without calcination. Ultrasonics Sonochemistry. 2010;**17**:878-883. DOI: http://dx.doi.org/10.1016/j.ultsonch.2010.03.010

[54] Dittmann R, Wintermantel E, Graule T. Sintering of nano-sized titania particles and the effect of chlorine impurities. Journal of the European Ceramic Society. 2013;**33**:3257-3264. DOI: http://dx.doi.org/10.1016/j.jeurceramsoc.2013.05.028

[55] Khatun N, Rini E, Shirage P, Rajput P, Jha S, Sen S. Effect of lattice distortion on bandgap decrement due to vanadium substitution in TiO_2 nanoparticles. Materials Science in Semiconductor Processing. 2016;**50**:7-13. DOI: http://dx.doi.org/10.1016/j.mssp.2016.04.002

[56] Vranješ M, Šaponjić ZV, Živković LS, Despotović VN, Šojić DV, Abramović BF, Čomor MI. Elongated titania nanostructures as efficient photocatalysts for degradation of selected herbicides. Applied Catalysis B-Environmental. 2014;**160**:589-596. DOI: http://dx.doi.org/10.1016/j.apcatb.2014.06.005

[57] Farbod M, Khademalrasool M. Synthesis of TiO_2 nanoparticles by a combined sol–gel ball milling method and investigation of nanoparticle size effect on their photocatalytic activities. Powder Technology. 2011;**214**:344-348. DOI: http://dx.doi.org/10.1016/j.powtec.2011.08.026

[58] Wetchakun N, Incessungvorn B, Wetchakun K, Phanichphant S. Influence of calcination temperature on anatase to rutile phase transformation in TiO_2 nanoparticles synthesized by the modified sol–gel method. Materials Letters. 2012;**82**:195-198. DOI: http://dx.doi.org/10.1016/j.matlet.2012.05.092

[59] Mutuma BK, Shao GN, Kim WD, Kim HT. Sol–gel synthesis of mesoporous anatase–brookite and anatase–brookite–rutile TiO_2 nanoparticles and their photocatalytic properties. Journal of Colloid and Interface Science. 2015;**442**:1-7. DOI: http://dx.doi.org/10.1016/j.jcis.2014.11.060

[60] Ngo PD. Energy dispersive spectroscopy. In: Lawrence C, editor. Failure Analysis of Integrated Circuits. 1st ed. USA: Springer; 1999. p. 205-215. DOI: 10.1007/978-1-4615-4919-2_12

[61] Dwivedi C, Raje N, Nuwad J, Kumar M, Bajaj PN. Synthesis and characterization of mesoporous titania microspheres and their applications. Chemical Engineering Journal. 2012;**193**:178-186. DOI: http://dx.doi.org/10.1016/j.cej.2012.04.029

[62] Martins AC, Cazetta AL, Pezoti O, Souza JR, Zhang T, Pilau EJ, Asefa T, Almeida VC. Sol-gel synthesis of new TiO_2/activated carbon photocatalyst and its application for degradation of tetracycline. Ceramics International. 2016. DOI: http://dx.doi.org/10.1016/j.ceramint.2016.12.088

[63] Black DL, McQuay MQ, Bonin MP. Laser-based techniques for particle-size measurement: A review of sizing methods and their industrial applications. Progress in Energy and Combustion Science. 1996;**22**:267-306. DOI: http://dx.doi.org/10.1016/S0360-1285(96)00008-1

[64] Song D, Hatami M, Wang Y, Jing D, Yang Y. Prediction of hydrodynamic and optical properties of TiO_2/water suspension considering particle size distribution. International Journal of Heat and Mass Transfer. 2016;**92**:864-876. DOI: http://dx.doi.org/10.1016/j.ijheatmasstransfer.2015.08.101

[65] Lenggoro IW, Xia B, Okuyama K, de la Mora JF. Sizing of colloidal nanoparticles by electrospray and differential mobility analyzer methods. Langmuir. 2002;**18**:4584-4591. DOI: 10.1021/la015667t

[66] Abbas Z, Holmberg JP, Hellström AK, Hagström M, Bergenholtz J, Hassellöv M, Ahlberg E. Synthesis, characterization and particle size distribution of TiO_2 colloidal nanoparticles. Colloids and Surfaces A. 2011;**384**:254-261. DOI: http://dx.doi.org/10.1016/j.colsurfa.2011.03.064

[67] Jiang Z-Z, Gu D-M, Wang Z-B, Qu W-L, Yin G-P, Qian K-J. Effects of anatase TiO_2 with different particle sizes and contents on the stability of supported Pt catalysts. Journal of Power Sources. 2011;**196**:8207-8215. DOI: http://dx.doi.org/10.1016/j.jpowsour.2011.05.063

Chapter 7

Synthesis of Titanium Dioxide Nanorod Arrays Using a Facile Aqueous Sol-Gel Route for Ultraviolet Photosensor Applications

Marmeezee Mohd. Yusoff,
Mohamad Hafiz Mamat and
Mohamad Rusop Mahmood

Additional information is available at the end of the chapter

http://dx.doi.org/10.5772/intechopen.68460

Abstract

In this chapter, we review the state of the art of aqueous sol-gel route for ultraviolet (UV) photosensor applications based on the nanostructured TiO_2. The performance of UV photosensor is associated with the high surface-to-volume ratio, porosity, surface defects, light trapping, and the intensity of the UV radiation. One-dimensional (1D) growth of TiO_2 nanorod arrays (TNAs) offers an enhance charge carrier mobility to overcome the photocurrent loss due to the existence of multiple grain boundaries in granular-like and porous nanostructures. Photoelectrochemical cell (PEC)-based device structure is preferred in TNA-based UV photosensor due to its low cost, facile fabrication process, high contact area, low recombination of the excitonic charge carriers, high photocurrent gain, and fast response and recovery times. It also could work in applied bias mode, as well as in "self-powered" mode. Our study has introduced a new one-step method to deposit a thin film TNA on an FTO-coated glass substrate at low temperature and a rapid process using a facile glass container. The fabricated PEC-based UV photosensor using the deposited TNAs has successfully shown its potential in the application of UV photosensor.

Keywords: TiO_2, nanorod arrays, UV photosensor

1. Introduction

Titanium dioxide (TiO_2) nanostructures emerged as one of promising materials for electronic device applications such as diodes, transistors, memristors, sensors, and solar cells. TiO_2 with numerous nanostructures yield a unique, expedient, and novel size and shape-dependent characteristics that are significant for high performance devices. TiO_2 nanorod array (TNA) is one of the favorable nanostructures that become tremendously significant in optoelectronic applications due to its large surface area, high electron mobility, and efficient light scattering ability. The physical tunings of its active surface area by controlling the diameter size and length of the nanorod, for instance, improve its functionality and sensitivity, thus enhancing the performance of the device. In ultraviolet (UV) photosensor application, these great characteristics of the TNAs enable the fabrication of a device with high responsivity and reliability. Although there are many methods available to synthesis the TNAs, aqueous sol-gel-based method compromise numerous advantages including a facile low temperature and low-time deposition, uniform and large-scale deposition-compatible, and producing excellent TNA crystalline properties. In this chapter, synthesis of TNAs via aqueous sol-gel methods will be discussed for the application of UV photosensor. The structural, optical, and electrical properties of deposited TNAs will be investigated. The performance of the fabricated TNA-based UV photosensor will also be discussed.

2. Ultraviolet photosensor using TiO_2 nanomaterials

Recent developments in the field of semiconductor have led to a renewed interest in TiO_2 nanostructures due to their unique and excellent properties in optic, electronic, and photonic devices. TiO_2 is basically a nontoxic material, physically stable and chemically robust to the environments that has extensive applications, for instance in photocatalytic activities, detecting devices, biomedical devices, and solar cell. In general, TiO_2 is an n-type semiconductor in natural, intrinsically prone to UV radiation and distinguish from visible light region due to its wide band gap energy of around 3.2 eV at an ambient temperature [1, 2]. Thus, TiO_2 has an excellent UV absorption, particularly in thin film based on its optical property. The downsize of TiO_2 particle in the nanoscale region also received excessive attention due to its distinct characteristics exhibited in electrical, mechanical, chemical, and others due to its quantum confinement effects, compared to their bulk component. Significant part of nanostructured TiO_2 material is due to high surface-to-volume ratio and high porosity, which increases the reactive surface area in particular reaction cycles in definite volume to achieve the requirement of high performance and high efficiency device applications. In addition to high surface-to-volume ratio, the number of atoms on the surface area plays the most important role of the overall number of atoms in the nanostructured materials, which enhance the thermal stability due to low surface energy generated compared to their bulk. This property is considered unique for certain electronics applications [3].

UV photosensor for detecting UV radiation is of interest and an important aspect in modern industries and environmental monitoring. In medical sector, it could be used to monitor the UV radiation treatment on patient with diseases such as rickets, psoriasis, eczema and jaundice, as well as slimming treatment in beauty and health sector, from having a prolonged exposure which may be attributed to chronic and acute health side effects. The fundamental of UV photosensor is basically explained by the generation of photocurrent from the exposure of incident UV radiation on few semiconductor materials with wide energy band gap. This phenomenon comprises a number of mechanisms, including the light absorption, photogeneration of charge carriers, and charge carriers transport [4]. The concepts of conductivity in this phenomenon are central to the amount of photogenerated charge carriers (electrons and holes) per absorbed photon or quantum yield and the transportation of generated charge carriers. Overall, the gain of the photogenerated current under the exposure of UV radiation typically comprises of few elements such as lifetime of the charge carriers, light trapping, charge carriers trapping and mobility, and the defects on the surface of the materials. In addition, the nanostructure and electrical properties of the materials such as conductivity and resistivity also play a major role in optimizing the performance of the UV photosensor.

Extensive research has been performed regarding UV photosensor based on TiO_2 nanostructures as the detecting components. Okuya et al. developed TiO_2 thin film-based UV sensors using a spray pyrolysis deposition (SPD) technique [5]. TiO_2 film was prepared using the sol-gel consisted of 0.05 M of titanium (IV) oxy acetylacetonate, ethanol solution, and aluminum (III) acetylacetonate as an additive. They produced TiO_2 thin films at 200 nm in $CuO/TiO_2/SnO_2$ p-i-n junction on a glass substrate and revealed that the used solvent with upper boiling limit has benefited for a particle growth due to wettability to a substrate and thus produced higher porosity. High porosity exhibited the maximum intensity due to improvement of light absorption from light scattering by small particles within the porous nanostructures. The doping of aluminum to TiO_2 layer, to some extent, was also attributed to increase the intensity by the reduction of carrier density, which enhanced the light absorption in correlation with low scattering of photogenerated carriers. The maximum intensity achieved was 12.5×10^{-4} A/W from 2 at % aluminum-doped TiO_2 films under the light intensity of 50 $\mu W/cm^2$ at the wavelength of 345 nm.

A UV photosensor based on metal-semiconductor-metal (MSM) structure using nanocrystalline TiO_2 thin films with Au Schottky contact was demonstrated through a sol-gel spin-coating technique [6]. An Au film was deposited by radio frequency (RF) magnetron sputtering to construct interdigitated $Au/TiO_2/Au$ structure with 260 nm thickness of Au film. Their study reported that the highest photoresponsivity at 199 A/W under 260 nm irradiation at 5 V bias, with the dark current at 1.9 nA. They also suggested that the high responsivity was primarily due to the reduction of Schottky barrier introduced by surface defects, which attributed to more carriers getting across the barrier and subsequently increased the measured photogenerated current. Moreover, the existence of neutral photoconductor region from the wide gap between the Au electrodes introduced high gain was also responsible for the high photoresponsivity of fabricated device. At the same time, the slow photoresponse, which occurred in MSM structure mostly, was primarily due to the defect in the nonocrystalline neutral region and could be further improved by enhancements on material and structure.

Han et al. introduced a UV photosensor based on a p-n junction consisting of n-type TiO_2 nanocrystals and p-type poly(9,9-dihexylfluorene) (PFH) [7]. The sol-gel consisted of oleic acid, triethylamin, NH_4HCO_3, cyclohexane, and $Ti(OBu)_4$ was mixed together at room temperature for the synthesization of TiO_2 nanocrystals via solvothermal method at 180°C for 20 hours. The synthesized TiO_2 nanocrystals were then mixed with the PFH solution to form a hybrid dispersion, before spin coated on the substrate. The sensitivity of the fabricated device was 2.94 mA/W under UV irradiation and 6.92 mA/W under 365 nm UV light. The work suggested that TiO_2 content was in correlation with the photogenerated current density, which is related to the deviation of the size of the interface between the two components that attributed to the agglomeration of TiO_2 nanoparticles. It was also mentioned that the high crystallinity of TiO_2 nanocrystals and a small size distribution were very significant for chemicophysical property, which enhanced the electron-transporting mechanism in the fabricated device.

One study by Vigil et al. examined the trend in a sealed two-electrode photoelectrochemical cell (PEC) using a double-layer of nanocrystalline TiO_2 [8]. A double layer of TiO_2 (**Figure 1**) was introduced through a "doctor-blading" technique for a mesoporous TiO_2 (~5 μm thickness) layer on top, while the bottom TiO_2 film (~100 nm thickness) obtained via microwave-assisted chemical bath deposition (MW-CBD) on transparent conducting oxide (TCO)-coated glass as contact electrodes. The sol-gel was consisted of the aqueous solutions of $(NH_4)_2TiF_6$ solution and H_3BO_3. The fabricated device showed an *I*–*V* curve of a junction diode characteristic. The measured short circuit photogenerated current values was 0.3–0.4 mA under UV radiation (λ = 340 nm) and linearly dependent on radiation intensity. The chapter suggested that the mesoporous structure with nanocrystallites (20–30 nm in size) allowed the penetration of electrolyte, which rapidly captured the holes through the iodide ions in the electrolyte from the high surface area of TiO_2 during the irradiation and thus hindered the recombination of electron-hole pairs for longer diffusion of electrons to TCO anode contact. They also proposed that the existence of the dense thin film deposited on top of the TCO inhibited the recombination of electrons in the TCO with species in the electrolyte.

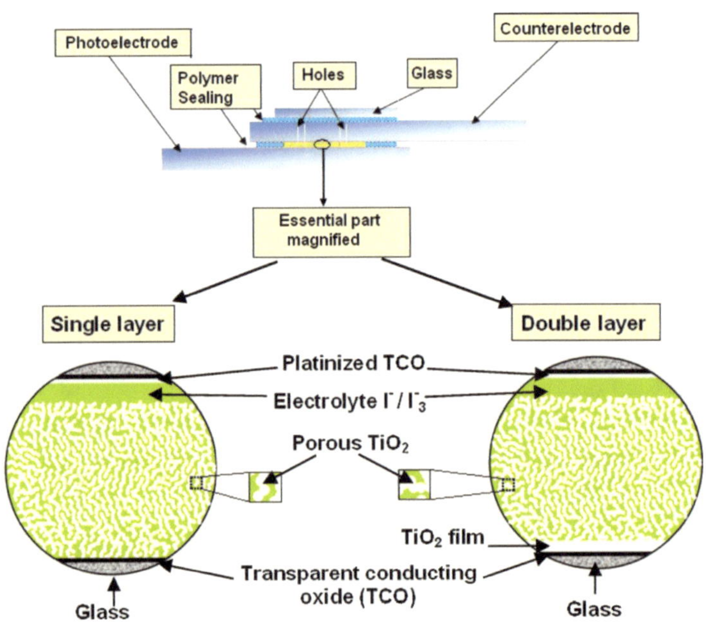

Figure 1. Schematic of UV photosensor based on photoelectrochemical cell (PEC) using a double-layer of nanocrystalline TiO_2 [8].

Another study of two-electrode PEC-based self-powered UV photosensor using a nanocrystalline TiO_2 was reported by Li et al. [9]. The performance of the fabricated self-powered photosensor was described to be isolated from the surrounding environment through the unique sandwich structure of PEC. Nanocrystalline TiO_2 was deposited through a spin-coating technique for a mesoporous structure (450 nm thickness) on an FTO-coated glass substrate as a contact electrode and followed by annealing at 500°C in air for 2 hours. The sol-gel was consisted of tetrabutyltitanate, acetic acid, ethanol, poly(vinyl pyrrolidone) (PVP), *N,N*-dimethylformamide (DMF), and tri-ethanolamine (TEOA). The measured short circuit photogenerated current values was 54 mA/cm^{-2} under UV radiation (λ = 330 nm, 150 mW/cm^{-2}) at 0 V bias and linearly dependent on radiation intensity.

Xiang et al. proposed a porous and dense nanocrystalline TiO_2 films grown on the FTO-coated glass substrate using a spin-coating technique from nanosized TiO_2 particles synthesized via hydrothermal method for vertical geometry UV photosensor [10]. Both porous and dense films were prepared at the same thickness, and the porous film was consisted of particles with an average size of 30 nm. The sol-gel was consisted of titanium isopropoxide, acetic acid, DI water, and HNO_3. Polyethylene glycol was added in the sol-gel to produce the porous film. They suggested that the response time of vertical geometry photosensor was faster than the lateral geometry photosensor for MSM-based device. The *I–V* characteristic of the porous film showed better rectifying behavior than that of the dense film. The photoresponsivity measured in photovoltage of the porous film was higher than that of the dense film under the 248-nm laser pulse irradiation due to light-scattering effects, which improved the light harvesting efficiencies in the deposited film. However, the response time of the porous film was slightly longer than that of the dense film in correlation with higher defect density in the porous film. They also suggested that the dislocation direction, which was parallel to the vertical geometry of photosensor, reduced the response time than that of the lateral geometry due to the reduction of mobility fluctuation of the charge carriers.

Another report of MSM-based TiO_2 UV photosensor on the $LaAlO_3$ substrate was introduced by Lv et al. using a sol-gel spin coating technique with Pt-interdigitated electrodes on the top of TiO_2 film [11]. The sol-gel was consisted of tetrabutyltitanate ($Ti(OC_4H_9)_4$), ethanol (C_2H_5OH), CH_3COOH, acetylacetone, and DI water. A high responsivity of 31.7 A/W was achieved under the UV radiation (260 nm, 26.5 μW/cm^2) at 5 V bias. Very low dark current was reported at 50 pA, and the ratio of the photogenerated current to dark current reached 5 orders of magnitude. The device also showed fast response and recovery at the ON and OFF states, respectively. They suggested that the low dark current achieved was due to increase of Schottky barrier height between Pt and TiO_2, which attributed to a wide depletion region. The formation of a heterojunction and depletion region at the interface of $TiO_2/LaAlO_3$, and the formation of reverse-biased junction from one of the two back-to-back Schottky junctions at Pt/TiO_2 interfaces were also attributed to the reduction of the dark current. The decrease of rise time was also due to the depletion region at the interface of $TiO_2/LaAlO_3$, where a large number of electrons from the electron-hole pairs generated and separated during the irradiation of UV were drifted toward the anode electrode. Meanwhile, the significant decrease of fall time during the OFF state was due to the rapid recombination of the balanced electrons with the holes in the depletion region at the interface of $TiO_2/LaAlO_3$.

Selman et al. developed a p-n heterojunction-based UV photosensor consisted of n-type TiO_2 on the p-type silicon substrate [12]. The work involved a deposition of TiO_2 seed layer on the p-type (1 1 1)-oriented silicon wafer via radio-frequency (RF) magnetron sputtering. Chemical bath deposition (CBD) technique was performed to grow rutile TiO_2 nanorods (NRs) with an average diameter of 30–35 nm and an average length of 85 nm on the seed layer of silicon substrate. The sol-gel was consisted of titanium (III) chloride ($TiCl_3$), urea (NH_2CONH_2), and DI water. A Pt electrode (~100 nm thickness) was deposited as a front contact on the TiO_2 NRs via RF magnetron sputtering. Their study stated that the responsivity of the device was significantly improved in correlation with bias voltage, and the responsivity achieved were 164.8, 377.0, and 538.5 A/W at 3, 5, and 7 V bias, respectively, under the UV radiation (405 nm, 0.22 mW/cm^2). High responsivity was due to high density and larger and rougher surface area of the deposited TiO_2NRs, and good formation of p-n junction in the TiO_2NRs/p-Si(1 1 1) interface. In addition, high photogenerated current achieved was due to the high crystallinity structure of the deposited rutile TiO_2, with reduced grain boundaries and defects on the structure that created electron trap to hinder the transportation of the charge carriers. Moreover, high aspect ratio also increased the photogenerated current due to oxygen desorption at the surface of TiO_2NRs.

Another interesting study that utilized the MSM-based UV photosensor with a NH_4^+ modified TiO_2 film and interdigitated Au electrodes on quartz substrates was reported by Liu et al. [13]. TiO_2 was deposited using a sol-gel spin coating technique and subsequently immersed in ammonium sulfide [$(NH_4)_2S$] solution to generate NH_4^+ modified TiO_2 films. The sol-gel consisted of tetrabutyltitanate ($Ti(OC_4H_9)_4$), ethanol (C_2H_5OH), acetic acid (CH_3COOH), acetylacetone, and DI water. A high photogenerated current of 84.93 μA and responsivity of 361.07 A/W was achieved under the UV radiation (300 nm, 61.9 μW/cm^2) at 5 V bias. Surface modification of TiO_2 film also showed the decrease of rise time to 366.9 ms and dark current to 88.8 pA than that of the TiO_2 without the surface modification. They found that the decrease in the Schottky barrier height (SBH) between TiO_2 and Au was due to the generation of dipole between NH_4^+ and exposed oxygen ions, replacing the H^+ on the deposited TiO_2 surface, which increased the performance of the device.

Yang et al. fabricated an MSM-based UV photosensor with two-dimensional (2D) TiO_2 nanosheets (TiO_2NS) deposited on 150 nm thickness of Ag IDT-electrodes on a glass substrate [14]. TiO_2NS was prepared via a sol-gel hydrothermal technique and calcined at different temperatures for 2 hours. The sol-gel was consisted of tetrabutyltitanate (TBOT), HCl, polyethylene oxide-polypropylene oxide-polyethylene oxide (PEO_{20}-PPO_{70}-PEO_{20}), ethanol (EtOH), and ethylene glycol (EG). The thickness of deposited TiO_2NS was about 1–7 nm, in correlation with 2–10 stacking layers of the monolayer (0.62 nm thickness). The measured photogenerated current of 69.4 nA was achieved under the UV radiation (3 mW/cm^2) at 1 V bias, with ultralow dark current (~2.63 pA) and fast response and recovery times. Low dark current may partly attributed from the boundaries between the nanosheets, which hindered the mobility of the charge carriers.

In previous studies on the TiO_2-based UV photosensor, the application of nanostructure materials have a lot of advantages over bulk materials, comprising high photogenerated current, high responsivity, fast response and recovery times, low power consumption, and miniaturization of the device structure. Thus far, a number of studies have highlighted the factors that are associated with the high responsivity of nanostructured devices in correlation with a high surface-to-volume ratio, surface defects, light trapping, charge carriers trapping, and high porosity [4]. The performance of the device was also dependent on the device structure and the detection mechanism of photon through the photogenerated charge carriers. Thus far, several studies have used some instances of UV photosensor based on p-i-n photodiodes, p-n junction, Schottky barrier (SB), and MSM structure, and these types of photosensors required an external bias with potential difference to hinder the recombination of the excitonic charge carriers. Such mechanism for most conventional sensors generally required an external power supply sources from batteries, or additional supplies from the controller systems for instances, which against the concept of miniaturization of future device system and also prominently reduced the independency and mobility of the detection system. Furthermore, energy supply has becoming one of the great challenges for large-scale area applications of the sensors [15]. The techniques to synthesize and deposit the TiO_2 nanostructures on the substrate are also important to be considered in addition to produce a rapid production and low-cost device. Much of the previous research has focused on the synthesization and deposition of granular-like and porous nanocrystalline TiO_2 that contributed to the enlargement of reactive surface area of detecting component. Much of the literature has also emphasized the lifetime of photogenerated charge carriers in conjunction to reduce the recombination rate of electron-hole pairs. The charge carrier's transportation is also relatively important in maximizing the overall photocurrent gain, as the effective charge carrier mobility is desired for high efficiency devices.

3. TiO_2 nanorod arrays (TNAs) and TiO_2 nanowires (TNWs) in UV photosensor

A considerable amount of literature has been studied on controlling the nanostructures of TiO_2 to enhance the responsivity of the UV photosensor. Granular-like and porous nanocrystalline TiO_2, for instance, provides a high reactive surface area to enhance the detection. However, the granular-like structure has deficiency on charge carrier's mobility in attribution to multiple grain boundaries, which hinder the transportation of charge carriers. Recently, researchers have shown an increased interest in one-dimensional (1-D) nanostructure of TiO_2 nanorod arrays (TNAs) and TiO_2 nanowires (TNWs) for UV photosensor applications due to their unique and excellent characteristics on size and shape, as well as their optical, electrical, and physical and chemical properties. Structures of TNAs and TNWs show high aspect ratio (length/diameter) larger than 1, with outstanding charge carrier's mobilities that are extremely suitable for a UV-sensing device. The size and shape may influence the optical property and give different performance on the photosensing device. Therefore, control over the shape and size depending on the method and experimental parameters used to synthesize the nanorods' structure has been one of the vital and challenging tasks. TNAs or TNWs offer a direct carrier's pathway, which increased the transportation rate and enhanced the performance of photosensor, as well as a photovoltaic type of optoelectronic devices [16, 17].

Previous research has established that the single-crystal 1D TNAs or TNWs could be synthesized via physical vapor deposition (PVD) method [18, 19] and chemical vapor deposition (CVD) [20]. Others are such as sol-gel coating, anodic oxidation, metal-organic chemical vapor deposition (MOCVD), template-assisted and sol-gel hydrothermal-based technique. Conversely, the synthesization and the deposition on the substrate for most techniques have required advanced instrumentation, for example of having a high pressured chamber, high temperature condition, or both, and addition of multiple procedures. Among those techniques, a one-step sol-gel hydrothermal method is preferred and widely investigated for their facile, rapid and considerably low temperature of synthesization and deposition technique.

Cao et al. deposited highly oriented TNAs on an FTO-coated glass substrate (**Figure 2**) and implemented amperometric measurement in 0.5-M sodium sulfate aqueous solution to investigate the photogenerated current responses under UV radiation [17]. TNAs were synthesized via a hydrothermal technique in a Teflon-lined stainless-steel autoclave, with sol-gel containing 10 mL of deionized (DI) water, 10 mL of HCl, and 0.4–0.8 mL of titanium butoxide for 4–24 hours at 120–180°C. The result presented in this study suggested that the TNAs with 40–60 nm in diameter and 6.0 μm in length were deposited for 4 hours at 150°C. The TNAs however aggregated at the bottom and only detached at tips. The TNAs were also congregated at higher temperature (180°C), with size increased to 50–100 nm in diameter and 2.0 μm in length, and in contrast, decrease to 20–40 nm in diameter and 0.5 μm in length at lower temperature (120°C). The titanium precursor (titanium butoxide) was also influenced the growth of the TNAs, as the size increased in correlation with the concentrations. However, the alignment of the nanorods was reduced at low concentration than that of higher concentration. The diameter size of the TNAs was also increase in correlation with time, even though the growth in length was hindered due to collision among the nanorods and aggregated to bigger size of nanorods. The dark current was found to be negligible without any radiation and the measured photogenerated current of 47.69 $\mu A/cm^2$ was achieved under the radiation (100 mW/cm^2), and linearly dependent on radiation intensity. Sample with higher specific surface area with separated nanorods showed better performance, and the designed UV photosensor corresponded well to 365 nm light. They also suggested that the small dimension and rod-shaped TNAs was important parameters for better mobility of charge carriers and light harvesting.

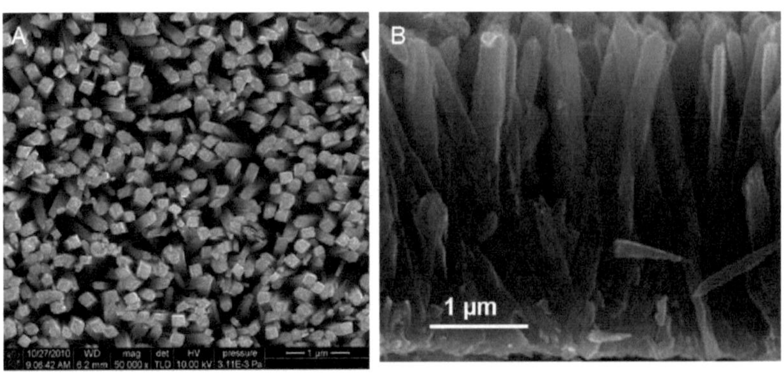

Figure 2. FESEM images of (A) top-view, and (B) cross-section of deposited TNAs film on FTO substrate at 150°C for 4 hours [17].

One study by Xie et al. examined the growth of TNAs on an FTO-coated glass substrate via a hydrothermal technique at low temperature (180°C, 2 hours) for the application of UV photosensor, using water electrolyte [2]. TNAs were synthesized via a hydrothermal technique in a Teflon-lined stainless-steel autoclave, with sol-gel containing deionized (DI) water, HCl, and titanium tetrachloride. A pure single-crystalline rutile phase of tetragonal-shaped TNAs was consistently deposited on the substrate with 100–150 nm in diameter. A high photogenerated current of 4.67 μA and open circuit voltage of 0.408 V were achieved under the UV radiation (365 nm, 61.9 μW/cm^2) at 0 V bias and linearly dependent on radiation intensity. The maximum responsivity of 0.025 A/W was achieved at 310-nm wavelength. The fabricated device also showed a fast response time of 0.15 s (rise) and 0.05 s (decay). In contrast to the proposed device structure, they initiated that the photoconductivity-based mechanism possessed much longer recovery time in behalf of the existence of depletion region at the surface of the material, which created electrons trap. It has been suggested that the formation of built-in Schottky barrier at TNAs/water interface separated the photogenerated charge carriers under the UV radiation without application of external bias and produced photogenerated current. Thus, the fabricated device works on both photodiode and photovoltaic modes.

Xie et al. also studied the growth of TNAs on an FTO-coated glass substrate via a hydrothermal technique with similar sol-gel at low temperature (180°C, 2 hours) for a PEC-based UV photosensor, using a liquid crystal-embedded electrolyte [21]. It has been shown that gap between TNAs was filled by the electrolyte, which contributed to the increase of charge separation and transportation. The purpose of having the TNAs to create quasi-solid-state electrolyte was to increase the pathway of the optical light within the PEC structure through the light-trapping mechanism, which suggested to enhance the light absorption efficiency and thus increased the photogenerated charge carriers. An increased photogenerated current of 175 μA/cm^2 and open circuit voltage of 0.46 V were achieved under the UV radiation (365 nm, 2.5 mW/cm^2) at 0 V bias and linearly dependent on radiation intensity. The maximum responsivity of 0.09 A/W was achieved at 383-nm wavelength. The fabricated device also showed fast response time (less than 0.03 s) for both rise and decay times.

Yanru et al. fabricated a PEC-based UV photosensor with nanobranched TiO_2 arrays on an FTO-coated glass substrate as a photoanode and deposited via a two-step chemical synthesis process [22]. TNAs were synthesized via a hydrothermal technique in a Teflon-lined stainless-steel autoclave, with sol-gel containing deionized (DI) water, HCl, and titanium tetrachloride for 2 hours at 180°C. Nanobranches were synthesized via aqueous chemical growth on the TNAs through immersion in titanium tetrachloride solution for several hours at 25°C. The nanobranched TNAs showed lower transmittance than that of the bare TNAs due to increase of light absorption, which attributed to increase of light scattering within the nanostructures. The photogenerated current of 31 μA/cm^2 and open circuit voltage of 0.32 V were achieved for bare TNAs, while photogenerated current of 373 μA/cm^2 and open circuit voltage of 0.44 V were achieved for TNAs with 18-hour deposition of nanobranches both under the UV radiation (365 nm, 2 mW/cm^2) at 0 V bias. The peak responsivity was increased linearly with the increased of the nanobranches. It has been described that the improved responsivity was due to increase of light scattering and contacting surface area between TNAs with nanobranches and the electrolyte. In addition, ultrathin nanobranches reduced the recombination of electron-hole pair through a very efficient hole transportation to the TiO_2/electrolyte interface.

However, further increased in length of the nanobranches disconnected the branches from the TNAs and thus reduced the active surface area interconnection with the electrolyte.

A hybrid UV photosensor using the TNWs and poly (9,9-dihexylfluorene) (PFH) was reported by Zhang et al. on an FTO-coated glass substrate [23]. TNWs were synthesized via a hydrothermal technique in a Teflon-lined stainless-steel autoclave, with sol-gel containing toluene, tetrabutyltitanate, titanium tetrachloride, and HCl for 12 hours at 150°C. PFH was spin coated onto the deposited TNWs, and 1 mm^2 of Au electrode deposited on top of the hybrids. It has been shown that the TNWs (22.6 nm diameter size) were uniformly deposited on the substrate with constant density and size distribution. The PFH was filled and concealed the whole TNWs to avoid a direct contact between the TNWs and Au electrode and thus effectively formed the TNWs/PFH heterojunction. The absorption region was formed at 320–390 nm through the spectrum overlapping of TNWs/PFH hybrid film. The *I–V* characteristics showed a typical rectifying behavior through the formation of the TNWs/PFH heterojunction, with low dark current (0.19 $\mu A/cm^2$) at 2 V reverse bias. The photogenerated current of 0.75 μA and maximum responsivity of 568 mA/W were achieved under the UV radiation (330 nm, 132 $\mu W/cm^2$) at 2 V reverse bias and linearly dependent on the voltage bias. The photogenerated carriers were instantly collected by both electrodes under the irradiation of UV owing to the direct pathway of the TNWs and rapid charge carrier's separation at the heterojunction. It has been demonstrated that the improved responsivity could be achieved through the energy level matching between the TiO_2 and PFH, which smoothed the charge transportation at the interface and reduced the recombination between electrons and holes. The energy level matching assisted the transportation of holes from the valance band of TiO_2 to the highest occupied molecular orbit (HOMO) of PFH and electrons from the lowest unoccupied molecular orbital (LUMO) of PFH to the conduction band of TiO_2.

Another hybrid structure of TNWs/polyaniline nanoflowers (TNWs/PANI NFs) on an FTO-coated glass substrate was used to fabricate a multilayer device of FTO/TNWs/PANI NFs/TNWs/FTO (**Figure 3**) by Zu et al. for the application of self-power UV photosensor [15]. TNWs were synthesized via a hydrothermal technique in a Teflon-lined stainless-steel autoclave, with sol-gel containing deionized (DI) water, HCl, and titanium butoxide for 20 hours at 150°C. PANI NFs were deposited on the TNWs by *in situ* wet chemical deposition method at about 0–5°C for 1 hour per layer, before functionalized by the immersion method in PSS solution for 12 hours. It has been shown that the tetragonal shape of TNWs (~140 nm diameter, ~2 μm length) were uniformly deposited on the substrate with very high density and high-aspect ratio. The nonlinear and asymmetric *I–V* characteristics showed a typical rectifying behavior through the formation of a p-n heterojunction of p-type PANI NFs and n-type TNWs. In addition, the used of PSS further enhanced the photogenerated current of the UV photosensor. The photogenerated current of 0.8 μA (1 layer), 2.7 μA (23 layer), and 8.0 μA (5 layer) was achieved under the UV radiation (365-nm wavelength) at 0 V bias and linearly dependent on the thickness of deposited PANI NFs. It has been described that the inner potential difference was formed at the interface of TNWs/PANI NFs under the UV irradiation, which contributed to the separation of the photogenerated electron-hole pairs and thus forming the photocurrent.

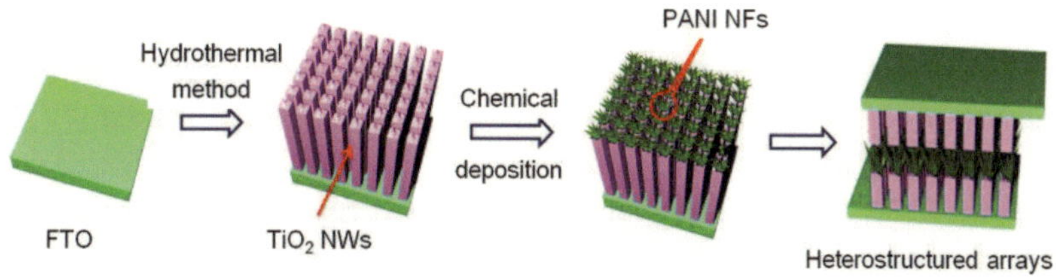

Figure 3. Schematic of the fabrication of TNWs/PANI NFs/TNWs heterostructured UV photosensor [15].

Yan et al. introduced an enhanced responsivity of self-powered UV photosensor using electrochemically reduced TNAs (R-TNAs) on an FTO-coated glass substrate [24]. TNAs were synthesized via a hydrothermal technique in a Teflon-lined stainless-steel autoclave, with sol-gel containing HCl aqueous solution and tetrabutyltitanate for 2 hours at 180°C. It has been shown that the TNAs (~100–150 nm diameter, ~3–3.5 μm length) were deposited nearly perpendicular to the substrate. The R-TNAs were prepared under −1.8 V (vs. SCE) in a conventional three-electrode system, and a disordered shell (~4 nm) with inconsistent distances between the adjacent lattice planes was formed at the outer layer of the TNAs. The disordered shell was suggested to be in correlation with bias voltage and reaction time and attributed to the electrochemical reduction. The photogenerated current of 0.24 mA/cm^2 was achieved for bare TNAs, while photogenerated current of 0.39 mA/cm^2 was achieved for R-TNAs, both under the UV radiation (385 nm, 4 mW/cm^2) at 0 V bias. The responsivity was increased linearly with the increased of the illumination intensity. High responsivity of 0.09 A/W was achieved under the UV intensity (385 nm, 1.6 mW/cm^2). The fabricated device also showed fast response time (less than 3 ms) for both rise and decay times than that of bare TNAs (4 ms) due to increase of charge transfer in the disordered shell. In addition, the open circuit voltage of the R-TNAs was higher than that of bare TNAs, which attributed to fast accumulation of holes in the disordered shell and decrease recombination of photogenerated charged carriers in the TNAs.

In previous studies on TNAs or TNW-based UV photosensor, the application of rod-like or wire-like structure of TiO_2 material has a lot of advantages over granular-like nanostructure, comprising high mobility of charge carriers and low carriers trapping at the grain boundaries, which contributed to high photogenerated current, high responsivity, and fast response and recovery times. The considerably high surface-to-volume ratio of the nanostructure also leads to miniaturization of the device and somehow reduces the power consumption for the passive device. Thus far, a number of studies have highlighted the factors that are associated with the high responsivity of nanostructured devices in correlation with high surface-to-volume ratio, surface defects, light trapping, charge carriers trapping, and high porosity.

Much of the current literature on TNAs or TNW-based UV photosensor has paid particular attention on the PEC-based structure due to their low cost, facile fabrication process, high contact area to the electrolyte, low recombination of the excitonic charge carriers, high photocurrent gain, and fast response and recovery times. Yanru et al. indicated that the PEC-based photosensor effectively evaded the utilization of complex epitaxial processes and the use of high-cost single-crystal substrates, which has been highly required for optoelectronic applications [22].

The existing literature on TNAs or TNWs-based UV photosensor also indicated that PEC-based devices worked in "self-powered" mode, which did not require any additional voltage supply from batteries or external supplies in order to operate. Thus, the PEC-based UV photosensor could be prominently reduced the size, independency and mobility of the whole detection system, and partly fulfilled the challenges for large-scale area applications of the sensors.

4. TiO_2 nanorod arrays in an ultraviolet photosensor via a facile one-step immersion method

There are a number of studies reported on the attempted hydrothermal method to prolong the synthesization and deposition processes of a preferred TNA film in attribution to the technical prospective of the TNAs, which is estimated to be remarkably enhanced if a fine-tuning of nanorods morphology is somehow achieved in the application of UV photosensor. Although the method has been extensively investigated by a number of researchers to enhance the deposition process on the substrate, the use of high-pressure vessel or autoclave even under low temperature condition was commonly reported. The usage of the tools was also difficult in attribution to the size and weight of the stainless steel materials and the Teflon line inside to hold the heat and the pressure inside the container. The time taken to heat the autoclave during the deposition process and cooling after the deposition procedure will somehow affect the deposited sample and included in the whole deposition process. Therefore, it is essential to improve the current method to a more simple, rapid, and consistent one-step sol-gel immersion method to synthesized and deposit the TNAs on the substrate.

In our process, we have synthesized and deposited the TNAs using an enhanced one-step sol-gel immersion method via a simple-modified Schott bottle with cap clamps as described in our former report [25]. The use of Schott bottle (**Figure 4**) as a glass container in our studies was intentionally to replace the use of stainless steel–based autoclave for the deposition of thin film TNAs at low temperature. The utilization of glass container was also to further

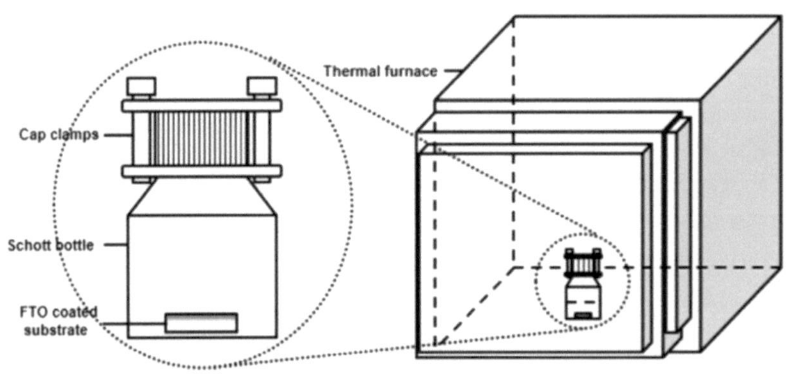

Figure 4. Setup for TNAs synthesis in a Schott bottle with cap clamps [25].

enhance the investigation on the deposition of the thin film TNAs at low temperature and low time growth parameters for the purpose of optimization, simplification, and miniaturization of the device. In addition, the condition of sol-gel solution in the glass container also could be monitored and controlled manually, or by implementation of automation system to maintain the quality of the deposited film, which is highly essential in product manufacturing, especially during mass production.

In general, most of the sol-gel methods for synthesizing nanomaterials involve precursor, solvent, and stabilizer. The precursor provides the core composition of atoms or ions of the intended nanostructures, while the stabilizer component is added to control the growth rate of the nanoparticles in the solution, or epitaxially growth on the surface substrate. The presence of stabilizer in sol-gel minimized the reaction rate from developing instant precipitation or large aggregation, to some extent at room temperature. In our method, TNAs were prepared by a sol-gel immersion method as described in our previous work [25]. The FTO-coated glass substrate was used as the catalyst for the growth of the TNAs. The substrate was initially cleaned with acetone, ethanol, and DI water for 10 minutes subsequently in an ultrasonic water bath (Hwasin Technology Powersonic 405, 40 kHz). Sol-gel containing HCl (37 %, Merck) and DI water in a 1:1 volume ratio was prepared in a Schott bottle. After constant stirring for 10 minutes, 0.07 M of titanium (IV) butoxide (97%, Sigma-Aldrich) was slowly added in the prepared solution and stirred vigorously for another 30 minutes. The cleaned substrate was immersed in the sol-gel inside the Schott bottle. Once the conductive side of the substrate was placed at the bottom and facing upward, the bottle was then tightly sealed with a cap and clamped to attain the internal pressure throughout the deposition process at 150°C in an oven. After the TNAs were deposited at the elevated temperature for 2 hours, the bottle was immediately cooled down until the sample could be taken out from the bottle. The sample was then rinsed with DI water and dried at room temperature. Finally, the sample was annealed in a furnace at 450°C for 30 minutes to improve crystallinity.

The surface morphologies and topography of the TNAs were observed via field-emission scanning electron microscopy (FE-SEM, ZEISS Supra 40VP) and atomic force microscopy (AFM, Park System), respectively. The crystallinity of the synthesized TNAs was characterized via X-ray diffraction (XRD, Shimadzu XRD-6000, Cu K-alpha radiation, wavelength 1.54 Å). The optical property was characterized using a UV-Vis-NIR spectrophotometer (Cary 5000). The synthesized and deposited TNA on FTO-coated glass substrate was used as an electrode in the proposed fabrication of UV photosensor based on PEC structure. PEC structure was selected for our intention to design a UV photosensor to operate at low-powered and self-powered modes as previously described. The counter electrode was designed using platinum (Pt)-coated FTO-coated glass substrate. Both electrodes were compiled together, with a spacer (Surlyn, DuPont) in between the electrodes, with effective area designed approximately at 1.0 cm^{-2} in the centre of the device. The prepared electrolyte comprising 0.5 M of sodium sulfate (Na_2SO_4) was filled in the inner side of the spacer between the two electrodes.

5. Performance of synthesized TNAs-based UV photosensor

The proposed sol-gel immersion method is used for the development of a one-dimensional growth of uniform and densely array of nanorods structure. TNAs (120 nm diameter, 1.52 μm in length) were successfully synthesized and deposited on the FTO-coated glass substrate, as shown in **Figure 5**. The topography images of AFM (**Figure 6**) exhibited the comparable image structure of the top view as observed using FESEM. The measured root mean square roughness, R_{rms}, and average roughness, R_a, of the synthesized TNAs were approximately 38.52 and 30.37 nm, respectively. In general, the growth of TNAs is basically formed via the chemical reactions throughout the nucleation and growing processes through the hydrolysis of titanium(IV) butoxide in acidic solution. The proposed reaction mechanisms were emphasized on two steps as follows [25, 26]:

$$\mathrm{Ti}\left(\mathrm{OCH_2\,CH_2\,CH_2\,CH_3}\right)_4 + 4\,\mathrm{H_2O} \xrightarrow{\mathrm{HCl}} \mathrm{Ti\,(OH)_4} + 4\mathrm{ROH}, \quad (1)$$

$$\mathrm{Ti}\,(OH)_4 \rightarrow \mathrm{TiO_2} + 2\,\mathrm{H_2O}, \quad (2)$$

where $R = C_4H_9$.

At a regular state, the process of hydrolysis for titanium yields a definite amount of heat and occurs immediately at room temperature. The HCl was used as a stabilizer in the sol-gel to minimize the reaction rate from instantly developing precipitation or aggregation during the preparation stage, in attribution to the presence of free H^+ [27, 28]. At the same time, the 1D growth of TNAs at the preferred [001] direction was due to the presence of Cl^-, which bonded to the (1 1 0) plane and preventing the growth of TiO_2 crystal structure at its direction, to some extent [16, 29].

The crystal clear condition of the prepared sol-gel at room temperature showed the sufficiency of HCl content in the solution. Eq. (1) exhibits the hydrolysis process of titanium, which removes four carbon atoms from the precursor to form titanium hydroxyl and alkoxy group. The presence of H^+ through the addition of HCl as the stabilizer in this equation prevents the hydrolysis rate of titanium. The condensation process in Eq. (2) synthesizes TiO_2 and produces water. In order to produce highly aligned 1D growth of the TNAs from titanium alkoxide in the sol-gel, the ratio of DI water and HCl, and the growth rate through time and temperature need to be tuned and optimized. On heating during the sol-gel immersion process, the growth rate increases in correlation with synthesization and deposition time, which contributes to the

Figure 5. FESEM images of the TNAs at magnifications of (a) 20,000× and (b) 70,000×. (c) Cross-sectional images of the synthesized TNAs.

Figure 6. AFM images of the synthesized TNAs [25].

nucleation of TiO_2 on the seeds of FTO from the substrate. The nucleation of TiO_2 is basically based on the atomic structuring of TiO_6 octahedron as the fundamental unit, which plays an important role in the formation of tetragonal structure of TiO_2. The arrangement and interconnection of TiO_6 octahedral determine the phase and structure of the synthesized TiO_2. The rutile phase of TNAs synthesized in our method indicated that the edge sharing of TiO_6 octahedral occurred during the formation of tetragonal structure of TiO_2.

Figure 7 shows the XRD pattern of the synthesized and deposited TNAs. The rutile peaks of TiO_2 were detected at (1 0 1), (2 1 0), (2 1 1), (3 1 0), (0 0 2), and (1 1 2) planes, and no peaks which indicated the anatase and brookite phases observed in the pattern (JCPDS No.01-072-1148). Intense peaks at (0 0 2) and (1 0 1) planes showed that the 1D growth of TNAs was at the [0 0 1] direction.

Figure 8(a) shows the spectrum of optical absorbance of the synthesized and deposited TNAs on the FTO-coated glass substrate in a range of 200–800 nm. The spectrum exhibits that the intensity of the absorbance is significantly increased below 400-nm wavelength, which is in the UV region. Therefore, the optical characteristic of the deposited TNAs to absorb the UV radiation could be utilized for the application of UV photosensor as predicted. The performance of the developed UV photosensor was measured under 365-nm UV radiation at a bias voltage of 1 V. **Figure 8(b)** exhibits the time-dependent photogenerated current of the developed UV photosensor using the deposited TNAs. The result shows an average increase of 0.1 mA of photogenerated current under UV radiation, which is attributed to the photogenerated electrons through the excitation from the valance band into the conduction band. The rise and fall times are negligible throughout the measurement. Thus, the fabricated UV photosensor from the deposited TNAs facilitates the potential application in a low-power UV detecting device.

The present study was designed to introduce a new one-step method to synthesize and deposit a thin film TNAs on the FTO-coated glass substrate at low temperature and rapid process using the facile glass container for the application of UV photosensor. Although the fabricated UV photosensor has successfully shown its potential application in sensing the UV radiation, more research is required to determine the optimum efficiency of the device, through the variation of synthesization and deposition parameters, composition of electrolytes, composition of sol-gel, device structure, and many more. This research has opened up many opportunities of further investigation.

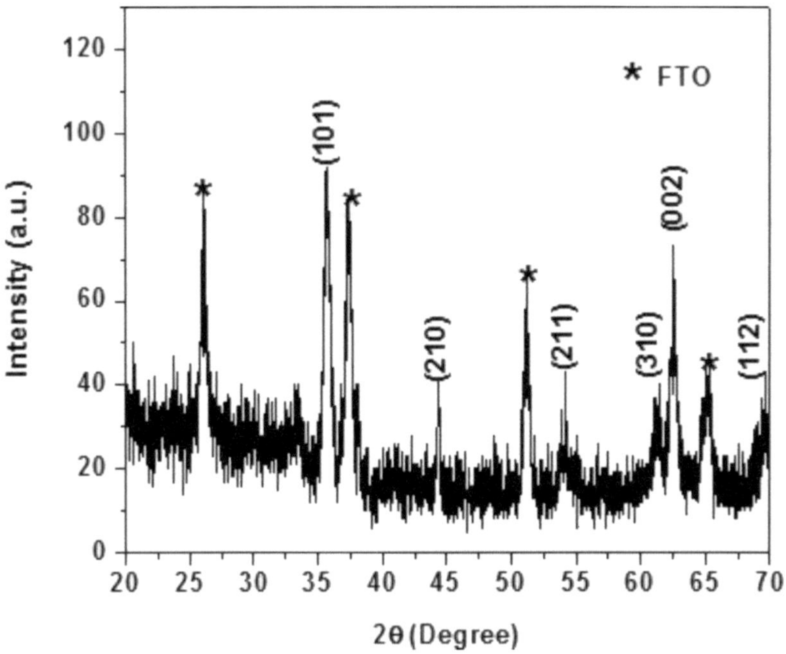

Figure 7. XRD pattern of the rutile-phased TNAs [25].

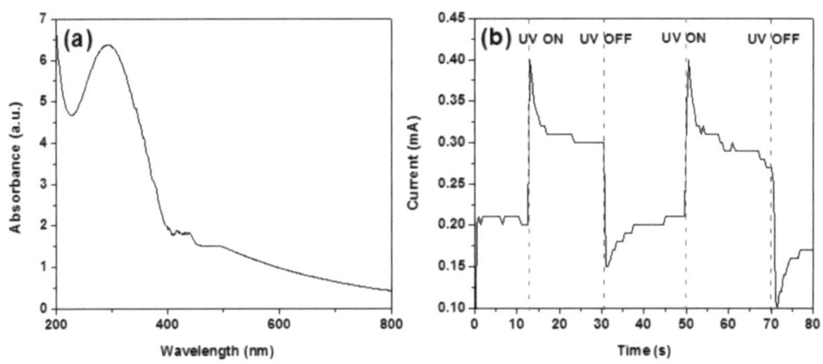

Figure 8. (a) UV-Vis absorbance spectrum, (b) photocurrent response under UV irradiation (365 nm, 1 V bias) of synthesized TNAs [25].

6. Conclusion

The aim of the present study was to examine the previous works on TiO_2 nanostructures in the application of UV photosensor and consequently open up new possibilities and opportunity in future works for enhancing the development of the device. This study has found that the nanostructured materials have a lot of advantages over bulk materials, and the high responsivity of the UV photosensor was mostly associated with the high surface-to-volume ratio, porosity, surface defects, light trapping, intensity of the UV radiation, and low-charge-carriers trapping. One-dimensional growth of TNAs or TNWs offers an enhance charge carriers mobility to overcome the photocurrent loss due to the existence of multiple grain boundaries in granular-like and porous nanostructures with comparatively high surface-to-volume ratio. The performance of the device was also dependent on the device structure and the detection mechanism of photon through the photogenerated charge carriers. PEC-based device structure is preferred the TNA-based UV photosensors due to their low cost, facile fabrication process, high contact area to the electrolyte, low recombination of the excitonic charge carriers, high photocurrent gain, and fast response and recovery times. It also could work in an applied bias mode, as well as in 'self-powered' mode, which did not require any additional voltage supply. Besides the high responsivity, the design of the device also needs to fulfill the requirement for small size, independency and mobility of the whole detection system. Our study was designed to introduce a new one-step method to synthesize and deposit a thin film TNAs on an FTO-coated glass substrate at low temperature and rapid process using a facile glass container. The fabricated PEC-based UV photosensor using the deposited TNAs has successfully shown its potential application in sensing the UV radiation. Our introduced method has opened up many opportunities of further investigation in the development of 1D growth of TNAs for various device applications.

Acknowledgements

This work was financially supported by the Fundamental Research Grant Scheme 600-RMI/FRGS 5/3 (57/2015)) from the Ministry of Education Malaysia. The authors also would like to thank the Faculty of Electrical Engineering (FKE), NANO-ElecTronic Centre (NET), NANO-SciTech Centre (NST), Research Management Centre (RMC) of University Teknologi MARA, Shah Alam, Malaysia (UiTM), the Ministry of Higher Education of Malaysia, and International Islamic University Malaysia (IIUM).

Author details

Marmeezee Mohd. Yusoff[1,3], Mohamad Hafiz Mamat[1,2]* and Mohamad Rusop Mahmood[1,2]

*Address all correspondence to: hafiz_030@yahoo.com

1 Faculty of Electrical Engineering, NANO-ElecTronic Center (NET), University Teknology MARA (UiTM), Shah Alam, Selangor, Malaysia

2 NANO-SciTech Centre (NST), Institute of Science (IOS), University Teknology MARA (UiTM), Shah Alam, Selangor, Malaysia

3 Manufacturing and Materials Engineering Department (MME), Kulliyyah of Engineering, International Islamic University Malaysia (IIUM), Malaysia

References

[1] Pfeifer V, Erhart P, Li S, Rachut K, Morasch J, Brötz J, Reckers P, Mayer T, Rühle S, Zaban A, Mora Seró I, Bisquert J, Jaegermann W, Klein A. Energy band alignment between anatase and rutile TiO_2. The Journal of Physical Chemistry Letters. 5 December 2013;**4**:4182–4187

[2] Xie Y, Wei L, Wei G, Li Q, Wang D, Chen Y, Yan S, Liu G, Mei L, Jiao J. A self-powered UV photodetector based on TiO_2 nanorod arrays. Nanoscale Research Letters. 2013;**8**:188

[3] Cao G, Wang Y. Nanostructures and Nanomaterials: Synthesis, Properties, and Applications. 2nd ed. New Jersey: World Scientific; 2011

[4] Soci C, Zhang A, Bao XY, Kim H, Lo Y, Wang D. Nanowire photodetectors. Journal of Nanoscience and Nanotechnology. March 2010;**10**:1430–1449

[5] Okuya M, Shiozaki K, Horikawa N, Kosugi T, Kumara GRA, Madarász J, Kaneko S, Pokol G. Porous TiO_2 thin films prepared by spray pyrolysis deposition (SPD) technique and their application to UV sensors. Solid State Ionics. 2004;**172**:527–531

[6] Xue H, Kong X, Liu Z, Liu C, Zhou J, Chen W, Ruan S, Xu Q. TiO_2 based metal-semiconductor-metal ultraviolet photodetectors. Applied Physics Letters. 2007;**90**:201118

[7] Han Y, Wu G, Wang M, Chen H. High efficient UV-a photodetectors based on monodispersed ligand-capped TiO_2 nanocrystals and polyfluorene hybrids. Polymer. 2010;**51**:3736–3743

[8] Vigil E, Peter LM, Forcade F, Jennings JR, González B, Wang H, Curbelo L, Dunn H. An ultraviolet selective photodetector based on a nanocrystalline TiO_2 photoelectrochemical cell. Sensors and Actuators A: Physical.2011;**171**:87–92

[9] Li X, Gao C, Duan H, Lu B, Pan X, Xie E. Nanocrystalline TiO_2 film based photoelectrochemical cell as self-powered UV-photodetector. Nano Energy. 2012;**1**:640

[10] Xiang WF, Yang PR, Wang AJ, Zhao K, Ni H, Zhong SX. Vertical geometry ultraviolet photodetectors with high photosensitivity based on nanocrystalline TiO_2 films. Thin Solid Films. 2012;**520**:7144–7146

[11] Lv K, Zhang M, Liu C, Liu G, Li H, Wen S, Chen Y, Ruan S. TiO_2 ultraviolet detector based on $LaAlO_3$ substrate with low dark current. Journal of Alloys and Compounds. 2013;**580**:614–617

[12] Selman AM, Hassan Z, Husham M, Ahmed NM. A high-sensitivity, fast-response, rapid-recovery p–n heterojunction photodiode based on rutile TiO_2 nanorod array on p-Si(1 1 1). Applied Surface Science. 2014;**305**:445–452

[13] Liu G, Tao C, Zhang M, Gu X, Meng F, Zhang X, Chen Y, Ruan S. Effects of surface self-assembled NH_4^+ on the performance of TiO_2-based ultraviolet photodetectors. Journal of Alloys and Compounds. 2014;**601**:107

[14] Yang J, Jiang Y-L, Li L-J, Muhire E, Gao M-Z. High-performance photodetectors and enhanced photocatalysts of two-dimensional TiO_2 nanosheets under UV light excitation. Nanoscale. 2016;**8**:8170–8177

[15] Zu X, Wang H, Yi G, Zhang Z, Jiang X, Gong J, Luo H. Self-powered UV photodetector based on heterostructured TiO_2 nanowire arrays and polyaniline nanoflower arrays. Synthetic Metals. February 2015;**200**:58–65

[16] Liu B, Aydil ES. Growth of oriented single-crystalline rutile TiO_2nanorods on transparent conducting substrates for dye-sensitized solar cells. Journal of the American Chemical Society. 25 March 2009;**131**:3985–3990

[17] Cao C, Hu C, Wang X, Wang S, Tian Y, Zhang H. UV sensor based on TiO_2 nanorod arrays on FTO thin film. Sensors and Actuators B: Chemical. 2011;**156**:114–119

[18] Meng L, Chen H, Li C, dos Santos MP. Preparation and characterization of dye-sensitized TiO_2 nanorod solar cells. Thin Solid Films. 2015;**577**:103–108

[19] Binetti E, El Koura Z, Bazzanella N, Carotenuto G, Miotello A. Synthesis of mesoporous ITO/TiO_2 electrodes for optoelectronics. Materials Letters.2015;**139**:355–358

[20] Chen JB, Wang CW, Kang YM, Li DS, Zhu WD, Zhou F. Investigation of temperature-dependent field emission from single crystal TiO_2 nanorods. Applied Surface Science. 2012;**258**:8279–8282

[21] Xie Y, Wei L, Li Q, Chen Y, Liu H, Yan S, Jiao J, Liu G, Mei L. A high performance quasi-solid-state self-powered UV photodetector based on TiO_2 nanorod arrays. Nanoscale. 2014;**6**:9116–9121

[22] Yanru X, Lin W, Qinghao L, Yanxue C, Shishen Y, Jun J, Guolei L, Liangmo M. High-performance self-powered UV photodetectors based on TiO_2 nano-branched arrays. Nanotechnology. 2014;**25**:075202

[23] Zhang M, Li D, Zhou J, Chen W, Ruan S. Ultraviolet detector based on TiO_2 nanowire array–polymer hybrids with low dark current. Journal of Alloys and Compounds. 2015;**618**:233–235

[24] Yan P, Wu Y, Liu G, Li A, Han H, Feng Z, Shi J, Gan Y, Li C. Enhancing photoresponsivity of self-powered UV photodetectors based on electrochemically reduced TiO_2 nanorods. RSC Advances. 2015;**5**:95939–95942

[25] Yusoff MM, Mamat MH, Malek MF, Suriani AB, Mohamed A, Ahmad MK, Alrokayan SAH, Khan HA, Rusop M. Growth of titanium dioxide nanorod arrays through the aqueous chemical route under a novel and facile low-cost method. Materials Letters. 2016;**164**:294–298

[26] Cozzoli PD, Kornowski A, Weller H. Low-temperature synthesis of soluble and processable organic-capped anatase TiO_2 nanorods. Journal of the American Chemical Society. 2003;**125**:14539–14548

[27] Li Y, Liu J, Jia Z. Morphological control and photodegradation behavior of rutile TiO_2 prepared by a low-temperature process. Materials Letters.2006;**60:**1753–1757

[28] Iraj M, Nayeri FD, Asl-Soleimani E, Narimani K. Controlled growth of vertically aligned TiO_2 nanorod arrays using the improved hydrothermal method and their application to dye-sensitized solar cells. Journal of Alloys and Compounds. 2016;**659**:44–50

[29] Feng X, Shankar K, Varghese OK, Paulose M, Latempa TJ, Grimes CA. Vertically aligned single crystal TiO_2nanowire arrays grown directly on transparent conducing oxide coated glass: Synthesis details and applications. Nano Letters. 12 November 2008;**8**:3781–3786

Chapter 8

Advancement of Sol-Gel–Prepared TiO_2 Photocatalyst

Robabeh Bashiri, Norani Muti Mohamed and
Chong Fai Kait

Additional information is available at the end of the chapter

http://dx.doi.org/10.5772/intechopen.68357

Abstract

This chapter elaborates a review of sol-gel–prepared TiO_2 photocatalyst for different photocatalytic applications. Among the semiconductors employed, TiO_2 is known as an attractive photocatalyst owing to its high photosensitivity, nontoxicity, easy availability, strong oxidizing power and long-term stability. Some research works related to the effect of sol-gel preparation parameters on physicochemical properties and different photocatalytic applications of prepared TiO_2 photocatalysts are reported. Furthermore, various sol-gel and related systems for modification of TiO_2 photocatalytic performance, including transition metals and co-doing of TiO_2, were considered. The results illustrated that doping TiO_2 with metal ions through sol-gel method usually resulted in an improved efficiency of TiO_2 photocatalyst. This method has all the advantages over other preparation techniques in terms of purity, homogeneity, felicity and flexibility in introducing dopants in a large concentration, stoichiometry control, ease of processing and composition control.

Keywords: TiO_2, metal doping, hydrolysis, physicochemical properties, photocatalytic activity

1. Introduction

An excessive deal of attention has recently been paid to the development of photocatalyst which being used in a variety of products and research areas especially for environmental and energy applications. Semiconductor-based photocatalyst have been extensively studied due to its electronic configuration to absorb applied solar spectrum for photocatalytic reaction. The semiconductor consists of a valence band (VB) and conduction band (CB). The energy difference between the top of the VB and the bottom of the CB levels is known as band gap energy (Eg) which is responsible for photoactivity of the semiconductor. **Figure 1** displays the general principle of photocatalytic reaction over the semiconductor as summarized in the following steps:

1. Harvesting photons with energy equivalent or larger than the bandgap energy to form electron-hole pairs.

2. Photogenerated electrons are injected to the CB, while photogenerated holes with a positive charge are left behind in the VB.

3. Surface chemical reactions consist of reducing an acceptor species ($A^+ \rightarrow A$) and oxidizing a donor species ($D \rightarrow D^+$) [1].

Semiconducting oxides such as TiO_2, $SrTiO_3$, ZnO, α-Fe_2O_3, WO_3, Ta_2O_5, VO_2, $KTaO_3$ and Fe_2O_3 appear to be the most considered materials for photocatalysts due to the fact that their properties can be modified over wide ranges through changing their semiconducting properties. Moreover, these materials are economically promising because their processing technologies are relatively simple [2, 3]. Among different metal oxides, TiO_2 is one of the most significant photocatalyst with a large variety of applications in energy and environment (**Figure 2**),

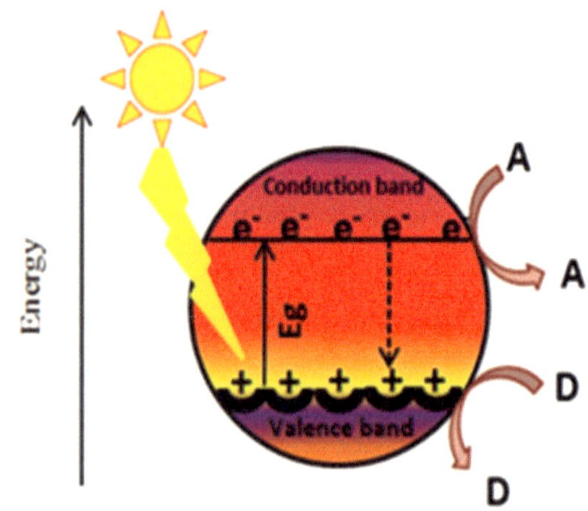

Figure 1. A schematic diagram of general principle of the photocatalytic reaction.

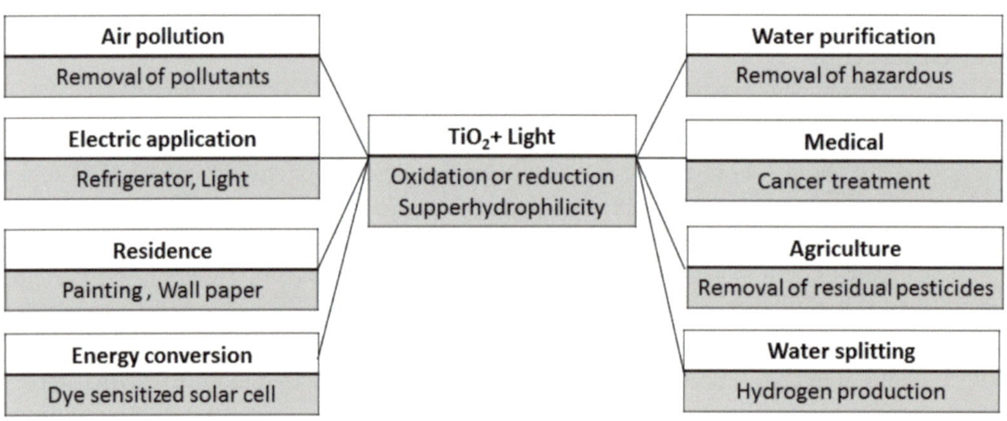

Figure 2. Different photocatalytic applications of TiO_2.

regardless of its limitations. The following section is dedicated to the discussion of the intrinsic properties of TiO_2 to fully understand its function as a photocatalyst and its synthesis process.

2. Titanium dioxide (TiO_2) photocatalyst

TiO_2 has high resistance to corrosion and photocorrosion in an aqueous medium, cheaper than many other photosensitive materials, easily available, environmentally friendly. Furthermore, its electronic properties can be varied by just changing the defect chemistry or oxygen stoichiometry [9]. The photocatalytic applications of TiO_2 are restricted because of its large bandgap energy (3.0–3.2 eV) and small electron mobility of 1 cm^2/V s [4–6]. TiO_2 has three crystalline phases, including anatase, rutile and brookite. Anatase and rutile are the most common structures for photocatalytic studies. **Figure 3a** and **b** describes the lattice structures of anatase and rutile, respectively, in the form of distorted TiO_6 octahedral with six O^{2-} ions around each Ti^{+4} ion. The structure of rutile represents an irregular octahedron with slightly orthorhombic distortion compared to octahedron in anatase [34]. The structural difference between anatase and rutile explains the significant difference in their electronic band properties.

However, rutile with band gap energy of 3.0 eV has a threshold absorption edge of 415 nm in the visible region compared to anatase with band gap energy of 3.2 eV and threshold absorption of 390 nm. Researchers have reported that anatase is the appropriate crystalline phase for photocatalytic hydrogen production [36]. It has more negative CB position (~0.2 eV) compared to rutile as demonstrated in **Figure 4**, while their VB positions (O2p) are almost in the similar position.

2.1. Sol-gel synthesis of TiO_2 nanoparticles

Photocatalytic activity of TiO_2 is strongly influenced by the synthesis condition and methodology. Various methods have been applied for synthesis of TiO_2 photocatalyst: electrochemical [7, 8], combining inverse micelle and plasma treatment [9], dip coating [10], two step wet

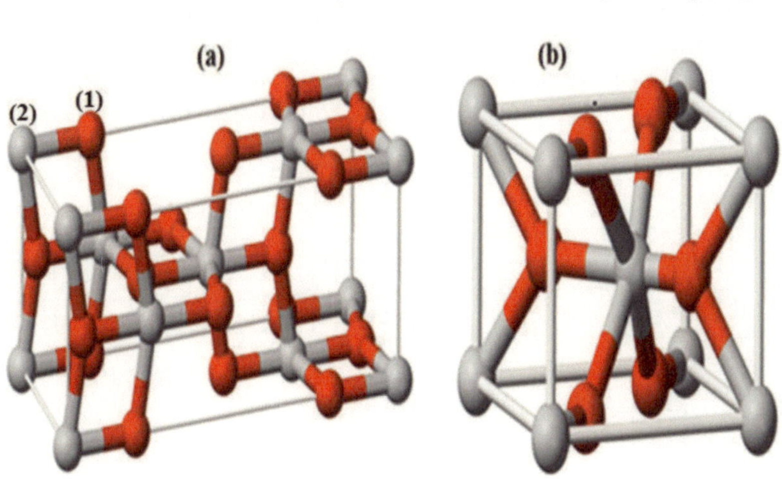

Figure 3. The bulk structural model of: (a) anatase and (b) rutile (ball (1) and (2) represent oxygen and titanium atoms, respectively) [35].

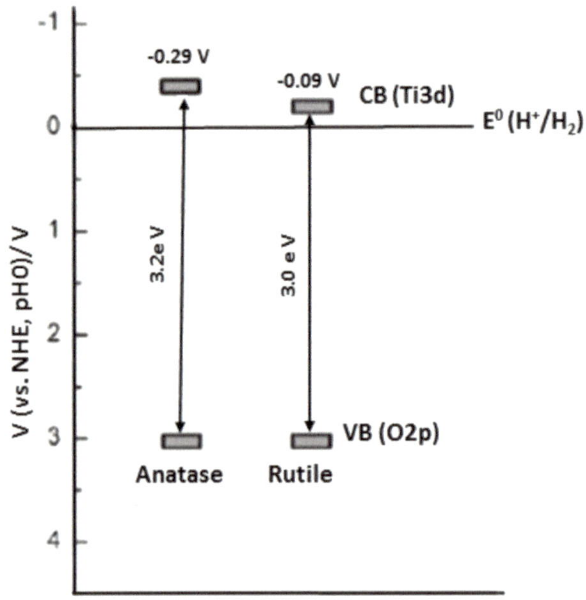

Figure 4. Electronic bands of anatase and rutile [36].

chemical [11], precipitation [12], thermal (hydrothermal and solvothermal) [13, 14], chemical solvent and chemical vapor decomposition (CSD and CVD) [15, 16], ultrasonic irradiation [17], two-route sol-gel [18], and sol-gel [19–21]. Sol-gel method is a technique to improve the physicochemical and electrochemical properties of TiO_2 nanocrystalline. It provides a simple synthesis process of nanoparticle at ambient temperature under atmospheric pressure, and this technique does not require a complicated set up. The benefits derived from preparing TiO_2 by sol-gel method (process flow chart shown in **Figure 5**) such as purity, homogeneity, and flexibility of the growth of TiO_2 can be effectively controlled by hydrolysis and condensation of titanium alkoxide in the aqueous medium [22]. Typically, sol-gel–derived precipitates are amorphous in nature, requiring further heat treatment to induce crystallization. Based on literatures, the sol-gel method has been modified using various techniques like calcination, ultrasonic [23], hydrothermal [24, 25], or surfactant [26] to obtain better properties of the synthesized powders.

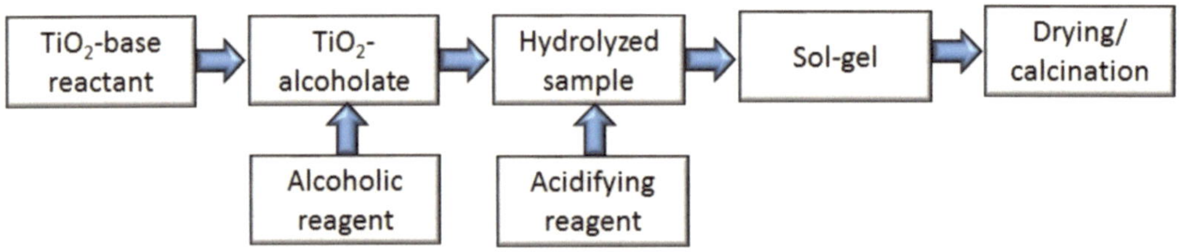

Figure 5. A schematic diagram of sol-gel process to synthesize TiO_2-based photocatalysts [22].

In normal sol-gel processing, gelation and calcintion are necessary. If the calcination temperature for crystallization is not high enough, some organic molecules will remain in the product, leading to the incomplete crystallization of titania. However, high temperature calcination causes high aggregation rate and/or phase transformation [27]. Hydrothermal synthesis provides an easy route to prepare a well-crystalline and phase-pure oxide in a tightly closed vessel. The hydrothermal-assisted sol-gel method helps to increase the number of hydroxyl groups (OH^-) on the surface of the photocatalyst and requires a lower calcination temperature. The prepared TiO_2 by this method has high thermal stability, well-crystalline phase, small particle sizes, and large surface area, which are beneficial to improve the photocatalytic activity [28–30]. The sol-gel process is typically based on the formation of inorganic polymer by hydrolysis and condensation of metal precursor like titanium alkoxide to oxopolymers, which are transformed into a metal alkoxide in aqueous solutions or organic solvents [31–34]. A number of parameters can influence on the hydrolysis process like water to alkoxide ratio, pH, and solvents. Water plays a significant role during the hydrolysis process to determine the reaction mechanism, the number of active sites on the surface of the photocatalyst and the physical properties of the photocatalyst during photocatalytic reaction [35].

2.2. Effect of hydrolysis process on sol-gel method

The water to alkoxide molar ratio (R_W) plays an important role in the structure of TiO_2 due to controlling of the formation of nucleus and the growth of crystallites. The total sol-gel synthesis can be described by an alcoholic permutation reaction, hydrolysis, and condensation reactions, which are competitive with alkoxolation, oxolation and olation as follows:

1. Alcoholic permutation

$$M(OR)Z + x\,R'(OH) \leftrightarrow M(OR)Z - x(OR') + x ROH \quad (1)$$

2. Hydrolysis

$$M(OR)n + nH_2O \rightleftarrows M(OH)n + nROH \quad (2)$$

3. Condensation reactions

A. Alcoxolation

$$\text{M-OH} + \text{M-OR} \rightleftarrows \text{M-O-M} + \text{R-OH} \quad (3)$$

B. Oxolation

$$\text{M-OH} + \text{M-OH} \rightleftarrows \text{M-O-M} + H_2O \quad (4)$$

The water to alkoxide molar ratio is investigated in two main groups: low R_W and high R_W. Hydrolysis of alkoxide with low water content less than stoichiometric ratio ($R_W < 4$) causes an incomplete hydrolysis. Therefore, unhydrolyzed alkoxide can be absorbed on the surface of TiO_2 and increase particle size with irregular shape and low surface area [36]. A high R_W ($R_W > 100$) causes the completion of nucleation and growth during the hydrolysis of TiO_2

within nanosecond; thus, TiO_2 particle is unstable, and the white suspension is immediately formed due to precipitation of a large aggregate [37]. Furthermore, adding external solvent like simple alcohols can significantly affect the controlling hydrolysis and condensation rate.

Venkatachalam et al. [38] investigated the influence of water to alkoxide ratio, pH, and type of solvent (methanol, isopropyl alcohol, acetic acid) on the physicochemical properties of TiO_2 and consequently its photocatalytic activity by using Bisphenol-A (BPA) as a model pollutant. **Table 1** illustrates that the prepared photocatalysts in the presence of acetic acid had smaller particle size and higher surface area compared to prepared TiO_2 in the presence of isopropanol and methanol. It was explained that acetic acid as a catalyst causes a rapid hydrolysis process, formation of titanium hydroxide, and their condensation to form TiO_2 nanoparticles. In addition, the variation of pH from 3 to 9 during hydrolysis process illustrates that the prepared TiO_2 photocatalyst at pH = 9 has very low surface area due to rapid hydroxylation of titanium precursor and agglomeration of TiO_2 particles.

It was reported that high water: alkoxide ratio improves the nucleophilic attack of water on titanium (IV) isopropoxide; however, it had negative effect on the surface area beyond the optimum level (350) due to more agglomeration. Furthermore, the presence of residual alkoxy groups decreases the crystallization rate of TiO_2, which favor the formation of less dense anatase phase exclusively. At low water content, the hydrolysis rate is slow, and the existence of excess titanium alkoxide in the solvent favors the development of Ti–O–Ti chains and formation of high ratio of rutile phase. They reported that the synthesized TiO_2 photocatalyst under the optimum molar ratio of alkoxide, acetic acid and water is 1:10:350 had maximum photocatalytic mineralization of BPA over nano-sized TiO_2 photocatalystcompered to P25 (commercial TiO_2) and other prepared photocatalysts.

Bashiri et al. [29] reported synthesis a series of 10 mol% Cu/TiO_2 photocatalysts by varying H_2O:alkoxide molar ratios (8, 16, 32, and 64) using sol-gel–associated hydrothermal method. The influence of hydrolysis rate on the physicochemical properties (**Table 2**) and photocatalytic

Photocatalyst	BET surface area ($m^2 g^{-1}$)	Catalyst anatase crystallite size (nm)	Anatase:rutile ratio (%)	Band gap energy (eV)
Hydrolyzing agent				
Methanol	69	17	69:31	3.19
Isopropyl alcohol	84	12.6	74.26	3.21
Glacial acetic acid	107	8.3	82.18	3.27
Source:solvent:water				
1:10:150	86	12	73:27	3.19
1:10:250	94	10.1	78:22	3.26
1:10:350	110	8.1	83:17	3.28
1:10:450	91	8.6	74:26	3.21

Table 1. Physicochemical properties of nano-sized TiO_2 photocatalysts [38].

Photocatalysts	H_2O:alk xide	Crystallite size (nm)	Average particle size (nm)		BET surface area (m^2/g)	Band gap (eV)	Hydrogen production (μmol)
			TEM	FESEM			
TiO_2	0	13.55	13.89	27.54	82.69	3	24.5
CuT8	8	9.63	11.47	22.194	85.57	2.96	1228.8
CuT16	16	11.22	12.08	19.72	83.5	2.9	3926.8
CuT32	32	12.88	13.34	16.835	89.19	2.72	10571.0
CuT64	64	10.54	12.53	20.185	84.57	2.83	3010.3

Table 2. Physicochemical properties of all prepared photocatalysts [29].

hydrogen production by water photosplitting in an aqueous NaOH-glycerol solution were investigated. It was demonstrated that the amount of water has strong influence on the hydrolysis and polymerization rate, changing the physicochemical properties and photocatalytic activity of the prepared photocatalysts.

The average crystal sizes ranged from 9.63 (CuT8) to 13.55 nm (TiO_2), which confirmed that the presence of copper and variation of water contents can strongly control crystal size. The low water content (CuT8) causes the incomplete hydrolysis, leads to aggregates, and surpasses the crystal growth. However, adding more water showed negative impact on the crystallinity due to formation of aggregates rather than the growth of crystals, strong nucleophilic reaction and fast hydrolysis rate. In this case, more alkoxy groups in the alkoxide are substituted by hydroxyl groups from H_2O, and the quantity of unhydrolyzed alkyl in the precursor is reduced hence reduction in steric hindrance by the residual alkyls preventing crystallization to anatase phase [37, 39]. The better crystallinity and proper crystal size of CuT32 compared to other photocatalysts can be explained by the moderate hydrolysis rate during preparation. The variation of the BET surface area of mesoporous photocatalysts from 82.69 to 89.19 m^2/g illustrates that the grain size was decreased in CuT8 and CuT64 but the surface area did not increase due to the agglomeration during hydrolysis process. The photocatalyst with the H_2O:alkoxide molar ratio of 32 produced the highest cumulative hydrogen production of 10571 μmol in 300 min compared to TiO_2 and other synthesized Cu/TiO_2 photocatalysts in the aqueous NaOH-glycerol solution. The mesoporous nanoparticles with larger specific surface area, lower bandgap energy, more absorbance in the visible region, presence of Ti^{3+} with higher photocatalytic activity and the coexistence of Cu_2O and CuO are responsible for better photocatalytic performance of CuT32 photocatalyst.

In the other work, Behnajady et al. [40] synthesized titanium dioxide nanoparticles by sol-gel method. Various precursors under different synthesis conditions such as solvent and water percent, reflux temperature and time, sol drying method and calcination temperature are studied as shown in **Table 3**.

The photocatalytic activity of prepared TiO_2 was studied by photodegradation of C.I. Acid Red 27 as a model contaminant from textile industry under UV light irradiation. Results of characterization revealed that the type of the precursor and solvent is effective on the particle size

Parameter	Variations
Precursor	Titanium tetraisopropoxide (TTIP), titanium n-butoxide (TBOT)
Solvent	Methanol, ethanol, isopropanol
Solvent molar percent (%)	1, 5, 10, 15, 20
Water molar percent (%)	10, 20, 40, 65, 80, 100
Reflux temperature (°C)	50, 65, 80
Reflux time (h)	1, 3, 6
Sol drying method	Thermal drying, freeze drying
Calcination temperature (°C)	350, 450, 750

Table 3. Experimental parameters [40].

and crystalline structure of the synthesized TiO_2 nanoparticles, which can strongly control the photocatalytic performance of TiO_2. Optimal conditions for synthesis of TiO_2 nanoparticles with excellent photocatalytic activity were obtained from titanium isopropoxide precursor in methanol solvent with molar ratio of 1:65:1(precursor:water:solvent) under reflux for 3 h at reflux temperature of 80°C, employing thermal drying method for sol drying and calcination temperature of 450°C. Titanium dioxide nanoparticles produced under optimal conditions show higher photocatalytic activity than commercial TiO_2-P25. In the following section, the various sol-gel and related systems of doping TiO_2 including noble metal, co-doping, transition metal doping, and their photocatalytic performance in degradation of pollutants in aqueous solutions and solar hydrogen production are discussed.

3. Doped-TiO_2 photocatalysts by sol-gel method

Many researchers have attempted to modify TiO_2 surface to overcome its limitation for photocatalytic reaction. The aim of these modifications is to extend the absorbance edge into the visible region, reducing charge carrier recombination and decreasing fast backward reaction. The modification of TiO_2 mainly was conducted through the following strategies:

1. Doping metal ions with a $\mathbf{d}^n$ ($0 < n < 10$) electronic configuration
2. Valence band control using an anion's **p** orbitals or the **s** orbital of p-block metal ions
3. Spectral sensitization [41]

The metal dopant can strongly influence on the number of surface defects by changing the morphology such as crystal structure, crystallinity and particle size [1, 42]. Type of metal dopants (electronegativity and affinity), total metal loading, preparation method, and chemical state of metal are crucial parameters to control the effect of metal loading in TiO_2 [28]. In this

http://dx.doi.org/10.5772/intechopen.68357

section, we discuss various sol-gel and sol-gel–related systems of doping TiO_2, including co-doping, transition metal and their photocatalytic performance.

3.1. Noble metal–doped TiO_2 by sol-gel method

Loading noble metals (platinum (Pt), rhodium (Rh), palladium (Pd), silver (Ag) and gold (Au) with a low overpotential have been investigated as the effective materials to enhance photocatalytic activity in terms of hydrogen production on the surface of TiO_2 [48–56]. **Figure 6** displays that their Fermi energy levels are lower than TiO_2 and the formation of Schottky barrier helps (retarding the electron/hole recombination) to pass photo-excited electrons from the CB of TiO_2 to the deposited metal particles on the surface of TiO_2 until achieving a thermodynamic equilibrium and their Fermi level energy are aligned [43]. Hence, accumulation of electrons in the noble metal particles is caused that their Fermi levels shift to more negative and closer to the CB level of TiO_2, which is suitable for the reduction of water to hydrogen. Pt and Au have been used intensively as a co-catalyst compared with other noble metals due to their lower overpotential, desirable electron affinity and work function compared to other noble metals [44].

Rosseler et al. described a comparison between sol-gel–prepared Au and Pt-doped TiO_2. They showed that the optimum photocatalyst of 3 wt%-Au/TiO_2 produced maximum amount of hydrogen up to 120 μmol/min in methanol (1, v/v%) solution under 150 W metal halide lamp (large portion of the visible light range) with intensity of 30 mW/cm^2. Low photocatalytic activity of Pt/TiO_2 was corresponded to low extension of absorption edge to the visible and great activity of Pt toward backward reaction:($H_2 + 1/2\ O_2 \rightarrow H_2O$) even at room temperature. Furthermore, they pointed out that the photocatalytic activity of a photocatalyst can be tuned by following considerable parameters:

1. The type and content of the metallic co-catalyst and metal-support interactions
2. The surface crystallographic, anatase/rutile ratio, porosity properties of the TiO_2
3. The relative amount of methanol added as sacrificial agent [38]

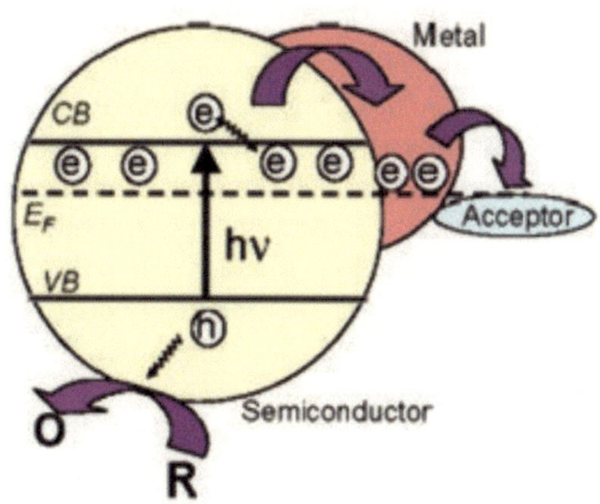

Figure 6. Fermi level equilibration in a semiconductor-metal nanocomposite [43].

Naseri et al. [45] synthesized sol-gel-derived Ag/TiO_2 thin films with various Ag:Ti molar ratios of 0, 1, 2, 5, 10 and 20 mol%. Ag/TiO_2 photocatalyst was immersed in 1 M KOH as the photoanode, Pt wire as the cathode, and Ag/AgCl as the reference electrode (RE) under the illumination of Xe short arc lamp with intensity of 1000 W/m^2 to measure photoelectrochemical activity. The 1 mol% Ag doped TiO_2 photocatalyst (crystallite size: 11 nm) had maximum photocurrent density of 0.8 mA/cm^2 and hydrogen production rate of 197 μmol/h. The modification of TiO_2 with noble metals is the promising method, but their high cost and relatively low availability strongly limit their application for large-scale photocatalytic systems. Therefore, largely available and cheap transition metals with acceptable photocatalytic activity can replace with the noble metals [46–49].

3.2. Transition metal–doped TiO_2 by sol-gel method

Transition metal (Cr, Mn, Fe, Co, Ni, Cu, and Zn) doped TiO_2 nanoparticles were synthesized by the sol-gel method using 2-hydroxylethylammonium format as an ionic liquid by Ghasemi et al. [50]. The performance of prepared photocatalysts was evaluated by degradation of Acid Blue92 (AB92) in aqueous medium under UV light. The results illustrated that transition metal (TM) doped TiO_2 nanoparticles was significantly enhanced photodegradation of AB92 in water compared to pure TiO_2. The studies revealed that transition metal (TM) doped TiO_2 nanoparticles have smaller crystalline size and higher surface area than pure TiO_2. Dopant ions in the TiO_2 structure caused significant absorption shift into the visible region. They explained that this better photocatalytic degradation may correspond to high electron-hole generation and low charge carrier recombination rate. The Fe/TiO_2 photocatalyst displayed maximum efficiency rate constant for AB92 degradation. The reason for the highest activity of Fe/TiO_2 could be the lowest crystalline size, the highest surface area and the minimum band gap energy. A decrease in crystallite size can give rise to larger surface area, which can increase the available surface active site and consequently leads to a higher adsorption of dye, separationof electron-hole generation, and interfacial charge carrier transfer rate for degradation. The photocatalytic hydrogen production of prepared NiO/TiO_2 photocatalysts with sol-gel and conventional incipient wetness impregnation methods was investigated by Sreethawong et al. [51]. They observed that sol-gel–prepared NiO/TiO_2 under 300 W high-pressure Hg lamp in an aqueous methanol solution had markedly higher hydrogen production rate of 162.6 μmol/h compared to TiO_2 photocatalyst (87.2 μmol/h). Furthermore, they pointed out that the extending absorption edge of NiO/TiO_2 to the longer wavelengths between 600 to 800 nm and combination of single-step sol-gel process with pore-controlling surfactant caused high photocatalytic performance. Singh et al. [52] studied the photoelectrochemical performance of synthesized anatase Fe/TiO_2 photoanode with sol-gel spin coating method. The photoelectrochemical behavior of Fe/TiO_2 photoanode was explored under 150 W Xenon Arc lamp and reaction condition; electrolyte: NaOH with pH = 13, cathode: Pt and reference electrode (RE): saturated calomel electrode (SCE). The four layers of 0.2 at % Fe/TiO_2 photoanode with thickness of 1.30 μm showed better shift of the absorption edge toward the visible region with bandgap energy of 2.89 eV. In addition, the measured photocurrent density (jp was ~0.92 mA/cm^2 at zero external bias. Many researchers have been reported that co-doped TiO_2 is more effective technique to improve the photocatalytic performance of TiO_2 compared to single-doped TiO_2 due to the synergistic effect of the presence of two metals.

3.3. Co-doped TiO_2 by sol-gel method

The photocatalytic properties of bimetallic nanoparticles are directly related to their composition structure and catalyst preparation processes. Kim et al. [53] synthesized Cu-Ag/TiO_2 photocatalyst with the sol-hydrothermal method (to split water under UV irradiation with intensity of 18 W/cm^2 in the presence of methanol. The $Cu_{(0.03)}$-$Ag_{(0.07)}$ $Ti_{(0.9)}O_2$ photocatalyst (crystallite size: 15.28 nm) markedly increased hydrogen production rate up to 1093 mmol/h compared to Ag/TiO_2 (312 mmol/h), Cu/TiO_2 (900 mmol/h), and TiO_2 (200 mmol/h). They suggested that the Cu or Ag components in the TiO_2 framework were reduced by attracting the excited electrons from the valence band of TiO_2, because of the greater reduction potential of CuO or AgO than that of pureTiO_2. This hinders the recombination of charge carriers because the CuO or AgO component captures electrons, thereby increasing the number of holes over the valence band and allowing methanol decomposition to continue. Sun et al. [28] synthesized the single (Fe and Ni) and bimetallic (Fe-Ni) doped TiO_2 using the sol-hydrothermal method. They found out that 5% Fe-4% Ni/TiO_2 photocatalyst (crystallite size:13.8 nm, surface area: 98.35 m^2/g, and bandgap energy: 2.41 eV) had the maximum hydrogen production rate of 361.64 μmol/h.gcat compared to other metal loading ratios under a 500 W Xenon lamp equipped with a 400 nm cut-off glass filter in an aqueous ethanol solution. This result was corresponded to better ability of charge carrier separation, restriction of their recombination and red shifting absorption edge to 514 nm compared to TiO_2. In the other works, Sun et al. [54] successfully synthesized Ag and Fe doped TiO_2 using sol-hydrothermal method. The water photosplitting was conducted in an aqueous ethanol solution with two light sources: A 16-W high-pressure inner irradiation Hg lamp (λ = 254 nm) with intensity of 11.7 mW/cm^2 as the UV light source and 500 W Xe lamp equipped with a 400 nm cut-off glass as the visible light source. 4.5% Fe-4.5% Ag/TiO_2 photocatalyst (crystallite size: 12.1 nm, bandgap energy: 2.03 eV, and particle size: 12 nm) had a higher hydrogen production rate of 515.45 μmol/h.gcal compared to monometallic doped and undoped TiO_2. They concluded that the interaction of Fe and Ag with TiO_2 reduces particle size (~12 nm), shifts absorption edge into the visible region compared to anatase phase of TiO_2 ($\lambda \leq$ 367 nm), reduces charge carrier recombination, and enhances the photocatalytic performance. **Tables 3** and **4** show a summary of reported related work to modify TiO_2 with metal dopants.

Cocatalyst	Finding	Ref	Year
Ni	The extending absorption edge of NiO/TiO_2 to the long wavelengths between 600 and 800 nm and combination of single-step sol-gel process with pore-controlling surfactant caused high photocatalytic performance	[51]	2005
Fe	0.2 at.% Fe/TiO_2 had the lowest bandgap energy of 2.89 eV, maximum jp ~0.92 mA/cm^2 at zero external bias, donor density (N_D):4.3 × 1019 cm^{-3} and more negative flat band potential (Vfb): −0.92 compared with TiO_2	[52]	2008
Cr, Mn, Fe, Co, Ni, Cu, and Zn	The most active photocatalyst was Fe/TiO_2 with maximum photocatalytic degradation rate for AB92 degradation. The reason for the highest activity could be the lowest crystalline size, the highest surface area and the minimum band gap energy	[50]	2009

Cocatalyst	Finding	Ref	Year
Ag	1 mol% Ag/TiO_2 compared with other ratios: 0, 1, 2, 5, 10 and 20 mol% had the maximum hydrogen production rate of 196 μmol/h, N_D: 16.3 × 10^{21} cm^{-3}, V_{fb}: −1.14 V and jp ~8 mA/cm^2.	[45]	2012
Cu-Ag	The Cu-Ag/TiO_2 improved the H_2 production rate up to1093 mmol/h	[53]	2012
Fe-Ni	4wt% Fe–5%Ni/TiO_2 with low bandgap energy and more red shifting of absorption edge had a high hydrogen production rate of 361.64 μmol/h	[28]	2012
Ag-Fe	The higher photocatalytic hydrogen production with the rate of 515.45 μmol/h. gcat was belong to 4.5%Ag–4.5% Fe/TiO_2 compared to hydrogen production rate single metal doped and undoped TiO_2	[54]	2013

Table 4. Summary of transition metal–doped TiO_2 (single and co-doped).

4. Conclusion

Here, the previous research works confirm that sol-gel process is a simple and easy means of synthesizing nanoparticles at ambient temperature under atmospheric pressure, and this technique does not require complicated setup. The results of the investigation conclude that optimization of preparation conditions is essential for obtaining nanocrystalline TiO_2 materials with notably higher activity than Degussa (P-25) TiO_2. Doping TiO_2 with metals is favorable to improve the photocatalytic efficiency of the catalyst in sol-gel method compared to other methods. Furthermore, from the results available, it can be concluded that co-doping of TiO_2 generally enhances the photocatalytic efficiency of the catalyst compared to single doped TiO_2.

Acknowledgements

The authors would like to thank Centre of innovative Nanostructures & Nanodevices (COINN) and Universiti Teknologi PETRONAS for financial and technical support in making this research work feasible.

Author details

Robabeh Bashiri[1]*, Norani Muti Mohamed[1,2] and Chong Fai Kait[2]

*Address all correspondence to: susanbashiri@gmail.com

1 Centre of Innovative Nanostructures & Nanodevices (COINN), Universiti Teknologi PETRONAS, Bandar Seri Iskandar, Perak, Malaysia

2 Fundamental and Applied Sciences Department, Universiti Teknologi Petronas, Bandar Seri Iskandar, Perak, Malaysia

References

[1] Kudo A, Miseki Y. Heterogeneous photocatalyst materials for water splitting. Chemical Society Reviews. 2009;**38**(1):253–278

[2] Zhu J, Zäch M. Nanostructured materials for photocatalytic hydrogen production. Current Opinion in Colloid & Interface Science. 2009;**14**(4):260–269. doi:10.1016/j.cocis.2009.05.003

[3] Bak T, Nowotny J, Rekas M, Sorrell CC. Photo-electrochemical hydrogen generation from water using solar energy. Materials-related aspects. International Journal of Hydrogen Energy. 2002;**27**(10):991–1022. doi:10.1016/s0360-3199(02)00022-8

[4] Dholam R, Patel N, Adami M, Miotello A. Physically and chemically synthesized TiO_2 composite thin films for hydrogen production by photocatalytic water splitting. International Journal of Hydrogen Energy. 2008;**33**(23):6896–6903. doi:10.1016/j.ijhydene.2008.08.061

[5] Yu J, Wang B. Effect of calcination temperature on morphology and photoelectrochemical properties of anodized titanium dioxide nanotube arrays. Applied Catalysis B: Environmental. 2010;**94**(3–4):295–302. doi:10.1016/j.apcatb.2009.12.003

[6] Hendry E, Wang F, Shan J, Heinz TF, Bonn M. Electron transport in TiO_2 probed by THz time-domain spectroscopy. Physical Review B. 2004;**69**(8):081101

[7] Karuppuchamy S, Suzuki N, Ito S, Endo T. A novel one-step electrochemical method to obtain crystalline titanium dioxide films at low temperature. Current Applied Physics. 2009;**9**(1):243–248. doi:10.1016/j.cap.2008.02.004

[8] Fan L, Ichikuni N, Shimazu S, Uematsu T. Preparation of Au/TiO_2 catalysts by suspension spray reaction method and their catalytic property for CO oxidation. Applied Catalysis A: General. 2003;**246**(1):87–95. doi:10.1016/S0926-860X(03)00002-4

[9] Arimitsu N, Nakajima A, Kameshima Y, Shibayama Y, Ohsaki H, Okada K. Preparation of cobalt–titanium dioxide nanocomposite films by combining inverse micelle method and plasma treatment. Materials Letters. 2007;**61**(11–12):2173–2177. doi:10.1016/j.matlet.2006.08.044

[10] Kluson P, Luskova H, Cajthaml T, Solcova O. Non-thermal preparation of photoactive titanium (IV) oxide thin layers. Thin Solid Films. 2006;**495**(1–2):18–23. doi:10.1016/j.tsf.2005.08.275

[11] Gu D-E, Yang B-C, Hu Y-D. V and N co-doped nanocrystal anatase TiO_2 photocatalysts with enhanced photocatalytic activity under visible light irradiation. Catalysis Communications. 2008;**9**(6):1472–1476. doi:10.1016/j.catcom.2007.12.014

[12] Dvoranová D, Brezová V, Mazúr M, Malati MA. Investigations of metal-doped titanium dioxide photocatalysts. Applied Catalysis B: Environmental. 2002;**37**(2):91–105. doi:10.1016/S0926-3373(01)00335-6

[13] Feng X, Wang Q, Wang G, Qiu F. Preparation of nano-TiO_2 by ethanol-thermal method and its catalytic performance for synthesis of dibutyl carbonate by transesterification. Chinese Journal of Catalysis. 2006;**27**(3):195–196. doi:10.1016/S1872-2067(06)60011-9

[14] Zhu J, Deng Z, Chen F, Zhang J, Chen H, Anpo M, Huang J, Zhang L. Hydrothermal doping method for preparation of Cr^{3+}-TiO_2 photocatalysts with concentration gradient distribution of Cr^{3+}. Applied Catalysis B: Environmental. 2006;**62**(3–4):329–335. doi:10.1016/j.apcatb.2005.08.013

[15] Babelon P, Dequiedt AS, Mostéfa-Sba H, Bourgeois S, Sibillot P, Sacilotti M. SEM and XPS studies of titanium dioxide thin films grown by MOCVD. Thin Solid Films. 1998;**322**(1–2):63–67. doi:10.1016/S0040-6090(97)00958-9

[16] Kim B-H, Lee J-Y, Choa Y-H, Higuchi M, Mizutani N. Preparation of TiO_2 thin film by liquid sprayed mist CVD method. Materials Science and Engineering: B. 2004;**107**(3):289–294. doi:10.1016/j.mseb.2003.12.010

[17] Peng F, Cai L, Yu H, Wang H, Yang J. Synthesis and characterization of substitutional and interstitial nitrogen-doped titanium dioxides with visible light photocatalytic activity. Journal of Solid State Chemistry. 2008;**181**(1):130–136. doi:10.1016/j.jssc.2007.11.012

[18] Huang M, Xu C, Wu Z, Huang Y, Lin J, Wu J. Photocatalytic discolorization of methyl orange solution by Pt modified TiO_2 loaded on natural zeolite. Dyes and Pigments. 2008;**77**(2):327–334. doi:10.1016/j.dyepig.2007.01.026

[19] Liao DL, Badour CA, Liao BQ. Preparation of nanosized TiO_2/ZnO composite catalyst and its photocatalytic activity for degradation of methyl orange. Journal of Photochemistry and Photobiology A: Chemistry. 2008;**194**(1):11–19. doi:10.1016/j.jphotochem.2007.07.008

[20] Zhang X, Liu Q. Preparation and characterization of titania photocatalyst co-doped with boron, nickel, and cerium. Materials Letters. 2008;**62**(17–18):2589–2592. doi:10.1016/j.matlet.2007.12.061

[21] Crişan M, Brăileanu M, Răileanu M, Zaharescu M, Crişan D, Drăgan N, Anastasescu M, Ianculescu A, Niţoi I, Marinescu VE, Hodorogea SM. Sol–gel S-doped TiO_2 materials for environmental protection. Journal of Non-Crystalline Solids. 2008;**354**(2–9):705–711. doi:10.1016/j.jnoncrysol.2007.07.083

[22] Akpan UG, Hameed BH. The advancements in sol–gel method of doped-TiO_2 photocatalysts. Applied Catalysis A: General. 2010;**375**(1):1–11. doi:10.1016/j.apcata.2009.12.023

[23] Wei X, Wang H, Zhu G, Chen J, Zhu L. Iron-doped TiO_2 nanotubes with high photocatalytic activity under visible light synthesized by an ultrasonic-assisted sol-hydrothermal method. Ceramics International. 2013;**39**(4):4009–4016. doi:10.1016/j.ceramint.2012.10.251

[24] Huang C-H, Yang Y-T, Doong R-A. Microwave-assisted hydrothermal synthesis of mesoporous anatase TiO_2 via sol–gel process for dye-sensitized solar cells. Microporous and Mesoporous Materials. 2011;**142**(2–3):473–480. doi:10.1016/j.micromeso.2010.12.038

[25] Norani MM, Bashiri R, Chong FK, Sufian S, Kakooei S. Photoelectrochemical behavior of bimetallic Cu–Ni and monometallic Cu, Ni doped TiO_2 for hydrogen production. International Journal of Hydrogen Energy. 2015;**40**(40):14031–14038. doi:10.1016/j.ijhydene.2015.07.064

[26] Liu GQ, Jin ZG, Liu XX, Wang T, Liu ZF. Anatase TiO_2 porous thin films prepared by sol-gel method using CTAB surfactant. Journal of Sol-Gel Science and Technology. 2007;**41**(1):49–55. doi:10.1007/s10971-006-0122-9

[27] Chen-Chi Wang JYY. Sol-gel synthesis and hydrothermal processing of anatase and rutile titania nanocrystals. Chemistry of Materials. 1999;**11**:3113–3120

[28] Sun T, Fan J, Liu E, Liu L, Wang Y, Dai H, Yang Y, Hou W, Hu X, Jiang Z. Fe and Ni co-doped TiO_2 nanoparticles prepared by alcohol-thermal method: Application in hydrogen evolution by water splitting under visible light irradiation. Powder Technology. 2012;**228**:210–218. doi:10.1016/j.powtec.2012.05.018

[29] Bashiri R, Mohamed NM, Kait CF, Sufian S. Hydrogen production from water photosplitting using Cu/TiO_2 nanoparticles: Effect of hydrolysis rate and reaction medium. International Journal of Hydrogen Energy. 2015;**4**(18):6021–6037. doi:10.1016/j.ijhydene.2015.03.019

[30] Bashiri R, Mohamed NM, Fai Kait C, Sufian S. Influence of hydrolysis rate on properties of nanosized TiO_2 synthesized via sol-gel hydrothermal. In: Advanced Materials Research. 2015. Trans Tech Publications.

[31] Su C, Tseng CM, Chen LF, You BH, Hsu BC, Chen SS. Sol-hydrothermal preparation and photocatalysis of titanium dioxide. Thin Solid Films. 2006;**498**(1–2):259–265. doi:10.1016/j.tsf.2005.07.123

[32] Su C, Hong BY, Tseng CM. Sol–gel preparation and photocatalysis of titanium dioxide. Catalysis Today. 2004;**96**(3):119–126. doi:10.1016/j.cattod.2004.06.132

[33] Yu JC, Zhang L, Yu J. Direct sonochemical preparation and characterization of highly active mesoporous TiO_2 with a bicrystalline framework. Chemistry of Materials. 2002;**14**(11):4647–4653. doi:10.1021/cm0203924

[34] Chen Z, Zhao G, Li H, Han G, Song B. Effects of water amount and pH on the crystal behavior of a TiO_2 nanocrystalline derived from a sol–gel process at a low temperature. Journal of the American Ceramic Society. 2009;**92**(5):1024–1029. doi:10.1111/j.1551-2916.2009.03047.x

[35] Bessekhouad Y, Robert D, Weber JV, Synthesis of photocatalytic TiO_2 nanoparticles: optimization of the preparation conditions. Journal of Photochemistry and Photobiology A: Chemistry. 2003;**157**(1):47–53. doi:10.1016/s1010-6030(03)00077-7

[36] Sergio Valencia, Marín JM, Restrepo G. Study of the bandgap of synthesized titanium dioxide nanoparticules using the sol-gel method and a hydrothermal treatment. The Open Materials Science Journal. 2010;**4**(1874-088X/10):9–14

[37] Oskam G, Nellore A, Penn RL, Searson PC. The growth kinetics of TiO_2 nanoparticles from titanium(IV) alkoxide at high water/titanium ratio. The Journal of Physical Chemistry B. 2003;**107**(8):1734–1738. doi:10.1021/jp021237f

[38] Venkatachalam V, Palanichamy M, Murugesan V. Sol–gel preparation and characterization of nanosizeTiO_2: Its photocatalytic performance. Materials Chemistry and Physics. 2007;**104**(2):554–559

[39] Yu JC, Yu J, Zhang L, Ho W. Enhancing effects of water content and ultrasonic irradiation on the photocatalytic activity of nano-sized TiO_2 powders. Journal of Photochemistry and Photobiology A. 2002;**148**(1–3):263–271. doi:10.1016/s1010-6030(02)00052-7

[40] Behnajady MA, Eskandarloo H, Modirshahla N, Shokri M. Investigation of the effect of sol–gel synthesis variables on structural and photocatalytic properties of TiO_2 nanoparticles. Desalination. 2011;**278**(1–3):10–17. doi:10.1016/j.desal.2011.04.019

[41] Kazuhiko M. Photocatalytic water splitting using semiconductor particles: History and recent developments. Journal of Photochemistry and Photobiology C: Photochemistry Reviews. 2011;**12**(4):237–268. doi:10.1016/j.jphotochemrev.2011.07.001

[42] Kudo A. Photocatalysis and solar hydrogen production. Pure and Applied Chemistry. 2007;**79**(11):1917–1927

[43] Subramanian V, Wolf EE, Kamat PV. Catalysis with TiO_2/gold nanocomposites. Effect of metal particle size on the fermi level equilibration. Journal of the American Chemical Society. 2004;**126**(15):4943–4950. doi:10.1021/ja0315199

[44] Leung DYC, Fu X, Wang C, Ni M, Leung MKH, Wang X, Fu X. Hydrogen production over titania-based photocatalysts. ChemSusChem. 2010;**3**(6):681-694. DOI: 10.1002/cssc.201000014

[45] Naseri N, Kim H, Choi W, Moshfegh AZ. Optimal Ag concentration for H_2 production via Ag:TiO_2 nanocomposite thin film photoanode. International Journal of Hydrogen Energy. 2012;**37**(4):3056-3065. http://dx.doi.org/10.1016/j.ijhydene.2011.11.041

[46] Lalitha K, Sadanandam G, Kumari VD, Subrahmanyam M, Sreedhar B, Hebalkar NY. Highly stabilized and finely dispersed Cu_2O/TiO_2: A promising visible sensitive photocatalyst for continuous production of hydrogen from glycerol:water mixtures. The Journal of Physical Chemistry C. 2010;**114**(50):22181–22189. doi:10.1021/jp107405u

[47] Malato S, Fernández-Ibáñez P, Maldonado MI, Blanco J, Gernjak W. Decontamination and disinfection of water by solar photocatalysis: Recent overview and trends. Catalysis Today. 2009;**147**(1):1–59. doi:10.1016/j.cattod.2009.06.018

[48] Ni M, Leung MKH, Leung DYC, Sumathy K. A review and recent developments in photocatalytic water-splitting using TiO_2 for hydrogen production. Renewable and Sustainable Energy Reviews. 2007;**11**(3):401–425. doi:10.1016/j.rser.2005.01.009

[49] Gombac V, Sordelli L, Montini T, Delgado JJ, Adamski A, Adami G, Cargnello M, Bernal S, Fornasiero P. CuOx–TiO_2 photocatalysts for H_2 production from ethanol and glycerol

solutions. The Journal of Physical Chemistry A. 2009;**114**(11):3916–3925. doi:10.1021/jp907242q

[50] Ghasemi S, Rahimnejad S, Setayesh SR, Rohani S, Gholami MR. Transition metal ions effect on the properties and photocatalytic activity of nanocrystalline TiO_2 prepared in an ionic liquid. Journal of Hazardous Materials. 2009;**172**(2–3):1573–1578. doi:10.1016/j.jhazmat.2009.08.029

[51] Sreethawong T, Suzuki Y, Yoshikawa S. Photocatalytic evolution of hydrogen over mesoporous supported NiO photocatalyst prepared by single-step sol–gel process with surfactant template. International Journal of Hydrogen Energy. 2005;**30**(10):1053–1062

[52] Singh AP, Kumari S, Shrivastav R, Dass S, Satsangi VR. Iron doped nanostructured TiO_2 for photoelectrochemical generation of hydrogen. International Journal of Hydrogen Energy. 2008;**33**(20):5363–5368. doi:10.1016/j.ijhydene.2008.07.041

[53] Kim A-Y, Kang M. Effect of Ag-Cu bimetallic components in a TiO_2 framework for high hydrogen production on methanol/water photo-splitting. International Journal of Photoenergy. 2012;**2012**(2012):1–9. doi:10.1155/2012/618642

[54] Sun T, Liu E, Fan J, Hu X, Wu F, Hou W, Yang Y, Kang L. High photocatalytic activity of hydrogen production from water over Fe doped and Ag deposited anatase TiO_2 catalyst synthesized by solvothermal method. Chemical Engineering Journal. 2013;**228**:896–906. doi:10.1016/j.cej.2013.04.065

Chapter 9

Sol-Gel-Derived Doped ZnO Thin Films: Processing, Properties, and Applications

Asad Mahmood and Abdul Naeem

Additional information is available at the end of the chapter

http://dx.doi.org/10.5772/67857

Abstract

Sol-gel-derived zinc oxide (ZnO)-based materials with an improved microstructure are consumed in electronics and electrical frameworks owing to their crystal structure dependent properties, which can be exploited for optical, electrical, and photocatalytic applications. Despite research articles published each year on the strategies to improve the optoelectronic properties of ZnO, the topic is still actively pursued in literature. This chapter provides an insight into the recent developments for the sol-gel-derived processing of the pure and doped ZnO thin films. It also highlights the challenges and opportunities surrounding the processing of these devices. The recent developments in the synthesis of pure, doped ZnO, and corresponding applications of these films will be discussed in detail. Consequently, the aim of this chapter is to provide an overview of the novel developmental strategies to improve ZnO-based thin films by a sol-gel route with enhanced optical properties for practical applications ranging from optical and electrical circuits to sensing.

Keywords: sol-gel processing, zinc oxide, thin films, conductivity, optical properties.

1. Introduction

Zinc oxide, characterized as a direct wide band gap (Eg ≈ 3.37 eV @ °T-Room) semiconductor, is a potential material for photonics and optoelectronic applications [1]. The large excitonic energy (60 mV), which renders an effectual excitonic emission in ZnO, is exploited for applications such as ultraviolet light-emitting and laser diodes. Moreover, the pure and doped ZnO has been investigated for applications in the solar cells, photoelectrochemical cells (PECs), thin film transistors, gas sensors and nanogenerators. ZnO can be used for all of these applications due to its chemical and physical stability, abundance, economical feasibility and environment friendly. Also, due to its characteristic Eg and good optical transmittance, ZnO is well thought out in organic and solar hybrid cells, as a cathode buffer layer or, a transparent electrode. Crystalline ZnO exhibits superior electrical properties, both in the form of bulk and thin films. Due to factors such as small size, light weight, stability, and ability to configure in different frameworks, thin films are promising structures studied compared to their bulk counterparts. Undoped ZnO is seldom used, and the optoelectronic properties of ZnO are tailored for specific applications by various processing techniques, such as synthesis procedures, doping mechanism optimization, an introduction of impurities, controlling the microstructure, and the thickness of the films. Recent development in the nanoscience and nanotechnology has led to the miniaturization and diversification of the electronic and optical devices, which endeavor researchers to develop new synthetic strategies in order to process semiconductor-based thin films and powders on a viable scale [2].

ZnO films are useful for the transparent conductive layer applications in the LEDs, flat panel displays (FPD), and solar cells due to its high transmittance properties in the visible region and good electrical conductivity [3]. The electrical properties of ZnO are associated with the presence of interstitial Zn atoms and crystal defects, that is, oxygen vacancies due to the non-stoichiometry and defects generated in crystals during grain growth. To improve the conductivity in ZnO films compared to the metallic films, various elements in the form of impurities are introduced in the ZnO crystal lattice to replace Zn. The elements from group III (B, Al, Ga, and In) and group IV (Ti, Zr, Sn, and Hf) are commonly used to alter the properties of ZnO. The electronegativity and ionic radius are important parameters, which affect the dopant efficiency. Another approach is to modify the processing techniques to develop ZnO films with modified microstructure. The particle morphology such as from particles, tubes, and wires (in the nm or μm range), and film surface morphology influenced the properties of the aimed devices.

Zinc oxide exhibits a wurtzite hexagonal crystal structure ($a \approx 3.249$ Å; $c \approx 5.205$ Å ICDD PDF no. 36-1451) [4]. The undoped ZnO structure is composed of alternating planes of tetrahedrally coordinated O^{2+} and Zn^{2+} ions along the c-axis (**Figure 1(a)**). The characteristic non-central symmetric structure and properties such as piezoactivity and pyroelectricity are associated with the tetrahedral coordination of ZnO. Further, ZnO exhibits polar surfaces, and the asel plane is the most common surface. The positively and negatively charged surfaces (Zn-$_{(0001)}$ and O-$_{(0001)}$, respectively) are created due to the oppositely charged ions, which generate spontaneous polarization and normal dipole moment along the c-axis as well as a divergence in the surface energy. The lower valence band (VB) and the upper conduction band (CB) comprise the Eg of ZnO (**Figure 1(b)**). The given band structure for the pure ZnO was calculated using a VASP code [5]. At 0 K, the CB remains empty, while electrons only occupy the VB energy levels. Discrete energy levels are present in an isolated atom. However, the split in the energy level occurs during the crystal formation. In such a state, the closely spaced levels due to atomic interactions result in a continuous energy band. Various properties of the pure ZnO are presented in **Table 1**.

ZnO thin films are processed using methods such as atomic layer deposition (ALD) [7], chemical vapors deposition (CVD) [8], pulsed laser deposition (PLD) [9], RF magnetron sputtering

http://dx.doi.org/10.5772/67857

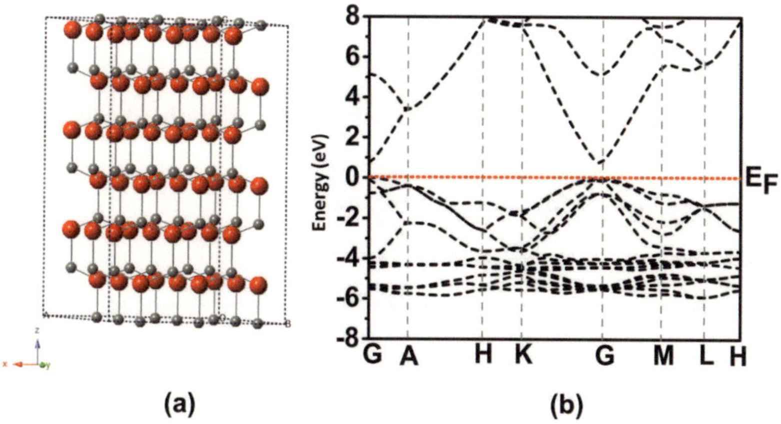

Figure 1. (a) Crystal structure ZnO. The small spheres represent zinc atoms, while the large spheres represent oxygen atoms; (b) calculated band structure of pure ZnO. The Fermi level (E_F) is set to 0 eV.

[10], epitaxial growth [11], self-assembly [12], sputtering technique [13], co-precipitation [14], electrodeposition [15], spray pyrolysis [16], and sol-gel method [17]. The sol-gel technique is also used for the fabrication of membranes [18], chemical sensors [19], optical gain media [20, 21], electrochemical devices [22], photochromic and non-linear applications [23], and nano-materials [24]. The sol-gel method is advantageous over other synthetic procedures. These advantages include as follows:

Properties	Characteristics
Crystal structure	Wurtize
Lattice constant (*a*)	0.325 nm
Lattice constan (*c*)	0.521 nm
Density (kg/m^3)	5.6 gcm^{-3}
Exciton binding energy	60 mV
Static dielectric constant (ε_s)	7.9
Optical dielectric constant	3.7
Optical band gap energy (E_g)	3.2 eV
Flat band potential (E_{fb})	−0.5 V versus saturated calomel electrode (SCE)
Effective electron mass (M)	0.24–0.3 $m_e m_e = 9.11 \times 10^{-31}$ kg
Effective hole mass (m_h)	0.45–0.6 $m_e m_e = 9.11 \times 10^{-31}$ kg
Electron mobility (μ_e)	200 $cm^2\ V^{-1} s^{-1}$
Point of zero charge (Pzc)	8–9 pH

Table 1. Various properties of pure ZnO [6].

- The sol-gel product exhibits better homogeneity and purity.
- The sol-gel process can be conducted at low temperatures.
- The sol-gel process offers control over stoichiometry of multiphase systems, particle size, shape and physiochemical properties.
- The procedure can be used for thin film fabrication.
- Sol-gel can be used for the synthesis of various inorganic and organic hybrid materials.
- Sol-gel process is easier to control the microstructure during sintering (densification process).

These advantages allow for commercial processing compared to other techniques such a ALD, PLD, and magnetron sputtering, which are costly, inefficient for large production purposes and have lengthy processing times. While recent research provides novel approaches in the development of ZnO-based thin films, this chapter presents a review with a goal to help ensure the device fabrication based on the sol-gel-derived ZnO thin films with current understanding and future perspectives. This chapter provides a plan to improve the approaches to semiconductor material applications, made available to a wider community of academics and practitioners, and present techniques for optimizing the parameters for better device performance. The chapter concludes with a summary of the contribution of the doped ZnO-based materials as the potential candidate for gas sensing, photocatalytic, and thermoelectric applications. This chapter will also provide researchers currently working in the field a useful literature in the form of articles and reviews.

2. Processing

The sol-gel method was initially used in 1800s. Later, around 1900s, this technology was implemented by the Schott Glass Company (Jena, Germany). The record of publications in the literature (patents and journal articles) related to the processing of materials via the sol-gel route from 1980 to 2010 is given in **Figure 2**.

A schematic view of the sol-gel process for the development of ZnO-based films is given in **Figure 3**. The sol-gel method can be modified for different purposes, and the parameters discussed here are modified continuously. Generally, the first step in a sol-gel process is the selection of suitable precursors, which will react through various steps and finally converting to colloidal particles (sol) or polymeric gels. A stable sol is required for the thin film deposition, which can be deposited by spin coating, dip coating, or drop casting techniques, while the sol is converted to a polymeric gel in order to synthesize powders. Metal ions or other reactive elements surrounded by ligands are largely used as a precursor for the sol-gel reaction. The most important precursors utilized in this method are metal alkoxides M(OR)n and alkoxysilanes due to their radial reaction mode. Metal alkoxides are derivatives of alcohols, ROH, which are weak acids, economical (inexpensive) and results in high purity hydrated oxides [25, 26]. Alcohols, such as absolute ethanol and isopropanol, are used as solvents for alkoxides, because they are immiscible with water. In the case of ZnO, acetates (zinc acetate dihydrate; $Zn(CH_3COO)_2.(H_2O)_2$) are commonly used as the Zinc (Zn) precursor [27]. Other

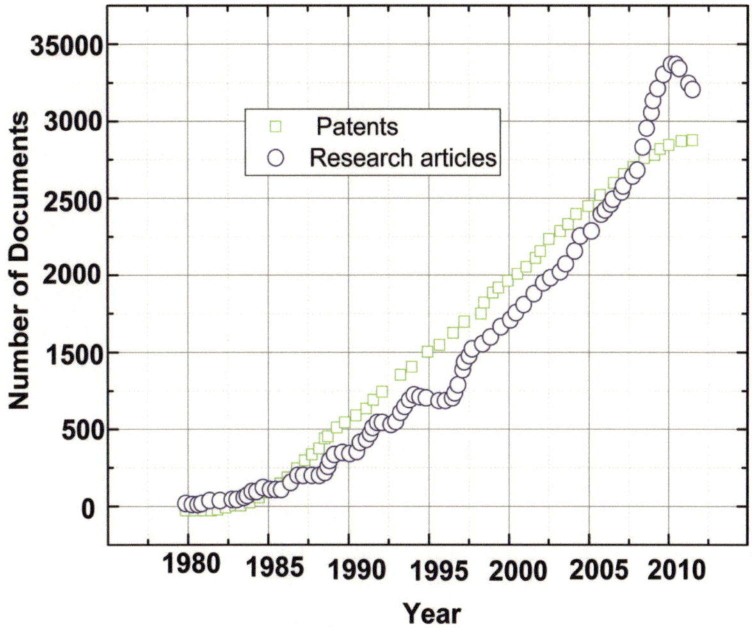

Figure 2. Publications in the sol-gel materials field, 1980–2010.

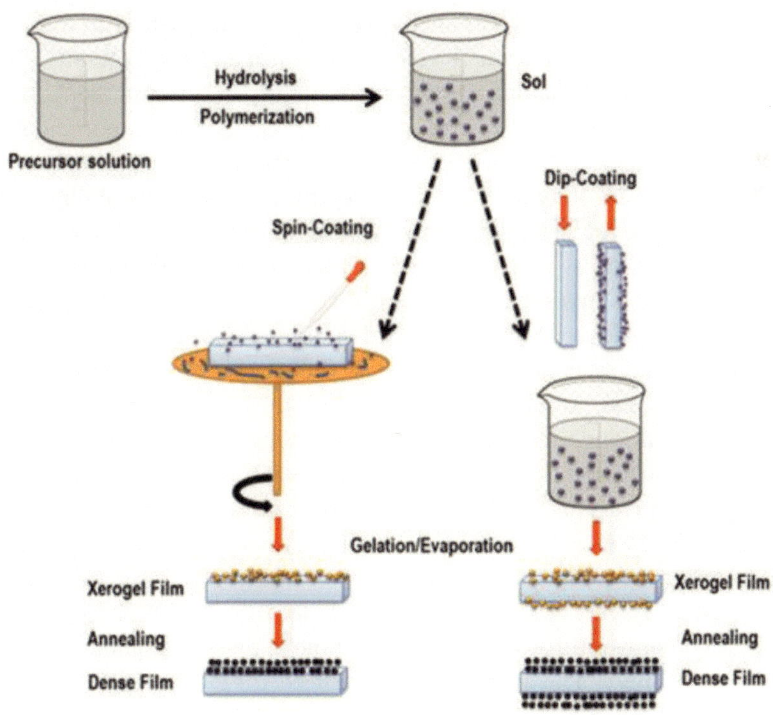

Figure 3. A schematic view of sol-gel process for the ZnO thin film processing.

Zn sources such as nitrate, acetylacetonate, perchlorate, and alkoxides (i.e., ethoxide, peroxides) are also used [28]. The alkoxide has disadvantages related to its high cost and sensitivity to moisture among others. On the contrary, metal salts are favored for large-scale production purposes. The Zn-salt plays an important role in developing a stable and clear colloidal solution. Acetates and nitrates are preferred compared to perchlorates, which might result in turbidity; however, the nitrate ions may result in the ionic impurities in the film annealing process. Further, the choice of precursor salt affects the morphology and hence the final properties of the films. The solvent used for the sol-gel process must exhibit a dielectric constant high enough to dissolve the metal salts. Generally, alcohols with low carbon number (i.e., methyl alcohol, ethyl alcohol, isopropanol, 1-butanol, and 2-methoxyethanol) are preferred as the potential solvents [29–32] and references therein. Ethylene glycol (boiling point; b.pt = 194.4°C) has also been reported exhibiting a dielectric constant of 40.61 (at 25°C) [33, 34].

During the sol-gel process, the precursor materials, that is, Zn source and the dopant elements (if applicable) in the predetermined stoichiometric ratios are first dissolved in a suitable solvent. In the second step, a stabilizing agent is added to the initial solution for the precursor metal(s). Monoethanolamine is a common stabilizer, which is used for the stabilization of Zn solution in sol-gel process. Acetylacetone, diethanolamine, tetramethylammonium hydroxide, and ethylenediamine tetraacetates are also used as stabilizers [27, 35]. A stabilizing agent is critical in sol-gel reactions to avoid both the premature precipitations and inhibiting the rapid conversion of the sol into a gel. In some cases, a catalyst is concurrently used for catalyzing the hydrolysis and condensation reactions. Some of the most common catalysts employed in sol-gel reactions are hydrochloric acid, potassium hydroxide, sodium hydroxide and ammonia [36]. The precursor solution is kept bathing on the magnetic hot plate with continued stirring until a clear and stable sol is achieved. The temperature is maintained (25–80°C) in an oil bath for homogeneous heating. The initial concentrations of the precursor materials, solvent to precursor ratios, and ratio of the stabilizing agents to the precursor are important as it can affect the solubility of precursors and stability of the sol. Furthermore, the concentration of the sol is an important factor, which defines the microstructure and thickness of films. Once a stable sol is achieved, it is aged for 24 h approximately at room temperature to confirm its stability. The stability is also checked by the Tyndall effect; laser light is applied to the solution, and the scattering phenomenon is investigated that is directly associated with the presence of small particles in the sol. The stable sols can be applied to different kinds of substrates using a drop casting, dip coating, and spin coating techniques. The speed of the spinning (revolution per minute; rpm) is controlled, which affects the film thickness and morphology. The spinning process is completed through the repetition of these steps: (a) spinning the sol (~3000 rpm/min for 30 s) on the substrate, (b) drying at low temperature (120°C for 10 min), (c) drying at relatively high temperature (350°C for 10 min), (d) drying in a furnace in the temperature range ~ 500–600°C for 10 min. The steps from (a) to (d) are reported as the preheat treatment process in literature. The parameters, such as drying time and temperatures, can be varied. After applying various layers by the same procedure, the films are annealed (referred as post-heating) at selected temperatures for longer dwell time to develop homogeneous, crystalline, and single phase ZnO films. The annealing regime affects the grain growth and the crystalline microstructure of the ZnO films [37]. **Table 2** presents various sol-gel methodologies recently reported for the ZnO film processing.

Precursor (mol L^{-1})	Alcohol	Additive	Aging (h)	Substrate	Ph-T[a] (°C)	Po-T[b] (°C)	Thickness (nm)	Cryst. orientation	Ref.
ZAD (0.4)	MeOH EtOH PrOH 2-ME	2-Ma MEA DEA	–	$Si_{(100)}$	140–170	400	–	(100), **(002),** (101)	[27]
ZAD	EtOH PrOH	MEA	15	Quartz	300	500	175–200	(100), **(002),** (101), (102), (110), (103), (112), (004)	[29]
ZAD (0.6)	PrOH	MEA	24	Glass	300	500	–	(100), **(002),** (101), (102), (110)	[38]
ZAD (0.1)	EtOH	MEA	24	Soda-lime Silica Silicon $Pt_{(50\ nm)}$/ $Ti_{(5\ nm)}$/ SiO_2	250	400	140	(100), (002), **(101)**	[30]
ZNH (0.3)	PrOH	MMEA	120	Sapphire	700	700	300–1900	(100), (002), (101)	[28]
ZNH (0.3)	PrOH	MEA	168	Sapphire	300–600	600	–	(100), (002), **(101)** or (100), **(002),** (101)	[39]
ZAD (0.75)	EtOH	MEA	–	Glass	200	500	–	(100), **(002),** (101)	[40]
ZAD (0.5)	EtOH	MEA	–	Glass	100	450, 600	–	(100), **(002),** (101), (102)	[41]
ZAD	MeOH PrOH	MEA	24	SiO_2/Si	150	500	–	(100), **(002),** (101) or (100), (002), **(101)**	[42]
ZAD (0.2-0.1)	2-ME	MEA	24	Glass	300	500	180–270	(100), **(002),** (101), (102), (110)	[31]

Precursor (mol L^{-1})	Alcohol	Additive	Aging (h)	Substrate	Ph-T[a] (°C)	Po-T[b] (°C)	Thickness (nm)	Cryst. orientation	Ref.
ZAD	EtOH MeOH BD PrOH	MEA	840	Glass	190	550	–	(100), **(002)**, (101)/ (100), **(002)**, (101)/ (100), **(002)**, (101)/ (100), (002), **(101)**	[43]
ZAD (0.2)	EtOH	MEA	72	Glass	230	450	98–366	(100), **(002)**, (101)	[44]
ZAD (0.5)	PrOH	DEA	–	Glass	25 (RT)	450	311	(100), (002), **(101)**, (102), (110), (103), (200), (112), (201), (004), (202)	[32]
ZAD (0.1–0.5)	2-ME	MEA	–	Quartz	170–350	250–400	–	(100), **(002)**, (101)	[45]
ZAD (0.1)	PrOH	DEA		Glass	550	400	350	(100), **(002)**, (101)	[46]
ZAD (0.2)	PrOH	MEA DEA TEA TE EN	–	Soda lime Glass	120–300	500	300–370	(10$\bar{1}$0), **(0002)**, (10$\bar{1}$1)	[35]
ZAD	EtOH -water	AcA	–	SiO_2	200	450	–	(100), (002), **(101)**	[47]
ZAD (0.75)	2-ME	MEA	24	Sapphire	300	600–900	50–350	(100), **(002)**, (101), (102), (110), (103), (112)	[48]

Precursor (mol L^{-1})	Alcohol	Additive	Aging (h)	Substrate	Ph-T[a] (°C)	Po-T[b] (°C)	Thickness (nm)	Cryst. orientation	Ref.
ZAD (0.3)	2-ME / PEG	MEA	24	Glass	300	600	257–277	(100), **(002)**, (101), (102), (110), (103)	[49]
ZAD	2-ME	TEA	36	Glass	500	600	230–350	(100), **(002)**, (101), (102), (110), (103), (112)	[50]
ZAD (0.75)	EtOH	MEA	–	Glass	200	500	–	(100), **(002)**, (101)	[51]
ZAD (0.1–0.3)	PrOH	MEA	24	Soda lime glass	80	400	218–437	$(10\bar{1}0)$, **(0002)**, $(10\bar{1}1)$	[52]
ZAD	2-ME	MEA	–	SiO_2/Si	200	500	–	(100), **(002)**, (101), (102), (110), (103), (112)	[53]
ZAD (0.5)	2-ME	MEA	–	Si-wafer	300	500–900	–	**(002)**	[54]
ZAD (0.5)	2-ME	DEA	168	Glass	300	500	170	(100), (002), **(101)**, (102), (110), (103)	[3]
ZAD	MeOH	–	720	Glass	180	400–500	–	(100), (002), **(101)**, (102), (110), (103), (200)	[55]
ZAD (0.09–0.75)	2-ME	MEA		$Si_{(001)}$	300	500	24–94	–	[56]

Abbreviation	Product name	Formula	Abbreviation	Product name	Formula
ZAD	Zinc acetate dihydrate	$Zn(CH_3COO)_2.2H_2O$	MeOH	Methanol	CH_3OH
ZNH	Zinc nitrate hexahydrate	$Zn(NO_3)_2.6H_2O$	EtOH	Ethanol	C_2H_5OH

Abbreviation	Product name	Formula	Abbreviation	Product name	Formula
2-ME	2-Methoxyethanol	$CH_3O(CH_2)_2OH$	PrOH	*x*-Propanol	C_3H_7OH
TE	Triethylamine	$N(CH_2CH_3)_3$	ED	Ethylenediamine	$C_2H_4(NH_2)_2$
DEA	Diethanolamine	$(HOCH_2CH_2)_2NH$	AcA	Acetic acid	CH_3COOH
MEA	Monoethanolamine	$(HOCH_2CH_2)_2NH_2$	PEG	Polyethyleneglycol	$C_{2n}H_{4n+2}O_{n+1}$
TEA	Triethanolamine	$(HOCH_2CH_2)_3N$	BD	1,4-butanediol	$HOCH_2CH_2CH_2CH_2OH$
2-Ma	2-(methylamino) ethanol	C_3H_9NO			

[a]Preheat treatment; [b]postheat treatment.
The bold entries are self explanatory which suggest the preferential crystallographic orientation.

Table 2. Recently published papers about the sol-gel-derived ZnO films.

3. Doping of ZnO

The term doping is used to introduce impurities in semiconductors. Similarly, doping mechanism is applied to improve the properties of ZnO (**Table 3**). Transition elements such as Cu^{2+}, Fe^{3+}, Co^{3+}, Mn^{2+} are used to alter the ZnO optical and electrical properties [57]. The dopants in ZnO crystal structure affect the crystal structure, volume, lattice parameters, hopping mechanism of electron, band gap, and particle morphology, which are efficient ways in which the properties of ZnO are improved [58]. It has been reported that 9 mol% Fe in ZnO decreases the d-spacing of ZnO crystal. The ionic radius of Fe^{3+} (0.68 Å) is smaller than Zn^{2+} (0.74 Å), which can possibly influence the crystal lattice due to the generation of tension in the crystal. Further, it decreases the crystal growth. The Eg also decreases with Fe due to the addition of more energy levels, consequently resulting in defects. Aydin et al. [59] reported a corresponding decrease in Eg for the samples with 5, 10, 15, and 20% Fe-doped ZnO as 3.08, 2.83, 2.79, and 2.75 eV, respectively. Xu and Li [60] processed sol-gel-derived Fe-ZnO on the Si substrates. The preferential orientations of the film perpendicular to the substrate surface along the c-axis and the crystalline nature were reported to increase for the 1% Fe doping concentration. However, with increasing Fe from 1% worsen the above-mentioned properties. The grain size was also decreased with increasing Fe content. The Eg was observed to increase. Aluminum (Al^{3+}; ionic radii: 0.053 nm) doping in Zn^{2+} (0.074 nm) decreases the interplanar d-spacing and lattice constants [61]. Chen et al. [62] investigated the effect of Ga-ZnO film thickness (230–480 nm), temperature (400–600°C) and atmosphere (air/argon) to improve the optical response and electronic microstructure of the ZnO. The films were processed by using the sol-gel dip coating technique, where thick films showed high crystalline nature. The annealing atmosphere resulted in the low carrier concentration, while an increase in the particle size promoted the career concentration. The argon atmosphere possibly produced films exhibiting resistivity $1.18 \times 10^{-2}\ \Omega$ cm, carrier density ($3.376 \times 10^{19}\ cm^{-3}$), and mobility (15.74 cm^2 (Vs)-1). Dubey et al. [63] used Mn and Li as the co-dopants to optimize the optical and magnetic behavior of ZnO thin films, which were processed through a sol-gel route according to the stoichiometric formula $Zn_{1-y-x}Mn_yLi_xO$ ($y = 0$, 0.02| $x = 0$–0.06. The lattice parameters were observed to increase linearly and were associated with the distortion

http://dx.doi.org/10.5772/67857

Dopant	Conc. (%)	Substrate	Annealing (°C)	Atm.	ϱ (Ω cm)	E_g (eV)	Application	Ref.
Al	24.71	p-type Si	500	H_2			Solar cells	[65]
Ti	0.5	Glass	400	Air	3.8– × –10^4	3.346	Transparent cond. layer	[66]
Fe	1–5	borofloat glass	650	Air		3.371		[67]
Cu	0.00, 0.03, 0.06, 0.10	$Si_{(100)}$	750	Air		3.22–3.43	Optical	[68]
Sn-Al			500	Air/ Vacuum	0.52–575.25	3.28–3.32	Optoelectronic	[69]
Co	0–10	Corning glass	550	Air		3.26–3.31	Magnetic	[70]
Na	3–30	Quartz glass	800	Air		3.25–3.293	Optical	[71]
Ru	0–6	Si/Quartz	600			3.278–3.372	Transistors	[72]
Li	0–0.2		500	Air			Sensors	[73]
Al-Ni	1.5	Corning glass	450–600	N_2/H_2	1.05– × –10^{-3}– 6.53 × 10^{-3}		Optoelectronic	[74]
Al B	0.25–5.0 0.25–1.25	Glass	550	Air/ 5% N_2/95% H_2	1.99– × –10^{-3} 4.01		Optoelectronic	[75]
Al		Quartz	600	Air	3.22–1.42	3.21–3.14	Gas sensing	[76]
Fe	1–4	FTO	450	–		3.26–3.21	Optical	[77]
In	3	Quartz	600	Air			Sensing	[78]
Mn	1–5	Glass	400	Air		3.43	Optical	[79]
Cu	1–4	Glass	500	–			Sensing	[80]
Ga	0–1	Alkali free glass	550			3.27–3.29	Sensing	[81]
Ag	0–3	Glass	500	Air		3.31–3.69	Optical	[82]
Al	0–12	Glass	300–500	Air			Optoelectronic	[83]
Fe	0–20	Glass	500			3.07–3.38	Optoelectronic/ Magnetic	[84]
Li	0.25–1.25	Glass	450	Air			Optoelectronic Thermal	[85]
Pb	1–4	Glass	500			3.19–3.25	Optoelectronic	[86]
Al-Ti	(1)Al–(0.1) Ti	Glass	550	Air	900 × 10^6–13 × 10^6	3.23–3.26	Optical	[87]
Co	0–0.09	Quartz glass	650	Air		3.26–3.28	Optical	[88]
Mn	0–12	Glass	450	Air		3.89–3.15	Optical	[89]
B	0–5	Alkali free glass	500		2.2 × 10^2	3.25–3.3	Optical	[90]
Ga	0–3	Glass	500	Vacuum	102–9	3.24–3.28	Photocatalytic	[91]

Table 3. Recently used impurities in the ZnO films for different applications.

of the Zn tetrahedron. The Mn K-edge XANES data revealed the Mn^{2+} in the 2% Mn-doped samples, while the oxidation state was reported as Mn^{2+} and Mn^{3+} in the co-doped samples. Furthermore, the methylene blue dye degradation was improved up to 90% with the co-doping mechanism. All the samples showed a preferred crystallographic orientation (100), (002), (101), (102), (110), (103), (112), and (210). The crystallite size calculated using a Scherer equation was observed to increase nonlinearly from 27.180 to 27.145 nm for the undoped to high doped samples, respectively. Vijayaprasath et al. [64] used various transition metals (Ni, Mn, Co = 0.03 mol%) as impurities in the ZnO crystal lattice by processing sol-gel-derived pure and doped ZnO films. The X-ray diffraction (XRD) studies confirmed the hexagonal wurtzite structure with (002) preferred orientation. Optical characterization showed a transparent character in the visible region while a d-d transition was reported in the violet region, which was associated with the crystalline defects and grain morphology. The crystalline size was reported as 28.70, 27.74, 26.00 and 27.74 nm for the pure ZnO, Ni-, Mn-, and Co-doped ZnO, respectively. An increase in the dislocation density (lines m^{-2}) and microstrain (lines^{-2} m^{-4}) was observed for all the transition elements. The resultant films exhibited well-defined ferromagnetic properties at room temperature. The coercivity of Ni-, Mn-, and Co-doped ZnO thin films was 41, 58 and 49 Oe (Oersted), respectively. It is important to further investigate the effect of elements in the ZnO both theoretically and experimentally. The new mechanism that can potentially improve the properties of ZnO is the co-doping mechanism rarely reported.

4. Application of ZnO

Zinc oxide has been utilized as a potential technological material for centuries. In the Bronze age, the smelting process for copper ore produced ZnO as a by-product. It was also used for the wounds healing and brass (Cu-Zn alloy) production. After industrialization in the mid-nineteenth century, ZnO was used in white paint, rubber vulcanization activation, and porcelain enamels. In the following, we will discuss comprehensive details of the selected electronic and optical applications of ZnO.

4.1. Gas sensing

ZnO nanostructures have been extensively studied for use in gas sensing applications. A gas sensor is designed to convert chemical information (concentration) of a particular gas present in a designated space into an electric (or optical) signal (**Figure 4**). The basic requirements of such a device configuration are: (1) compaction in order to install it in a commodity appliance and (2) must be economical. In principle, the gas sensors are designed by combining two key functions. The first one is the gas recognition, which is conducted through adsorption, reduction or electrochemical reactions to/by sensor materials or electrode (receptors). The second function is the transduction of this information in the form of signals (signal transduction). The interaction can be any physical or chemical effect on or around the receptors, such as a reaction product, adsorbed formations, generation of heat of reactions, changes in receptors mass, dimensions, surface/bulk properties and changes in the electrode potentials.

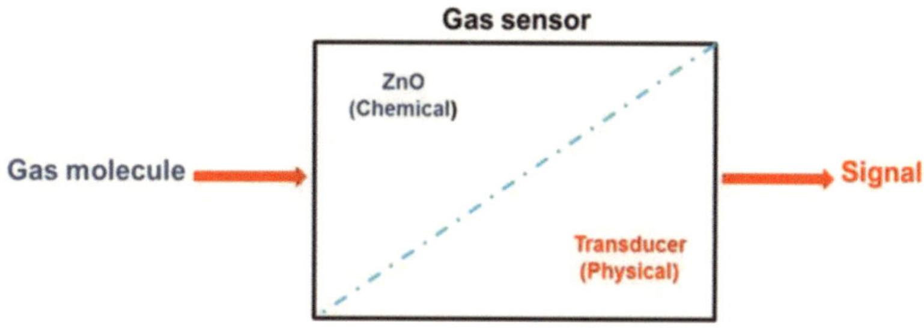

Figure 4. A typical ZnO-based gas sensor.

In the case of semiconductors, the changes in the work function are monitored, and consequently, the work function dependent resistance is monitored. The surface conductivity of the ZnO is recorded in response to the adsorbed gas. As mentioned in the previous text, the surface of the ZnO exhibits the point defects, which affect the conductivity of ZnO. The oxygen vacancies are dominant defects in ZnO that are generally generated in the films by annealing at high temperature. The resistance of ZnO-based sensors is monitored, which correspond to the adsorption and desorption of the target gas molecule on the surface of the sensors [92]. The surface of ZnO captures oxygen from the air, which is converted into oxygen ions by taking electrons from the conduction band. This phenomenon results in decreasing the charge carrier species, subsequently resulting in increasing the resistance across the circuit. For example, when ammonia molecules react with O_2 on the surface, the electrons flow back into the conduction band which decreases the circuit resistance. The oxygen and NH_3 reaction are as follows:

$$2NH_3(g) + 7O^- \rightarrow 2NO_2(g) + 3H_2O(g) + 7e^- \quad (1)$$

The gas sensing properties of ZnO depend on the surface areas, porosity, defect concentration, and working temperature. The thickness of the ZnO film by sol-gel route is controlled by varying the sol-gel concentrations and deposition cycles. Kumar et al. [44] reported that an increase in the film thickness from 98 to 366 nm resulted in particle size growth and surface roughness 5.8–47 nm, respectively. Further, the band gap was observed to decrease from 3.33 to 3.24 eV, respectively. Films with high surface roughness showed good gas sensing characteristics. Huo and Jayatissa [93] studied the effect of laser irradiation on the sol-gel-derived ZnO film thickness and H_2 sensing. A pulsed laser (£ = 532 nm, pulsed rate = 8 nm, pulsed f = 5 kHz, laser fluence range = 1.06–3.58 J/cm^2) was used in this experiment. The low irradiation promoted the grain growth and crystallinity, while the high-dose resulted in deterioration of the crystallinity. The H_2 sensing ability of the Al-doped ZnO was observed to be dependent on the film thickness and optimum dose of the laser. Liu et al. [76] reported the sensing and optical properties for the Al-doped ZnO films deposited on the quartz substrates and annealed with a different heating process. The films exhibited Eg = 3.21–3.14 eV, resistivity = 3.22–1.42 Ω cm, average transmittance 400–700 nm increased from 82.9 to 86.7%. The sensing properties were in ethanol atmosphere. The thin films showed high response and recovery toward ethanol gas sensing. Hou et al. [94] utilized two different Zn metal salts (Zinc acetate and Zinc nitrate) to develop Al-doped ZnO films. These precursors were selected as the controlling factors to investigate its effect on the films microstructure, conductivity, and gas sensing response for the Zinc acetate compared to the Zn-nitrate. Different concentrations of the hydrogen in the air were also investigated. Zinc acetate-derived films exhibited a skeletal wrinkles structure. The sample with 3.0% Al-doped ZnO showed conductivity, high tunability for a selective, and optimum operating temperatures.

4.2. Photocatalytic applications

The photocatalysis process can be exploited as a renewable energy source and an environmental technology for application ranging from solar cell to photoelectrochemical cells for water splitting reactions, purification of hazardous materials from industrial effluents, that is, removal/degradation of drugs, dye, and conversion of alcohols to fuel. The photocatalysis is an important electrochemical process, which occurs under light irradiations either from the sun or artificial light source that occurs at the surface of a photocatalyst (semiconductor material). The interaction of photon to the photocatalyst is a very important characteristic that facilitates/deteriorate the photocatalytic process. The photocatalysts have been under extensive investigations, and recently, a number of efforts have been given to tune the electronic microstructure of the semiconductor materials. For this purpose, efforts have been made not only to modify the synthesis procedures, but also the crystal orientations, impurities, tuning the defect chemistry such as TiO_2. However, there is still room to improve the photocatalysts for industrial applications and to understand the physical and chemical processes involved [95]. ZnO is one of the efficient photocatalysts that exhibit good photocatalytic activity, non-toxicity, and abundance [96]. The environmental contaminants are degraded by using ZnO as the photocatalysts. When the energy of the incident photons is equal to or greater than the Eg energy of the ZnO, reactive species such as H_2O_2, Radicles (i.e., anions ($^{\bullet}O^{2-}$) and hydroxyl ($^{\bullet}OH$)) are produced, which are strong oxidizing agents that facilitate the photocatalytic process of degrading the organic and pharmaceutical pollutants. Moreover, these reactive species can be used as antibacterial agents that can rupture the outer membrane of bacteria, inhibiting their growth. The antibacterial agents can be organic or inorganic; however, the inorganic antibacterial agents are more stable and safe. ZnO is categorized as the most promising inorganic antibacterial agent that showcase antibacterial activity, even in the absence of UV light. These properties can be exploited for various applications, such as wastewater treatment. Additionally, the thin ZnO films exhibiting antibacterial activity can be used to modify windows in buildings for protection purposes against certain microbes. Recently, the self-healing films with good antibacterial activity are under consideration to improve health care facilities for the public [97]. The ZnO is applied in the form of aqueous slurries as a photocatalyst; however, the process is technologically not feasible, and thus, it cannot be used to recycle and recover the catalyst after wastewater treatment. To overcome this issue, attempts have been reported to develop ZnO thin films on rigid substrates such as glass (beads, fibers), aluminum foil sheet, and stainless steel. The photocatalytic response of these films is lower compared to the nanopowders. A new approach is to develop immobilized nanostructured photocatalysts, exhibiting superior microstructure properties such as a high number of active sites and surface to volume ratio compared to thin films. Furthermore, mixed phases are fabricated to increase the photocatalytic activity. Georgakopoulos et al. [98] processed the TiO_2/

http://dx.doi.org/10.5772/67857

ZnO nanocomposites and studied the photoconductivity of various stoichiometric ratios both in vacuum and in air. The environment did not affect the photoconductivity amplitudes; however, a high decay rate was recorded in the air. Moreover, a composite ZnO (10%)/TiO_2 thin film degraded 78.1% of methyl orange (MO) compared to pure TiO_2 (49.3%) [99]. Despite absorption in the ultraviolet region of the solar spectrum by ZnO, for good photocatalytic properties, it is important that ZnO absorbs the visible region because the solar spectrum constitutes only 5–7% UV light, while the visible and IR percentage are higher 46 and 47%, respectively. Therefore, the Eg of the ZnO is narrowed by doping the transition elements to improve the absorption of the visible region of the solar spectrum [100]. Davar et al. [101] adopted a novel green chemistry route to develop ZnO for the degradation of various dyes such as methyl blue (MB), methyl red (MR), and methyl orange. The effect of sucrose solution was added in various ratios to investigate its effect on the lemon juice and zinc acetate solutions. The product materials studied by SEM and XRD showed an enhanced grain morphology and crystal phase. The as-synthesized ZnO showed a good photocatalytic response to the degradation of various dyes used in this work, and the photocatalyst can also be used for commercial applications in the degradation of dyes in the textile industry.

4.3. Thermoelectric properties

The demands for the generation of energy from sources that are free from carbon emission are increasing which has a great impact on the current social economic and political structure. The production of energy from sources such as fossil fuels and natural gas has increased the production of carbon dioxide emission in the atmosphere and sulfur compounds to an alarming rate, which poses a great threat to our environment and health. To overcome these issues, attempts have been made to produce energy from clean sources and utilize alternative means of energy. These include sources like hydrogen fuel economy, solar cells, piezoelectric materials, pyroelectric materials, and thermoelectric materials for energy harvesting either from sun radiations, vibrations and movements or, surrounding heat and heat generated in the automobiles and industries. Every year a tremendous amount of energy is wasted in the form of heat, which provides an opportunity to harness the energy for power generation. Techniques have been explored to recover waste heat, such as Organic Rankine Cycle (ORC), which utilizes an organic fluid, using superheated gases for running blades in a turbine for power generation, and thermoelectric generators, those can transfer the difference of heat energy between the two plates into a DC power by using Seebeck, Peltier, and Thomson effect, which can be exploited for the generation of electric powers from heat. In the past few years, many materials, both oxide and non-oxide-based, have been studied to improve the performance of the TEGs. The non-oxide-based TEGs have better ZT values; however, they contain lead and other heavy and expensive metals, which cannot be used for large-scale implementation. The oxide-based TEGs are cheap, abundant and have the potential to improve its properties by tuning the defect chemistry and conductivity in these materials. Further, these materials are more stable at high temperature compared to non-oxide based, which readily oxidizes at extreme temperature and deteriorate its properties. ZnO is considered to be one of the potential thermoelectric materials, which has recently gained attention due to its stability and economy. TEGs are based on p- and n-type semiconductor materials connected in series and parallel to deal with current and heat flow, respectively. The thermoelectric figure of merit, $ZT = S^2\sigma T/\kappa$ where T is the temperature, S is the Seebeck coefficient, σ is the electrical conductivity, and κ is the thermal conductivity, is the measure of the potential efficiency of a thermoelectric material [102]. The thermoelectric power factor (PF) of ZnO is comparatively higher and can be used as alternatives to the conventional thermoelectric materials. The PF is given as ZT ¼$(rS^2T)/k$, where r is defined as electrical conductivity. The doping mechanism can be adopted in ZnO to restructure the crystal architecture, which can possibly increase r. In such a way, the semiconductor behavior can be changed into a metallic one. The Al doping has been reported to increase the magnitude of r by the three order; however, the high k results in a low thermoelectric performance. The thermoelectric properties can be improved by improving the electronic transport mechanism [103]. Liang [103] reported the effect of Fe on the ZnO for thermoelectric applications. The XRD confirmed the ZnO solid solution and $ZnFe_2O_4$ spinal phases. The magnitude of ZT was observed to increase with increasing Fe concentration, which might be associated with phonon scattering, point defects, and electron transport mechanism.

5. Summary and future perspectives

This chapter reviewed the processing, effect of doping, and strategies to improve the development of the sol-gel derived pure and doped ZnO films, with an emphasis understanding the parameters that can influence the optical and electronic properties of these materials. Moreover, it has been discussed how these materials can be used for gas sensing, photocatalytic and thermoelectric applications. The number of publications on sol-gel chemistry is increasing every year, which shows its potentials in semiconductor-based films processing that can be exploited and applied on an industrial scale. The doping mechanism, film microstructure, and hence the final properties can be easily controlled by the sol-gel method; however, the parameters, which can influence the properties and durability of the sol-gel derived films, must be further investigated and optimized. For instance, very little attention has been given to understand the film growth and grain orientation on the nanoscale, and most of the studies generalize/overlook this approach. It is important to modify the sol-gel process to develop the homogeneous ZnO-based films with preferred crystal orientations and controlled microstructure, which can be proved as an alternative to atomic layer deposition and magnetron sputtering techniques. Few studies report the microstructure evaluation, which limits the usefulness of sol-gel for the sensitive and reliable applications. Future studies must focus on the grain formation in sol-gel during film processing and crystal orientation mechanism must be studied, which will open a new window in the sol-gel regime. Moreover, the grain-to-grain connection and density parameters must be investigated in detail, which strongly affect the electrical properties of the device. Although the sol-gel technology for the semiconductor processing has been used for decades, the methods have been overlooked. Despite hundreds of papers published each year, the sol-gel process has been little modified, that is, sol-precipitation and sol-hydrothermal method. The major issues, which are not only related to the current discussion, but also, in general, are the reproducibility of the results which are reported differently from lab to lab. The percentage of reproducibility of the experimental results reported is extremely low, which demolishes the adoption of such techniques.

The importance of the sol-gel technology and further modification in the techniques will increase in the near future. The new precursors and developing new sol-gel chemistry will change the film growth technology, and sol-gel has enormous potential for commercial applications. Sol-gel-derived films will generate unique properties to cope with the needs of future technological device applications.

Author details

Asad Mahmood and Abdul Naeem*

*Address all correspondence to: naeem@upesh.edu.pk

National Centre of Excellence in Physical Chemistry, University of Peshawar, Pakistan

References

[1] Pal D, Mathur A, Singh A, Singhal J, Sengupta A, Dutta S, Sudeshna SZ. Tunable optical properties in atomic layer deposition grown ZnO thin films. Journal of Vacuum Science & Technology A: Vacuum, Surfaces, and Films. 2017;35: 01B108. doi:10.1116/1.4967296

[2] Hoye RLZ, Muñoz-Rojas D, Nelson SF, Illiberi A, Poodt P, Roozeboom F, Judith L. Research Update: Atmospheric pressure spatial atomic layer deposition of ZnO thin films: Reactors, doping, and devices. APL Materials. 2015;3: 040701. doi:10.1063/1.4916525

[3] Tsay C-Y, Lee W-C. Effect of dopants on the structural, optical and electrical properties of sol-gel derived ZnO semiconductor thin films. Current Applied Physics. 2013;13: 60–65. doi:10.1016/j.cap.2012.06.010

[4] Wang ZL. Zinc oxide nanostructures: growth, properties and applications. Journal of Physics: Condensed Matter 1. 2004;16: R829–R858

[5] Kresse G, Hafner J. Ab initio molecular dynamics for liquid metals. Physical Review B. 1993;47: 558–561. doi:10.1103/PhysRevB.47.558

[6] Alias SS, Mohamad AA. ZnO nanocrystalline metal oxide semiconductor via sol gel method. SpringerBriefs in Materials. 2014: 1–8. doi:10.1007/978-981-4560-77-1_1

[7] Saha D, Misra P, Das G, Joshi MP, Kukreja LM. Observation of dopant-profile independent electron transport in sub-monolayer TiOx stacked ZnO thin films grown by atomic layer deposition. Applied Physics Letters. 2016;108: 032101. doi:10.1063/1.4939926

[8] Kaushik VK, Mukherjee C, Ganguli T, Sen PK. Material characterizations of Al:ZnO thin films grown by aerosol assisted chemical vapour deposition. Journal of Alloys and Compounds. 2016;689: 1028–1036. doi:10.1016/j.jallcom.2016.08.022

[9] Kumar V, Ntwaeaborwa OM, Swart HC. Deep level defect correlated emission and Si diffusion in ZnO:Tb(3+) thin films prepared by pulsed laser deposition. Journal of Colloid and Interface Science. 2016;465: 295–303. doi:10.1016/j.jcis.2015.12.007

[10] Suzuki T, Chiba H, Kawashima T, Washio K. Comparison study of V-doped ZnO thin films on polycarbonate and quartz substrates deposited by RF magnetron sputtering. Thin Solid Films. 2016;605: 53–56. doi:10.1016/j.tsf.2015.11.064.

[11] Demiroglu I, Bromley ST. Evidence for multi-polymorphic islands during epitaxial growth of ZnO on Ag(111). Journal of Physics Condensed Matter : An Institute of Physics Journal. 2016;28: 224007. doi:10.1088/0953-8984/28/22/224007

[12] Zhang B, Wang F, Zhu C, Li Q, Song J, Zheng M, Ma L, Shen W. A facile self-assembly synthesis of hexagonal zno nanosheet films and their photoelectrochemical properties. Nano-Micro Letters. 2016;8: 137–142. doi:10.1007/s40820-015-0068-y

[13] Kunj S, Sreenivas K. Near band edge emission characteristics of sputtered nano-crystalline ZnO films. Proceeding of International Conference on Condensed Matter and Applied Physics (ICC 2015). 2016;1728: 020520. doi:10.1063/1.4946571

[14] El-Shazly AN, Rashad MM, Abdel-Aal EA, Ibrahim IA, El-Shahat MF, Shalan AE. Nanostructured ZnO photocatalysts prepared via surfactant assisted co-precipitation method achieving enhanced photocatalytic activity for the degradation of methylene blue dyes. Journal of Environmental Chemical Engineering. 2016;4: 3177–3184. doi:10.1016/j.jece.2016.06.018

[15] Kıcır N, Tüken T, Erken O, Gumus C, Ufuktepe Y. Nanostructured ZnO films in forms of rod, plate and flower: Electrodeposition mechanisms and characterization. Applied Surface Science. 2016;377: 191–199. doi:10.1016/j.apsusc.2016.03.111

[16] Rajah S, Barhoumi A, Mhamdi A, Leroy G, Duponchel B, Amlouk M, Guermazi S. Structural, morphological, optical and opto-thermal properties of Ni-doped ZnO thin films using spray pyrolysis chemical technique. Bull Mater Sci. 2016;39: 177–186.

[17] Duan L, Zhao X, Zhang Y, Shen H, Liu R. Fabrication of flexible Al-doped ZnO films via sol-gel method. Materials Letters. 2016;162: 199–202. doi:10.1016/j.matlet.2015.10.023

[18] Yu Y, Pan W, Guo X, Gao L, Gu Y, Liu Y. A poly(arylene ether sulfone) hybrid membrane using titanium dioxide nanoparticles as the filler: Preparation, characterization and gas separation study. High Performance Polymers. 2017, 29(1) 26–35. doi:10.1177/0954008315626990

[19] Barczak M, McDonagh C, Wencel D. Micro- and nanostructured sol-gel-based materials for optical chemical sensing (2005–2015). Microchimica Acta. 2016;183: 2085–2109. doi:10.1007/s00604-016-1863-y

[20] Ong X, Zhi M, Gupta S, Chan Y. Wet-chemically synthesized colloidal semiconductor nanostructures as optical gain media. Chemphyschem : A European Journal of Chemical Physics and Physical Chemistry. 2016;17: 582–597. doi:10.1002/cphc.201500975

[21] Mikosch A, Ciftci S, Kuehne AJ. Colloidal crystal lasers from monodisperse conjugated polymer particles via bottom-up coassembly in a sol-gel matrix. ACS Nano. 2016;10: 10195–10201. doi:10.1021/acsnano.6b05538

[22] Lee S-W, Kim H, Kim M-S, Youn H-C, Kang K, Cho B-W, Roh KC, Kim KB. Improved electrochemical performance of $LiNi_{0.6}Co_{0.2}Mn_{0.2}O_2$ cathode material synthesized by citric acid assisted sol-gel method for lithium ion batteries. Journal of Power Sources. 2016;315: 261–268. doi:10.1016/j.jpowsour.2016.03.020

[23] Adachi K, Tokushige M, Omata K, Yamazaki S, Iwadate Y. Kinetics of coloration in photochromic tungsten(VI) oxide/silicon oxycarbide/silica hybrid xerogel: Insight into cation self-diffusion mechanisms. ACS Applied Materials & Interfaces. 2016;8: 14019–14028. doi:10.1021/acsami.6b04115

[24] Mohana Priya S, Geetha A, Ramamurthi K. Structural, morphological and optical properties of tin oxide nanoparticles synthesized by sol-gel method adding hydrochloric acid. Journal of Sol-Gel Science and Technology. 2016;78: 365–372. doi:10.1007/s10971-016-3966-7

[25] Ullah A, Ahn CW, Hussain A, Kim IW. The effects of sintering temperatures on dielectric, ferroelectric and electric field-induced strain of lead-free $Bi_{0.5}(Na_{0.78}K_{0.22})_{0.5}TiO_3$ piezoelectric ceramics synthesized by the sol-gel technique. Current Applied Physics. 2010;10: 1367–1371. doi:10.1016/j.cap.2010.05.004

[26] Kumar A, Gaurav, Malik AK, Tewary DK, Singh B. A review on development of solid phase microextraction fibers by sol-gel methods and their applications. Analytica Chimica Acta. 2008;610: 1–14. 10.1016/j.aca.2008.01.028.

[27] Guo D, Ju Y, Fu C, Huang Z, Zhang L. (002)-oriented growth and morphologies of ZnO thin films prepared by sol-gel method. Materials Science-Poland. 2016;34: pp. 555–563 doi:10.1515/msp-2016-0076

[28] Chia CH, Tsai WC, Chiou JW. Thickness effect on luminescent properties of sol-gel derived ZnO thin films. Journal of Luminescence. 2013;136: 160–164. doi:10.1016/j.jlumin.2012.11.019

[29] Popa M, Mereu RA, Filip M, Gabor M, Petrisor TJr, Ciontea L, Petrisor T. Highly c-axis oriented ZnO thin film using 1-propanol as solvent in sol-gel synthesis. Materials Letters. 2013;92: 267–270. doi:10.1016/j.matlet.2012.10.099

[30] Ayana DG, Ceccato R, Collini C, Lorenzelli L, Prusakova V, Dirè S. Sol-gel derived oriented multilayer ZnO thin films with memristive response. Thin Solid Films. 2016;615: 427–436. doi:10.1016/j.tsf.2016.07.025

[31] Malek MF, Mamat MH, Sahdan MZ, Zahidi MM, Khusaimi Z, Mahmood MR. Influence of various sol concentrations on stress/strain and properties of ZnO thin films synthesised by sol-gel technique. Thin Solid Films. 2013;527: 102–109. doi:10.1016/j.tsf.2012.11.095

[32] Das S, Bhattacharjee K, Maitra S, Das GC. Effect of oxygen partial pressure on the photoluminescence properties of sol-gel synthesized nano-structured ZnO thin films. Thin Solid Films. 2014;550: 65–70. doi:10.1016/j.tsf.2013.10.045.

[33] Zahn M, Ohki Y, Fenneman DB, Gripshover RJ, Gehman VH. J. Dielectric properties of water and water/ethylene glycol mixtures for use in pulsed power system design. Proceedings of the IEEE. 1986;74: 1182–1221.

[34] Naik GV, Bhat N. Poly-ols based sol-gel synthesis of zinc oxide thin films. Journal of The Electrochemical Society. 2011;158: H85. doi:10.1149/1.3515894

[35] Hosseini Vajargah P, Abdizadeh H, Ebrahimifard R, Golobostanfard MR. Sol-gel derived ZnO thin films: Effect of amino-additives. Applied Surface Science. 2013;285: 732–743. doi:10.1016/j.apsusc.2013.08.118

[36] Znaidi L. Sol-gel-deposited ZnO thin films: A review. Materials Science and Engineering: B. 2010;174: 18–30. doi:10.1016/j.mseb.2010.07.001

[37] Patil SL, Chougule MA, Pawar SG, Raut BT, Sen S, Patil VB. New process for synthesis of ZnO thin films: Microstructural, optical and electrical characterization. Journal of Alloys and Compounds. 2011;509: 10055–10061. doi:10.1016/j.jallcom.2011.08.030

[38] Mahroug A, Boudjadar S, Hamrit S, Guerbous L. Structural, optical and photocurrent properties of undoped and Al-doped ZnO thin films deposited by sol-gel spin coating technique. Materials Letters. 2014;134: 248–251. doi:10.1016/j.matlet.2014.07.099

[39] Chia CH, Tsai WC, Chou WC. Preheating-temperature effect on structural and photoluminescent properties of sol-gel derived ZnO thin films. Journal of Luminescence. 2014;148: 111–115. doi:10.1016/j.jlumin.2013.12.006

[40] Boudjouan F, Chelouche A, Touam T, Djouadi D, Khodja S, Tazerout M, Ouerdance Y, Hadjoub Z. Effects of stabilizer ratio on photoluminescence properties of sol-gel ZnO nano-structured thin films. Journal of Luminescence. 2015;158: 32–37. doi:10.1016/j.jlumin.2014.09.026

[41] Benramache S, Benhaoua B, Chabane F, Guettaf A. A comparative study on the nanocrystalline ZnO thin films prepared by ultrasonic spray and sol-gel method. Optik - International Journal for Light and Electron Optics. 2013;124: 3221–3224. doi:10.1016/j.ijleo.2012.10.001

[42] Foo KL, Kashif M, Hashim U, Ali ME. Sol-gel derived ZnO nanoparticulate films for ultraviolet photodetector (UV) applications. Optik-International Journal for Light and Electron Optics. 2013;124: 5373–5376. doi:10.1016/j.ijleo.2013.03.120

[43] Talebian N, Nilforoushan MR, Maleki N. Ultraviolet to visible-light range photocatalytic activity of ZnO films prepared using sol-gel method: The influence of solvent. Thin Solid Films. 2013;527: 50–58. doi:10.1016/j.tsf.2012.11.138

[44] Kumar V, Singh N, Mehra RM, Kapoor A, Purohit LP, Swart HC. Role of film thickness on the properties of ZnO thin films grown by sol-gel method. Thin Solid Films. 2013;539: 161–165. doi:10.1016/j.tsf.2013.05.088

[45] Guo D, Sato K, Hibino S, Takeuchi T, Bessho H, Kato K. Low-temperature preparation of (002)-oriented ZnO thin films by sol-gel method. Thin Solid Films. 2014;550: 250–258. doi:10.1016/j.tsf.2013.11.004.

[46] Nehmann JB, Ehrmann N, Reineke-Koch R, Bahnemann DW. Aluminum-doped zinc oxide sol-gel thin films: Influence of the sol's water content on the resistivity. Thin Solid Films. 2014;556: 168–173. doi:10.1016/j.tsf.2014.01.052.

[47] Heredia E, Bojorge C, Casanova J, Cánepa H, Craievich A, Kellermann G. Nanostructured ZnO thin films prepared by sol-gel spin-coating. Applied Surface Science. 2014;317: 19–25. doi:10.1016/j.apsusc.2014.08.046

[48] Cui L, Wang G-G, Zhang H-Y, Sun R, Kuang X-P, Han J-C. Effect of film thickness and annealing temperature on the structural and optical properties of ZnO thin films deposited on sapphire (0001) substrates by sol-gel. Ceramics International. 2013;39: 3261–3268. doi:10.1016/j.ceramint.2012.10.014

[49] Aydemir S, Karakaya S. Effects of withdrawal speed on the structural and optical properties of sol-gel derived ZnO thin films. Journal of Magnetism and Magnetic Materials. 2015;373: 33–39. doi:10.1016/j.jmmm.2014.01.077

[50] Goktas A, Mutlu IH, Yamada Y. Influence of Fe-doping on the structural, optical, and magnetic properties of ZnO thin films prepared by sol-gel method. Superlattices and Microstructures. 2013;57: 139–149. doi:10.1016/j.spmi.2013.02.010

[51] Khodja S, Touam T, Chelouche A, Boudjouan F, Djouadi D, Hadjoub Z, Fischer A, Boudrioua A. Effects of stabilizer ratio on structural, morphological, optical and waveguide properties of ZnO nano-structured thin films by a sol-gel process. Superlattices and Microstructures. 2014;75: 485–495. doi:10.1016/j.spmi.2014.08.010

[52] Thongsuriwong K, Amornpitoksuk P, Suwanboon S. Structure, morphology, photocatalytic and antibacterial activities of ZnO thin films prepared by sol-gel dip-coating method. Advanced Powder Technology. 2013;24: 275–280.doi:10.1016/j.apt.2012.07.002

[53] Li C-F, Hsu C-Y, Li Y-Y. NH_3 sensing properties of ZnO thin films prepared via sol-gel method. Journal of Alloys and Compounds. 2014;606: 27–31. doi:10.1016/j.jallcom.2014.03.120.

[54] Caglar M, Ruzgar S. Influence of the deposition temperature on the physical properties of high electron mobility ZnO films by sol-gel process. Journal of Alloys and Compounds. 2015;644: 101–105. doi:10.1016/j.jallcom.2015.04.167.

[55] Quiñones-Galván JG, Sandoval-Jiménez IM, Tototzintle-Huitle H, Hernández-Hernández d LA, Moure-Flores e FD, Hernández-Hernández A, Campos-González E, Guillén-Cervantes A, Zelaya-Angel O, Araiza-Ibarra JJ. Effect of precursor solution and annealing temperature on the physical properties of sol-gel-deposited ZnO thin films. Results in Physics. 2013;3: 248–253. doi:10.1016/j.rinp.2013.11.001.

[56] Guillemin S, Consonni V, Rapenne L, Sarigiannidou E, Donatini F, Bremond G. Identification of extended defect and interface related luminescence lines in polycrystalline

ZnO thin films grown by sol-gel process. RSC Adv. 2016;6: 44987–44992. doi:10.1039/c6ra04634g

[57] Janisch R, Gopal P, Spaldin NA. Transition metal-doped TiO_2 and ZnO—present status of the field. Journal of Physics: Condensed Matter. 2005;17: R657–R689. doi:10.1088/0953-8984/17/27/r01

[58] Ciciliati MA, Silva MF, Fernandes DM, de Melo MAC, Hechenleitner AAW, Pineda EAG. Fe-doped ZnO nanoparticles: Synthesis by a modified sol-gel method and characterization. Materials Letters. 2015;159: 84–86. doi:10.1016/j.matlet.2015.06.023

[59] Aydın C, Abd El-sadek MS, Zheng K, Yahia IS, Yakuphanoglu F. Synthesis, diffused reflectance and electrical properties of nanocrystalline Fe-doped ZnO via sol-gel calcination technique. Optics & Laser Technology. 2013;48: 447–452. doi:10.1016/j.optlastec.2012.11.004

[60] Xu L, Li X. Influence of Fe-doping on the structural and optical properties of ZnO thin films prepared by sol-gel method. Journal of Crystal Growth. 2010;312: 851–855. doi:10.1016/j.jcrysgro.2009.12.062

[61] Zhang Z, Bao C, Yao W, Ma S, Zhang L, Hou S. Influence of deposition temperature on the crystallinity of Al-doped ZnO thin films at glass substrates prepared by RF magnetron sputtering method. Superlattices and Microstructures. 2011;49: 644–653. doi:10.1016/j.spmi.2011.04.002.

[62] Chen S, Warwick MEA, Binions R. Effects of film thickness and thermal treatment on the structural and opto-electronic properties of Ga-doped ZnO films deposited by sol-gel method. Solar Energy Materials and Solar Cells. 2015;137: 202–209. doi:10.1016/j.solmat.2015.02.016

[63] Dubey DK, Singh DN, Kumar S, Nayak C, Kumbhakar P, Jha SN, Bhattacharya D, Ghosh AK, Chatterjee S. Local structure and photocatalytic properties of sol-gel derived Mn–Li co-doped ZnO diluted magnetic semiconductor nanocrystals. RSC Adv. 2016;6: 22852–22867. doi:10.1039/c5ra23220a

[64] Vijayaprasath G, Murugan R, Ravi G, Mahalingam T, Hayakawa Y. Characterization of dilute magnetic semiconducting transition metal doped ZnO thin films by sol-gel spin coating method. Applied Surface Science. 2014;313: 870–876. doi:10.1016/j.apsusc.2014.06.093

[65] Khan F, Baek S-H, Mobin A, Kim JH. Enhanced performance of silicon solar cells by application of low-cost sol-gel-derived Al-rich ZnO film. Solar Energy. 2014;101: 265–271. doi:10.1016/j.solener.2013.12.025

[66] Eshaghi A, Hakimi MJ, Zali A. Fabrication of titanium zinc oxide (TZO) sol-gel derived nanostructured thin film and investigation of its optical and electrical properties. Optik-International Journal for Light and Electron Optics. 2015;126: 5610–5613. doi:10.1016/j.ijleo.2015.09.243

[67] Ilican S, Özdemir Y, Caglar M, Caglar Y. Temperature dependence of the optical band gap of sol-gel derived Fe-doped ZnO films. Optik – International Journal for Light and Electron Optics. 2016;127: 8554–8561. doi:10.1016/j.ijleo.2016.06.074

[68] Yang S, Zhang Y, Mo D. Spectroscopic ellipsometry studies of sol-gel-derived Cu-doped ZnO thin films. Thin Solid Films. 2014;571: 605–608. doi:10.1016/j.tsf.2014.02.097

[69] Lee M-I, Huang M-C, Legrand D, Lerondel G, Lin J-C. Structure and characterization of Sn, Al co-doped zinc oxide thin films prepared by sol-gel dip-coating process. Thin Solid Films. 2014;570: 516–526. doi:10.1016/j.tsf.2014.04.051

[70] Dhruvashi, Shishodia PK. Effect of cobalt doping on ZnO thin films deposited by sol-gel method. Thin Solid Films. 2016;612: 55–60. doi:10.1016/j.tsf.2016.05.028

[71] Wang LW, Wu F, Tian DX, Li WJ, Fang L, Kong CY, Zhou M. Effects of Na content on structural and optical properties of Na-doped ZnO thin films prepared by sol-gel method. Journal of Alloys and Compounds. 2015;623: 367–373. doi:10.1016/j.jallcom.2014.11.055

[72] Jeng J-S. The influence of pH value and annealing temperature on the characteristics of ZnO–Ru composite films and their application in thin film transistors. Microelectronic Engineering. 2016;149: 1–4. doi:10.1016/j.mee.2015.08.014

[73] Zhao J, Xie C, Yang L, Zhang S, Zhang G, Cai Z. Enhanced gas sensing performance of Li-doped ZnO nanoparticle film by the synergistic effect of oxygen interstitials and oxygen vacancies. Applied Surface Science. 2015;330: 126–133. doi:10.1016/j.apsusc.2014.12.194

[74] Zhang XL, Hui KS, Bin F, Hui KN, Li L, Cho YR, Mane RS, Zhou W. Effect of thermal annealing on the structural, electrical and optical properties of Al–Ni co-doped ZnO thin films prepared using a sol-gel method. Surface and Coatings Technology. 2015;261: 149–155. doi:10.1016/j.surfcoat.2014.11.043

[75] Peng S, Tang Y, Jin L, Wang Y, Ma L, Cao F. Comparison of the electro-optical performance of ZnO:Al and ZnO:B thin films derived by sol-gel method. Surface and Coatings Technology. 2017;310: 251–255. doi:10.1016/j.surfcoat.2016.12.090

[76] Liu X, Pan K, Li W, Hu D, Liu S, Wang Y. Optical and gas sensing properties of Al-doped ZnO transparent conducting films prepared by sol-gel method under different heat treatments. Ceramics International. 2014;40: 9931–9939. doi:10.1016/j.ceramint.2014.02.090

[77] Kafle BP, Acharya S, Thapa S, Poudel S. Structural and optical properties of Fe-doped ZnO transparent thin films. Ceramics International. 2016;42: 1133–1139. doi:10.1016/j.ceramint.2015.09.042

[78] Pati S, Banerji P, Majumder SB. n- to p- type carrier reversal in nanocrystalline indium doped ZnO thin film gas sensors. International Journal of Hydrogen Energy. 2014;39: 15134–15141. doi:10.1016/j.ijhydene.2014.07.075

[79] Xin M, Hu LZ, Liu D-P, Yu N-S. Effect of Mn doping on the optical, structural and photoluminescence properties of nanostructured ZnO thin film synthesized by sol-gel technique. Superlattices and Microstructures. 2014;74: 234–241. doi:10.1016/j.spmi.2014.06.009

[80] Shewale PS, Patil VB, Shin SW, Kim JH, Uplane MD. H_2S gas sensing properties of nanocrystalline Cu-doped ZnO thin films prepared by advanced spray pyrolysis. Sensors and Actuators B: Chemical. 2013;186: 226–234. doi:10.1016/j.snb.2013.05.073

[81] Hou Y, Jayatissa AH. Low resistive gallium doped nanocrystalline zinc oxide for gas sensor application via sol-gel process. Sensors and Actuators B: Chemical. 2014;204: 310–318. doi:10.1016/j.snb.2014.07.082

[82] Khan F, Baek S-H, Kim JH. Influence of Ag doping on structural, optical, and photoluminescence properties of nanostructured AZO films by sol-gel technique. Journal of Alloys and Compounds. 2014;584: 190–194. doi:10.1016/j.jallcom.2013.09.055

[83] Eskandari F, Ranjbar M, Kameli P, Salamati H. Laser induced photoconductivity in sol-gel derived Al doped ZnO thin films. Journal of Alloys and Compounds. 2015;649: 35–45. doi:10.1016/j.jallcom.2015.07.093

[84] Ariyakkani P, Suganya L, Sundaresan B. Investigation of the structural, optical and magnetic properties of Fe doped ZnO thin films coated on glass by sol-gel spin coating method. Journal of Alloys and Compounds. 2017;695: 3467–3475. doi:10.1016/j.jallcom.2016.12.011

[85] Shinde SS, Bhosale CH, Rajpure KY. Photoelectrochemical properties of highly mobilized Li-doped ZnO thin films. Journal of photochemistry and photobiology B, Biology. 2013;120: 1–9. doi:10.1016/j.jphotobiol.2013.01.003

[86] Yilmaz M, Aydoğan S. The effect of Pb doping on the characteristic properties of spin coated ZnO thin films: Wrinkle structures. Materials Science in Semiconductor Processing. 2015;40: 162–170. doi:10.1016/j.mssp.2015.06.064

[87] Davoodi A, Tajally M, Mirzaee O, Eshaghi A. Fabrication and characterization of optical and electrical properties of Al–Ti Co-doped ZnO nano-structured thin film. Journal of Alloys and Compounds. 2016;657: 296–301. doi:10.1016/j.jallcom.2015.10.107

[88] Li B, Adjei R, Chen Z, Shen H, Luo J. Synthesis and characterization of highly preferred orientation polycrystalline Co-doped ZnO thin films prepared by improved sol-gel method. Journal of Sol-Gel Science and Technology. 2014;70: 19–23. doi:10.1007/s10971-014-3268-x

[89] Amoupour E, Ghodsi FE, Andarva H, Abdolahzadeh Ziabari A. Preparation and investigation of optical, structural, and morphological properties of nanostructured ZnO:Mn thin films. Pramana. 2013;81: 331–341. doi:10.1007/s12043-013-0566-8

[90] Tsay C-Y, Hsu W-T. Sol-gel derived undoped and boron-doped ZnO semiconductor thin films: Preparation and characterization. Ceramics International. 2013;39: 7425–7432. doi:10.1016/j.ceramint.2013.02.086

[91] Huang M-C, Lin J-C, Cheng S-H, Weng W-H. Influence of Ga dopant on photoelectrochemical characteristic of Ga-doped ZnO thin films deposited by sol-gel spin-coating technique. Surface and Interface Analysis. 2016. doi:10.1002/sia.6176

[92] Senthil T, Anandhan S. Structure-property relationship of sol-gel electrospun ZnO nanofibers developed for ammonia gas sensing. Journal of colloid and interface science. 2014;432: 285–296. doi:10.1016/j.jcis.2014.06.029

[93] Hou Y, Jayatissa AH. Effect of laser irradiation on gas sensing properties of sol-gel derived nanocrystalline Al-doped ZnO thin films. Thin Solid Films. 2014;562: 585–591. doi:10.1016/j.tsf.2014.03.089

[94] Hou Y, Soleimanpour AM, Jayatissa AH. Low resistive aluminum doped nanocrystalline zinc oxide for reducing gas sensor application via sol-gel process. Sensors and Actuators B: Chemical. 2013;177: 761–769. doi:10.1016/j.snb.2012.11.085

[95] Surendar T, Kumar S, Shanker V. Influence of La-doping on phase transformation and photocatalytic properties of $ZnTiO_3$ nanoparticles synthesized via modified sol-gel method. Physical chemistry chemical physics : PCCP. 2014;16: 728–735. doi:10.1039/c3cp53855a

[96] Habibi MH, Sheibani R. Nanostructure silver-doped zinc oxide films coating on glass prepared by sol-gel and photochemical deposition process: Application for removal of mercaptan. Journal of Industrial and Engineering Chemistry. 2013;19: 161–165. doi:10.1016/j.jiec.2012.07.019

[97] Thongsuriwong K, Amornpitoksuk P, Suwanboon S. Photocatalytic and antibacterial activities of Ag-doped ZnO thin films prepared by a sol-gel dip-coating method. Journal of Sol-Gel Science and Technology. 2012;62: 304–312. doi:10.1007/s10971-012-2725-7

[98] Georgakopoulos T, Todorova N, Pomoni K, Trapalis C. On the transient photoconductivity behavior of sol-gel TiO_2/ZnO composite thin films. Journal of Non-Crystalline Solids. 2015;410: 135–141. doi:10.1016/j.jnoncrysol.2014.11.034

[99] Chen Y, Zhang C, Huang W, Yang C, Huang T, Situ Y, Huang H. Synthesis of porous ZnO/TiO_2 thin films with superhydrophilicity and photocatalytic activity via a template-free sol-gel method. Surface and Coatings Technology. 2014;258: 531–538. doi:10.1016/j.surfcoat.2014.08.042

[100] Kant S, Kumar A. A comparative analysis of structural, optical and photocatalytic properties of ZnO and Ni doped ZnO nanospheres prepared by sol-gel method. Advanced Materials Letters. 2012;3: 350-354.

[101] Davar F, Majedi A, Mirzaei A, French R. Green Synthesis of ZnO Nanoparticles and Its Application in the Degradation of Some Dyes. Journal of the American Ceramic Society. 2015;98: 1739–1746. doi:10.1111/jace.13467

[102] Abutaha AI, Sarath Kumar SR, Alshareef HN. Crystal orientation dependent thermoelectric properties of highly oriented aluminum-doped zinc oxide thin films. Applied Physics Letters. 2013;102: 053507. doi:10.1063/1.4790644

[103] Liang X. Thermoelectric transport properties of Fe-enriched ZnO with high-temperature nanostructure refinement. ACS Applied Materials & Interfaces. 2015;7: 7927–7937. doi:10.1021/am509050a

Chapter 10

Photoluminescence Studies on ZnO Thin Films Obtained by Sol-Gel Method

Guadalupe Valverde Aguilar,
Mónica R. Jaime Fonseca,
Ángeles Mantilla Ramírez and
Antonio G. Juárez Gracia

Additional information is available at the end of the chapter

http://dx.doi.org/10.5772/intechopen.68529

Abstract

The sol-gel process is a friendly room temperature method to prepare transparent glasses in the form of monoliths, films, and fibers. The zinc oxide films have been obtained by sol-gel method, which are very important materials in the ceramic technology due to their piezoelectric properties and applications in various pressure transducers and acoustic-optic devices, surface and bulk acoustic wave devices, and solar cells. Structure and characteristic ultraviolet-blue emissions of amorphous and crystalline zinc oxide thin films–coated glass substrates by dip-coating deposition are explained by photoluminescence studies in this chapter.

Keywords: sol-gel, thin film, zinc oxide, photoluminescence

1. Introduction

Sol-gel method is a very attractive chemical route due to its simplicity and flexibility in the use of different source materials that allow to synthesize amorphous and polycrystalline materials at low cost. The focus of this chapter is to expand on that knowledge about this method as an efficient route to obtain amorphous and crystalline ZnO thin films particularly. It is well known that the ZnO thin films are very important materials in ceramic technology and thin films technology due to numerous properties, while the ZnO nanocrystals with high stability at low processing temperatures have applications in displays, emitters, and sensors. The optical properties of excitons confined in the ZnO nanocrystals were studied by optical absorption, infrared spectroscopy and photoluminescence studies. The photoluminescence mechanisms of the violet and blue emissions are discussed in detail.

2. Sol-gel process

Materials prepared by the sol-gel process have become the source of an important field of research in materials science since the 1970s. The sol-gel process is a room temperature method of preparing transparent inorganic glasses without melting [1, 2], which is much lower compared to the high temperatures needed by standard glass manufacturing processes. The fabrication of glasses or ceramic materials involves a set of reactions such as hydrolysis and condensation that convert an aqueous metal alkoxides [2] with a molecular formula of $M(OR)_n$ into different types of inorganic networks. The advantage of this process has the advantage of being able to yield high purity inorganic oxide glasses, besides versatile in creating different types of materials (thin films, spun fibers, particles, aerogels, and xerogels).

As the name implies, the sol-gel process is the conversion of a sol to a gel. The sol-gel process is therefore a series of hydrolysis and condensation reactions of the inorganic alkoxide monomers that forms colloidal particles (sol) and converts them into a continuous network (gel). Usually for the preparation of silicate materials, tetramethoxysilane (TMOS) or tetraethyl orthosilicate (TEOS) is the most popular alkoxides. TEOS can be hydrolyzed and condensed under relatively mild conditions to manufacture silica gels. The hydrolysis step of TEOS can be represented as the Equation: $Si(OEt)_4 + H_2O \rightarrow HO\text{-}Si(OEt)_3 + EtOH$. The first step in this equation is the generation of a silanol group (Si-OH) and the corresponding alcohol. The second step is the condensation of the silanol group. This step can occur in two different ways and is represented by the equation $HO\text{-}Si(OEt)_3 + HO\text{-}Si(OEt)_3 \rightarrow (EtO)_3\text{-}Si\text{-}O\text{-}Si(OEt)_3 + H_2O$ and $HO\text{-}Si(OEt)_3 + Si(OEt)_4 \rightarrow (EtO)_3\text{-}Si\text{-}O\text{-}Si(OEt)_3 + EtOH$.

The process continues as the silanol groups condense with other silonols groups or with an alkoxide. If the silonols condense with an alkoxide, they make siloxane bonds (Si-O-Si) that have water or an alcohol as a byproduct. By repeating these steps many times, a gel or a solid material is generated. The hydrolysis reaction can be greatly influenced by the presence of an acid or a base. Conducted in the presence of an acid catalyst, the hydrolysis step involves the electrophilic attack from the proton on the alkoxide oxygen atom resulting oxygen having a positive charge. This leads the bond between the silicon center and the oxygen that was attacked to become more polarized and facilitates the departure of the alcohol to form the silanol bond. In the case of a base catalyzed reaction, the hydroxyl attacks the silicon atom, and the alcohol group leaves to form $HO\text{-}Si(OEt)_3$. The hydrolysis reaction is much faster under acidic condition than under basic.

The pH of the solution also affects the condensation rates of the reaction which happens in two steps. The first step is the electrophilic attack of the proton on the oxygen of the silanol group. The second step is the formation of the siloxane bond after the loss of the hydronium cation. Similarly under basic conditions, the condensation reaction also has two steps. The hydrogen on the silanol group gets deprotonated by the hydroxide ion leaving a

http://dx.doi.org/10.5772/intechopen.68529

negatively charged oxygen on the silanol. This results in the formation of the siloxane bonds. The condensation rate under basic conditions is faster than that under acidic conditions.

Because the hydrolysis and condensation rates differ depending on the catalyst used, the structures can also be controlled by using the right catalyst. For acid catalysis, the hydrolysis reaction is faster than the condensation reaction. As a consequence, a network with less siloxane bonds and more silanol groups yields more linearly branched polymeric species. This makes this catalyst the preferred method for the formation of thin films. On the other hand, the condensation reaction is much faster using base as a catalyst. Therefore, there are fewer silanol groups in the network. The network then consists of highly branched clusters, which yields denser materials. This is the method used for the formation of particles.

The typical mixture involves TMOS or TEOS, water, a solvent as methanol or ethanol followed by the addition of a catalyst (hydrochloric acid, for example). During the sol-gel formation, the viscosity of the solution gradually increases as the sol, which is a colloidal suspension of small particles with sizes about 1–100 nm that are well dispersed in a liquid. It becomes interconnected through polycondensation reactions to form a rigid, porous network that contains a phase of continuous liquid, the gel. The gelation stage will take place depending of the conditions as Si:H_2O ratio, type and concentration of catalyst, alkoxide precursors, etc. This stage can spend from seconds to minutes, or days to months. In the final stage, drying, the alcohol and water evaporation from the pores produces the gel shrinks.

The advantages of using the sol-gel process are (1) the method is performed at room temperature so molecules that could easily denature at higher temperatures can be added to the glasses [3], (2) it allows the easy fabrication of the materials in various configurations such as monoliths, powders, fibers, and films (**Figure 1**), (3) the material is transparent so spectroscopic studies can be used to probe the chemistry going on in the material, (4) the chemical properties of the sol-gel-derived oxides can be manipulated by incorporating organic, organometallic, and inorganic functional groups into the gel framework [4], (5) the materials are chemically, photochemically, and electrochemically stable, and (6) the matrix can stabilize the entrapped reagent from photodegradation, oxidation process, or not friendly environments.

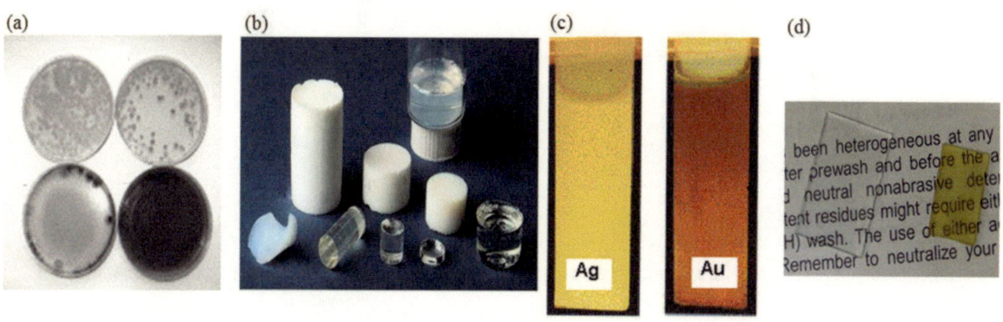

Figure 1. Picture of (a) monoliths doped with different concentrations of Ag, (b) monoliths of silica, TiO_2 and ZnO, (c) sol-gel films silica dip-coated onto glass wafers containing metallic nanoparticles, and (d) sol-gel films titania spin-coated onto glass wafers containing silver and iron nanoparticles.

Thin films are utilized as chemical sensors applications due to the short path length for diffusion. Bulk monoliths are often used for spectroscopic measurements due to their longer optical path length, while the powders are used in catalysis applications due to their high surface area.

As mentioned before, sol-gels can be made into powders, monoliths, films, and fibers. Monoliths can be prepared by casting the sol into a mold of any size or shape. The monolith undergoes gelation, aging, and drying to form the final xerogel. Films can be made by either spin coating [5] or dip coating [6] the sol onto a substrate, see **Figure 2**. Spin coating is a technique used since the 90s used to deposit thin films on different types of flat substrates. Placed a drop of liquid or semi-liquid precursor solution centered on the substrate (glass, silicon, etc.), which rotates at a high speed, around 3000 or 4000 rpm causing that the fluid is spread by the action of the centripetal acceleration. The thickness and properties of the final film depend on the precursor solution (viscosity, concentration, surface tension, etc.), as well as the spin-coater parameters as the rotational speed, time, and revolutions per minute (rpm). Any variation in the values of these parameters can alter the thickness and properties of the coated film. A precursor solution was placed on the glass wafers ($2.5 \times 2.5\ cm^2$) using a dropper and spun at a rate of 4000 rpm for 15 s, for example.

The technique of spin coating is widely used in the photolithography, in the manufacture of films with thicknesses lower than 10 nm, and to deposit photoresistances of near a micrometer wide.

Amorphous and crystalline ZnO thin films can be produced via a combined sol-gel route with spin-coating technique. First, a ZnO precursor solution can be prepared by sol-gel process. Then, this solution can be deposited on glass and silicon wafers by spin-coating technique to obtain amorphous thin films. Crystalline films are obtained by calcination process between 450 and 550°C to produce ZnO nanocrystals into the films [7].

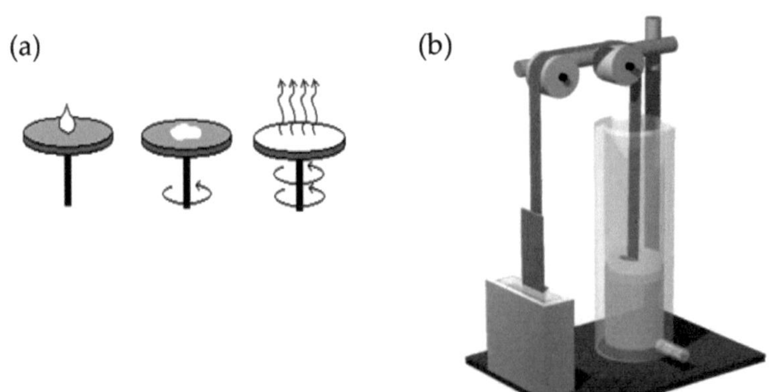

Figure 2. (a) Spin-coating technique [5] and (b) dip-coater device [6], both used to deposit sol-gel thin films on to silicon or glass substrates.

3. Photoluminescence

Luminescence is the process of emission of light by atoms or molecules excited. There are three types of luminescence procedures: *molecular fluorescence*, *phosphorescence*, and *chemiluminescence*. In these procedures, the molecules of the analyte are excited to give species whose emission spectrum provides information for qualitative or quantitative analysis.

A molecule in the ground state is raised to the excited state due to the absorption of photons that have enough energy. The excited molecules undergo a vibrational relaxation of the level of energy toward the lowest excited state by nonradiative processes and then return to ground state by emitting photons. The phenomena of fluorescence and phosphorescence fall within this definition, and therefore, the two phenomena are often referred by the general term *photoluminescence*. The fluorescence differs from phosphorescence by the electronic energy transitions which do not imply a change in electron spin. In terms of decay, the fluorescence is a radiative decay of the singlet state to the ground state occurring almost immediately, in about 1 ns. On the contrary, the time of half-life of triplet states is very large. This is why its decay to the ground state results in phosphorescence produced via one photon, since it can also decay by nonradiative processes. The chemiluminescence is based upon the emission spectrum of an excited species that is formed during a chemical reaction. Sometimes, the excited particles are produced by this reaction between the analyte and a suitable reagent (as a strong oxidant such as ozone or hydrogen peroxide), and a spectrum characteristic of the oxidation product of the analyte is observed instead of the analyte itself.

The main features of the luminescence methods are the sensitivity and selectivity. The luminescence tends to be three orders of magnitude smaller than those encountered in absorption. The selectivity of the luminescent is usually higher than that for absorption. There are many species capable of absorbing radiation, but the number of them that can reemit is much smaller; besides the excitation wavelength, as the emission can be selected to minimize interferences, while in spectrophotometry, only the wavelength of absorption is selectable.

Other advantage of photoluminescence method is their large linear concentration ranges, which are often significantly greater than those encountered in absorption methods. Finally, the selectivity of luminescence procedure is often better than that of absorption methods.

Luminescent materials have been used for more than a century for ionizing radiation detection, and the search for new and more efficient detector materials has been intense. Currently, a new promising direction of this research has emerged as the luminescence of nanocrystals. The basis for this is the observed luminescence intensity dependence on nanocrystal size [8]. The observed greater luminescence intensity for smaller size nanocrystals is explained by enhanced oscillator strength of the excitons [9].

3.1. Singlet/triplet excited states

A molecular electronic state in which all electrons spins are paired is called a *singlet* state, and no splitting of energy level occurs when the molecule is exposed to a magnetic field (the effects of the nuclear spin are negligible). On the other hand, the ground state for a free radical is a *doublet* state since the odd electron has two orientations in a magnetic field, resulting slightly different energies to the system.

A singlet or a *triplet* state is allowed if one of a pair of electrons of a molecule is excited to a higher energy level. In the excited single state, the spin of the promoted electron is still paired with the electron that is the ground state, while, in the triplet state, the spins of both electrons disappear. **Figure 3** shows a representation of both states.

3.2. Energy level diagrams for photoluminescent molecules

Figure 4 shows a partial energy level diagram for a typical photoluminescent molecule. The lowest heavy horizontal line represents the ground state energy of the molecule, which is a single state labeled as S_0. At room temperature, this state represents the energies of essentially all of the molecules in a solution.

The heavy lines above are energy levels for the ground vibrational states of three excited electronic states. The two lines on the left represent the first (S_1) and second (S_2) electronic singlet states. The one on the right (T_1) represents the energy of the first electronic triplet state. As usually happens, the energy of the first excited triplet state is lower than the energy of the corresponding singlet state. Several vibrational energy levels are associated with each of the four electronic states, as suggested by the lighter horizontal lines.

The excitation of this molecule can be brought about by absorption of two bands of radiation, one centered about the wavelength λ_1 ($S_0 \rightarrow S_1$) and the second around the shorter wavelength λ_2 ($S_0 \rightarrow S_2$). The direct excitation to the triplet state is not shown. This transition does not occur to any significant extent, because this process involves a change in multiplicity, and even that, it has a low probability of occurrence, it is called a forbidden transition.

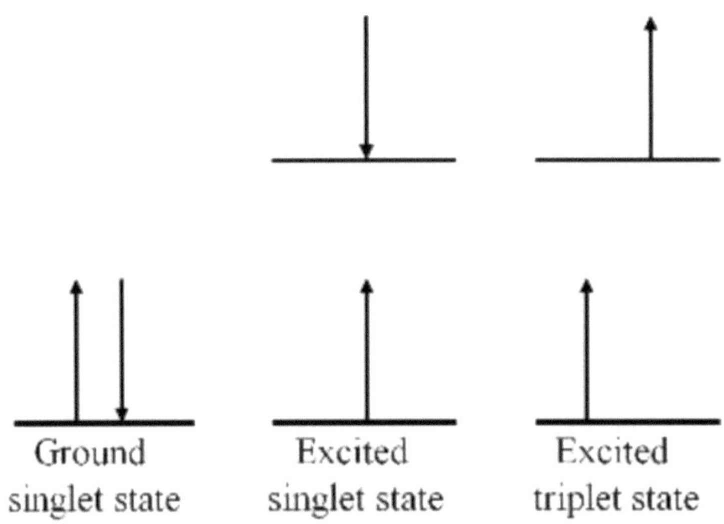

Figure 3. Schematic representation of different molecular electronic states.

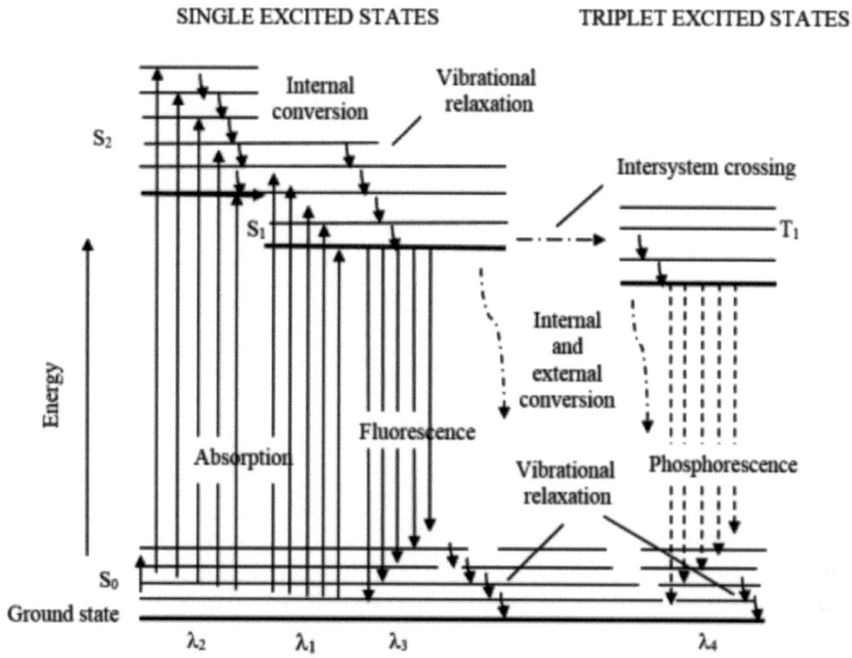

Figure 4. Partial energy diagram for a photoluminescent system.

The luminescent material ZnO exhibits two groups of luminescence bands: (i) in the region 1.8–2.4 eV due to defects and (ii) in the region 3.1–3.3 eV due to excitons [10]. The exciton luminescence lifetime is below 1 ns in ZnO [11], and therefore, it is possible to obtain a very fast scintillator. Luminescence of ZnO has been studied for thin films [12, 13], nanocrystals suspended in propanol [14], and nanocrystals embedded in solid matrices [15].

The luminescence can be influenced by the interface layer between a nanocrystal and the matrix [16]. ZnO nanocrystals (i.e., not embedded in any matrix) can modify the surface and luminescent properties.

4. Zinc oxide

Table 1 contains the main properties of the Zinc oxide (ZnO) [17–21], which is a semiconductor with a direct band gap of 3.37 eV at room temperature, and has a very stable even at room temperature which guarantee more efficient exciton emission at higher temperatures. The ionicity of this semiconductor resides at the borderline between covalent and ionic semiconductors. The difference of the electronegativities between the zinc and oxygen produces a high degree of ionicity in the bond. This produces an intense repulsion between the clouds of the charges from neighboring atoms with similar electrical loadings, making its crystalline structure more stable as hexagonal type, wurtzite. Due to its excellent chemical and thermal stability and specific optoelectronic properties, the ZnO has been used as

Property	Symbol (units)	Value
Lattice parameters	a, c, (Å)	3.253, 5.213
Coordination geometry	-	Tetrahedral
Fusion temperature	T_f (K)	>2250
Melting point	°C	2000
Density	ρ (g·cm^{-3})	5.675
Refractive index	nD	2.0041
Fractional ionic character	-	0.62
Std enthalpy of formation	ΔH (Jmol^{-1})	6.5×10^3
Std entropy of formation	ΔS (J·mol^{-1}·K^{-1})	100
Specific heat	C_p (J·mol^{-1}·K^{-1})	41
Thermal expansion coefficient	α_a (K^{-1}) α_o (K^{-1})	6.5×10^{-6} 3.0×10^{-6}
Thermal conductivity	λ (W·m^{-1}·K^{-1})	0.6
Elastic constants (300 K, 10 Gpa)	C_{11} (Pa) C_{12} (Pa) C_{13} (Pa) C_{33} (Pa) C_{55} (Pa) C_{66} (Pa)	20.70 11.77 10.61 20.95 4.48 0.45
Dielectric constants	$\varepsilon_{0\|\|}$, $\varepsilon_0\perp$ $\varepsilon_{\infty\|\|}$, $\varepsilon_\infty\perp$	8.75, 7.8 3.75, 3.70
Band gap (2 K) Band gap (300 K)	E_g (eV) E_g (eV)	3.42 3.35
Exciton binding energy	E_b (meV)	60
Effective mass of electrons	m_η	$0.28 \cdot m_0$
Effective mass of holes	m_ϱ	$0.58 \cdot m_0$

Table 1. Physical properties of ZnO.

photonic crystals, field emissions, photodetectors, photodiodes, light-emitting diodes, solar cells, varistors, gas sensors, and photocatalytic materials. Besides, the zinc oxide is a non-toxic material, it does not cause skin and eye irritation and there is no evidence of carcinogenicity, genotoxicity and reproduction toxicity in humans. There have been some concerns about the potential adverse effects on human health or the environment. Nevertheless, the current evidence shows that the ZnO particles or nanoparticles do not penetrate cells of the skin and remain on the outer layer of undamaged skin (the stratum corneum) with low systemic toxicity.

4.1. Technological applications

Zinc oxide has a wide variety of applications as luminescent material, such as vacuum fluorescent displays due to its room temperature ultraviolet emission and nonlinear optical properties. The combination of being a large-bandgap semiconductor and a luminescence material has allowed to study as nanostructured ZnO such as nanoparticles, nanowires, nanobelts, and nanotube [22]. Due to its crystalline structure, wurtzite, the ZnO has three exciton states called A, B, and C, whose energies are 3.3768 eV (367 nm), 3.3834 eV (366 nm), and 3.4223 eV (362 nm), respectively [21].

Specifically, the ZnO as thin film has relevant applications in ceramic technology and thin films technology [23–25]. ZnO nanocrystals with high stability at low processing temperatures have applications in displays, emitters, and sensors [23]. As films, exhibits a combination of piezoelectric, electrical, optical, and thermal properties, which have been already applied in devices as gas sensors, ultrasonic oscillators, and transparent electrodes in solar cells. Literature reports that pure and doped ZnO films have been produced by chemical vapor deposition, sputtering, spray pyrolysis, and the sol-gel process just to mention some methods of synthesis. At this point, our attention will be focus on the sol-gel route in combination with the dip-coating process to produce ZnO thin films, which represents an easy low cost and efficient route to coat large surfaces, permitting also the tailoring of the microstructure from the chemistry of the sol-gel synthesis [22, 24].

Special attention is deserved to the application as optoelectronic devices. On one hand, the fact of having the same crystal structure (with a difference of 2.0% and 0.5% for the parameters *a* and *c*, respectively) than the GaN [12], the material most studied so far for this application makes the ZnO an excellent substrate for heteroepitaxial growth. Its high exciton binding energy can permit the stimulated emission at temperatures higher than the environmental values (~550 K). All these properties have made that the study of ZnO has grown exponentially in recent years, especially after obtaining stimulated emission at room temperature in ZnO layers.

ZnO compounds doped with impurities are used as photoconductors in electrophotography, ceramic varistors, and sensor elements in the detection of combustible gases. As thin film, ZnO can exhibit piezoelectric properties, which can be used in pressure transducers and several acoustic-optical devices. ZnO doped with Al and I is used as transparent conductive electrodes to construct fluorescent screens because of their non-linear optical properties, and as solar cells made with ZnO, colloids show quantum size effects and luminescence. Furthermore, by it being ZnO oxide semiconductor material and the bandgap at room temperature, it becomes an emitter potential ultraviolet light at room temperature (such as ZnSe and GaN). Another important property of this material is the photocatalysis, which means that under illumination, this material can degrade organic pollutants, such as polluted water [25]. In this aspect, the ZnO shares this property with TiO_2 whose bandgap is 3.2 eV. But unlike the latter, the ZnO is not degraded by the photocorrosion. As gas sensor, it is known that ZnO is a good sensor of methanol, ethanol, propyl alcohol, LPG (Liquefied petroleum gas), carbon monoxide, and hydrogen.

The preparation as films has great advantages over other sensors. Since they can be manufactured in small sizes at large scale and at a lower cost, they are highly compatible with microelectronic and circuit technology, and additionally, incorporating impurities can give greater sensitivity ZnO film [22].

4.2. Crystal structure of ZnO: wurtzite

Under ambient conditions, the crystal structure of ZnO, wurtzite, is the most stable phase thermodynamically [26]. It has a hexagonal unit cell with two lattice parameters a and c, which ratio, and it is catalogue in the space group according to the Schöemflies notation and $P6_3mc$ in the Hermann-Mauguin notation. The parameters cell are a = b = 0, 381 nm and c = 0, 623 nm. Two hexagonal close-packed (hcp) sublattices interlace, each one consists of one type of atom displaced with respect to each other along the threefold c-axis by the amount of u = 3/8 = 0.375 in fractional coordinates. It consists of zinc and oxygen atoms tetrahedrally coordinated stacked according to the sequence ABABAB [27]. In such structure, the atoms are sufficiently far apart to compensate for the electrostatic repulsion. Each zinc atom is surrounded by a tetrahedron of four oxygen atoms and vice versa, which can be seen in **Figure 5**.

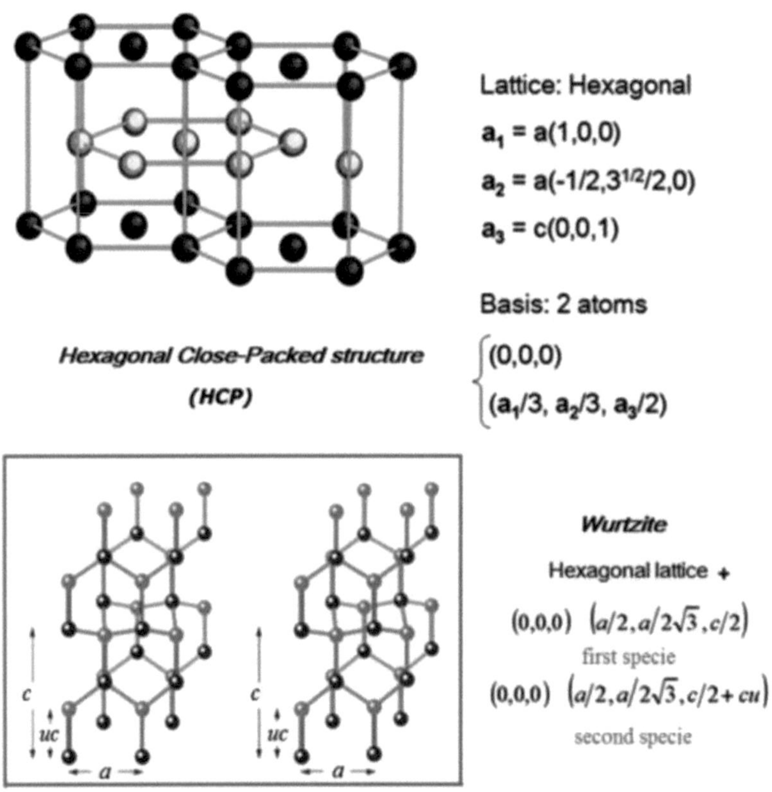

Figure 5. Structure of the primitive cell from ZnO. ● oxygen, ● zinc.

4.3. Optical properties of ZnO thin films

As mentioned above, the interest in ZnO has been motivated by possible optoelectronic applications. This is because it seems to be able to replace the compounds based on GaN that currently are used in optoelectronic devices as LEDs, laser diodes, and photodetectors that operate in the blue or UV range being cheaper and nontoxic. Other important factor that influences in this selection is the similarity of its band gap energy (3.37 eV at room temperature) with that of GaN (3.39 eV at room temperature) and the exciton binding energy of ZnO (60 meV) which is larger than that of GaN (18–28 meV). It can be useful in light-emitting devices. Emission properties of ZnO nanocrystals are influenced by many factors such as synthesis method, morphology of the crystals, dopants, and ligands used for surface coating.

4.3.1. Infrared spectra

Typical absorption spectra from amorphous and crystalline ZnO thin films taken at room temperature in the range of 300–900 nm are shown in **Figure 6**. The absorption spectrum of the amorphous film does not exhibit any band. After calcination process between 400 and 500°C [25, 26, 28], the spectrum of the crystalline film shows a narrow absorption peak A located around at 359 nm due to free excitons [21].

4.3.2. Optical absorption spectra

Other optical characterization on ZnO films is performance by infrared spectroscopy, shown in **Figure 7**. For amorphous and crystalline films, similar absorption bands were observed, and their respective assignment is described at **Table 2**. After the calcination treatment, the absorption bands located at 1342, 1410, 1557, 2900, 2945, and 3263 cm^{-1} disappeared.

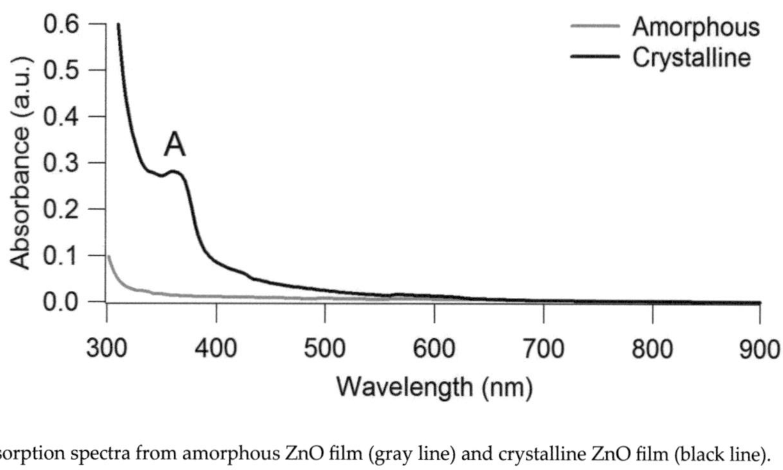

Figure 6. Absorption spectra from amorphous ZnO film (gray line) and crystalline ZnO film (black line).

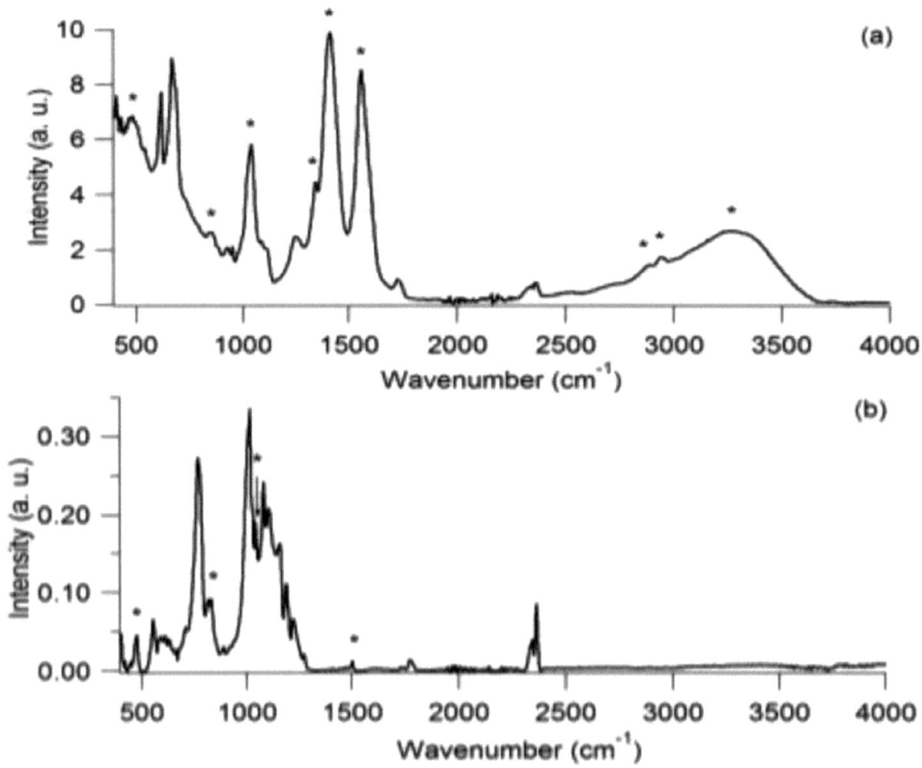

Figure 7. IR spectra of (a) amorphous ZnO film and (b) crystalline ZnO film.

ν_{exp} (cm^{-1}) Amorphous	ν_{exp} (cm^{-1}) Crystalline	Assignment	Ref.
478	476	ZnO stretching	[29]
854	842	Symm. bending ZnO	[29]
1042	1038	Asymm. stretching ZnO	[29]
1342	-	HOOC-R	[30]
1410	-	C-O	[31]
-	1502	Stretching mode of $ZnO{\cdot}H_2O$	[29]
1557		C=O	[31]
2900	-	C-H stretching	[30]
2945	-	C-H stretching	[30]
3263	-	H-O species	[29]

Table 2. IR frequencies [in cm^{-1}], of the amorphous and crystalline ZnO films.

4.4. Photoluminescence studies on ZnO thin films

Photoluminescence studies have been performed on ZnO thin films by different emissions as near ultraviolet, red, green, and blue. However, the recombination centers and possible mechanisms to explain the mentioned emissions are not clear totally. A crucial factor is the fact that the luminescence of the ZnO is very sensitive to its surface state determined and modulated by the synthesis method. Also, the kind of substrate can influence the luminescence of the film ZnO deposited on it. Although these factors can lead to different optical phenomena, in general, all published works coincide in the assignment of the emission bands mentioned.

Particularly, the sol-gel method can be an appropriated route to maximize the luminescent properties of the ZnO films. Typical room temperature luminescence emission spectra of ZnO nanocrystals (crystalline film) synthesized by sol-gel method are shown in **Figure 8**. The luminescent response of these nanocrystals, the amorphous film, and the silicon wafer was compared. The excitation wavelength of ZnO was 359 nm, which coincides with the maximum absorption (**Figure 6**). The D and E bands located at 561 nm and 626 nm correspond only to the substrate, silicon wafer. Three main luminescent bands from ZnO were identified as A, B, and C. **Table 3** summarizes the maximum emission peaks for amorphous and crystalline ZnO films. Their luminescence behavior is identified with a small violet emission A band and a narrow blue emission B and C bands.

The visible emission of the ZnO films is related to different intrinsic defects, such as oxygen vacancies, zinc vacancies, zinc interstitials, oxygen interstitials, and anti-site defect O_{Zn} [32]. The luminescence emission characteristics of ZnO films are dependent on both the crystal quality of the film and the film stoichiometry [32].

From **Figure 8**, among the most notable differences is the shape of all bands, which is more defined from crystalline film than amorphous film. The intensity of the photoluminescent peak arises for crystalline film in comparison to the amorphous one [33]. The crystallization of the ZnO film increases the visible emissions due to the increase of the excess oxygen

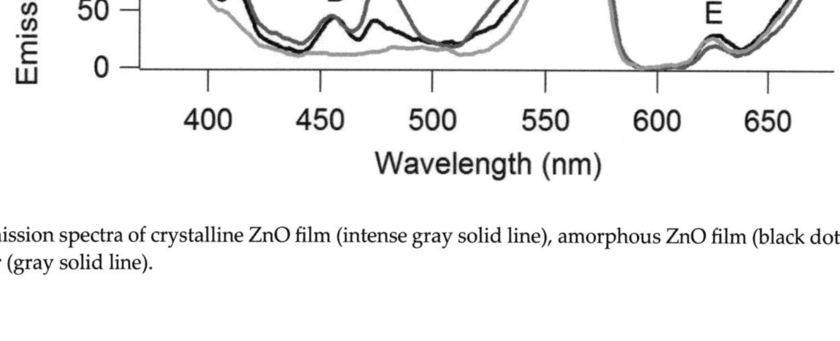

Figure 8. Emission spectra of crystalline ZnO film (intense gray solid line), amorphous ZnO film (black dotted line), and silicon wafer (gray solid line).

Film	Maximum emission peaks		
Amorphous	413 nm (3.0 eV)	456 nm (2.7 eV)	475 nm (2.6 eV)
Crystalline	411 nm (3.0 eV)	455 nm (2.7 eV)	474 nm (2.6 eV)

Table 3. Maximum emission peaks from amorphous and crystalline ZnO films.

concentrations and oxygen interstitials [4]. A slight blue shift means an improvement in the emissions [33, 34].

The violet emission peak located at 413–411 nm is due to the electron transition from the conduction band edge to the Zn vacancy level [33]. **Figure 9** shows the calculated energy levels of the intrinsic defects in ZnO applying full-potential linear muffin-tin orbital method [35]. Since it can be observed, the energy of the transition from conduction band to the Zn vacancies (V_{Zn}) corresponds to a violet emission with an energy of 3.0 eV [36]. Probably, these vacancies are on the grain boundaries. Other defects are the oxygen interstitial and oxygen antisite (O_{Zn}) which also favor the formation of an oxygen-rich environment.

The blue emission band observed at 456–455 nm (2.7 eV) could be produced by the electron transition from the shallow donor level of oxygen vacancies [28] and zinc interstitials [37, 38] to the valence band. Oba et al. report that the oxygen vacancies and zinc interstitials possess formation energies [37–39]. Oxygen vacancies produce two defect donor levels: One is the deep donor level located at 1.3–1.6 eV below the conductor band [40, 41] and the other is the shallow donor level below the conductor band in the range of 0.3–0.5 eV [41]. Valverde et al. calculated photon energy of the 2.7 eV from blue emission, which coincides with the value of 2.8 eV for the energy interval from the shallow donor level to the top of the valence band. Also, the energy level of zinc interstitials was located at 2.9 eV above the valence band for the samples with an energy gap of 3.36 eV [40]. For ZnO film with an energy gap of 3.4 eV, the energy level of zinc interstitials was around 2.72 eV above the valence band, which corresponds to a blue emission with a photon energy of the 2.7 eV (**Figure 9**). Finally, the blue band emission located at 475–474 nm (2.6 eV) from ZnO has been reported, but its exact origin has not been explained yet.

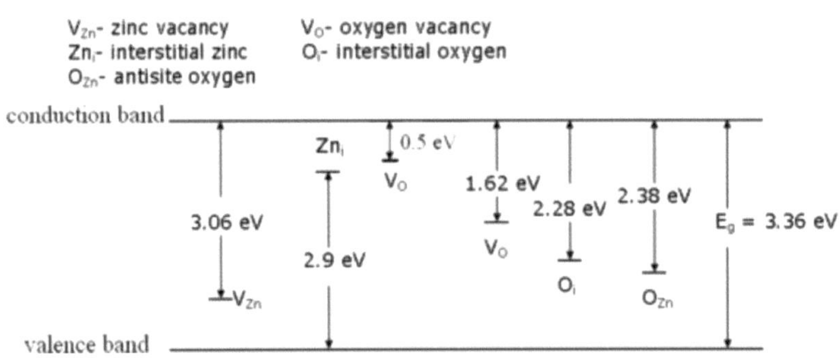

Figure 9. Schematic representation of calculated defect level in ZnO nanostructures.

Zinc oxide has been an important industrial material for centuries and is currently the subject of considerable new interest. As luminescent material, it has potential technological applications. Its different emissions as near ultraviolet, red, green, and blue can be substantially altered by the kind of substrate where the film is deposited and the synthesis method. Without a doubt, the sol-gel method is a friendly route to obtain uniform and stable zinc oxide films with high optical quality. Up to this time, the assignment of the emission bands is similar for different synthesis and substrates.

Acknowledgements

The authors thank the financial support to SIP 20170002, SIP 20170004, SIP 20170208, SIP 20170435, CONACyT 180379 and CONACyT 169157.

Author details

Guadalupe Valverde Aguilar*, Mónica R. Jaime Fonseca, Ángeles Mantilla Ramírez and Antonio G. Juárez Gracia

*Address all correspondence to: mvalverde@ipn.mx

CICATA-Legaria, Instituto Politécnico Nacional, Cd. de México, México

References

[1] Marci G, Augugliaro V, López-Munoz MJ, Martin C, Palmisano L, Rives V. Preparation, characterization and photocatalytic activity og polycrystalline ZnO/TiO_2 systems. Journal of Physical Chemistry B. 2001;**105**(5):1026-1032. DOI: 10.1021/jp003173j

[2] Hsu CC, Wu NL. Synthesis and photocatalytic activity of ZnO/ZnO_2 composite. Journal of Phochemistry and Photobiology A: Chemistry. 2005;**172**(3):269-274. DOI: 10.1016/j.jphotochem.2004.12.014

[3] Zhang DK, Liu YC, Liu YL, Yang H. The electrical properties and the interfaces of CuO/ZnO/ITO p-i-n heterojunction. Physica B: Condensed Matter. 2004;**351**(1):178-183. DOI: 10.1016/j.physb.2004.06.003

[4] Djurišić AB, Leung YH, Tam KH, Hsu YF, Ding L, Ge WK, Zhong YC, Wong KS, Chan WK, Tam HL, Cheah KW, Kwok WM, Phillips DL. Defect emissions in ZnO nanostructures. Nanotechnology. 2007;**18**(9):095702. DOI: 10.1088/0957-4484/18/9/095702

[5] Valverde-Aguilar G, García-Macedo JA, Juárez-Arenas R. Photoconductivity on nanocrystalline $ZnO\text{-}TiO_2$ thin films obtained by sol-gel. Proceedings of SPIE. 2008;**7041**:1-9. DOI: 10.1117/12.795461

[6] García-Macedo J, Valverde-Aguilar G, Gómez RW, Pérez- Mazariego JL, Marquina V. Optical properties of nanostructured sol-gel thin films doped with Fe_2O_3 and the study of their ferromagnetic properties by Mössbauer spectroscopy. Journal of Nanoscience and Nanotechnology. 2008;**8**(12):6491-6496. DOI: 10.1166/jnn.2008.019

[7] Yamamoto A, Atsuta S, Kanemitsu Y. Fabrication of ZnO nanocrystals dispersed in glass films for low-temperature optical studies. Physica E. 2005;**26**(1-4):96-99. DOI: 10.1016/j.physe.2004.08.031

[8] Pan Z, Zhang P, Tian X, Cheng G, Xie Y, Zhang H, Zeng X, Xiao C, Hu G, Wei Z. Properties of fluorine and tin co-doped ZnO thin films deposited by sol-gel method. Journal of Alloys Compounds. 2013;**576**:31-37. DOI: 10.1016/j.jallcom.2013.04.132

[9] Mohanta SK, Lee SH, Kong BH, Cho HK. Behavior of ultraviolet emission from nano-crystalline embedded ZnO film synthesized by solution-based route. Journal of Crystal Growth. 2009;**311**(6): 1539-1544. DOI: 10.1016/j.jcrysgro.2009.01.104

[10] Tai W-P, Oh J-H. Humidity sensing behaviors of nanocrystalline Al-doped ZnO thin films prepared by sol-gel process. Journal of Materials Science: Materials in electronics. 2002;**13**(7): 391-394. DOI: 10.1023/A:1016084309094

[11] Brinker CJ, Scherer GW. Sol-Gel Science: The Physics and Chemistry of Sol–Gel Processing. San Diego, CA: Academic Press; 1990. ISBN: 978-008057103-4

[12] Hench LL, West JK. The Sol-Gel Process. Chemical Reviews. 1990;**90**(1):33-72. DOI: 10.1021/cr00099a003

[13] Dave BC, Dunn B, Valentine JS, Zink JI. Sol-Gel Encapsulation methods for biosensors. Analytical Chemistry. 1994;**66**(22): 1120A-1127A. DOI: 10.1021/ac00094a001

[14] Dunn B, Zink JI. Optical properties of sol-gel glasses doped with organic molecules. Journal of Materials Chemistry. 1991;**1**(6): 903-913. DOI: 10.1039/JM9910100903

[15] Kachurin GA, Tyschenko IE, Zhuravlev KS, Pazdnikov NA, Volodin VA, Gutakovsky AK, Leier AF, Skorupa W, Yankov RA. Visible and near infrared luminescence from silicon nanostructures formed by ion implantation and pulse annealing. Nuclear Instruments and Methods in Physics Research B: Beam Interactions with Materials and Atoms. 1997;**122**(3):571-574. DOI: 10.1016/S0168-583X(96)00764-1

[16] Kayakuma Y. Quantum size effects of interacting electrons and holes in semiconductor microcrystals with spherical shape. Physical Review B. 1988;**38**(14-15):9797-9805. DOI: http://dx.doi.org/10.1103/PhysRevB.38.9797

[17] Urbieta A, Fernandez P, Hardalov Ch, Piqueras J, Sekiquchi T. Cathodoluminescence and scanning tunneling spectroscopy of ZnO single crystals. Materials Science and Engineering: B. 2002;**91-92**: 345-348. DOI: 10.1016/S0921-5107(01)01062-5

[18] Xiong G, Wilkinson J, Lyles J, Ucer KB, Williams RT. Luminescence and stimulated emission in zinc oxide nanoparticles, films, and crystals. Radiation Effects and Defects in Solids. 2003;**158**(1-6):83-88. DOI: 10.1080/1042015022000037607

[19] Zu P, Tang ZK, Wong GKL, Kawasaki M, Ohtomo A, Koinuma H, Segawa Y. Ultraviolet spontaneous and stimulated emissions from ZnO microcrystallite thin films at room temperature. Solid State Communications. 1997;103(8): 459-463. DOI:10.1016/S0038-1098(97)00216-0

[20] Li BS, Liu YC, Zhi ZZ, Shen DZ, Lu YM, Zhang JY, Fan XW. The photoluminescence of ZnO thin films grown on Si (100) substrate by plasma-enhanced chemical vapor deposition. Journal of Crystal Growth. 2002;**240**(3-4):479-483. DOI: 10.1016/S0038-1098(97)00216-0

[21] van Dijken A, Meulenkamp EA, Vanmaekkelberh D, Meijerink A. The luminescence of nanocrystalline ZnO particles: the mechanism of the ultraviolet and visible emission. Journal of Luminescence. 2000;**87-89**:454-456. DOI: 10.1016/S0022-2313(99)00482-2

[22] Liu YX, Liu YC, Shen DZ, Zhong GZ, Fan XW, Kong XG, Mu R, Henderson DO. The structure and photoluminescence of ZnO films prepared by post-thermal annealing zinc-implanted silica. Journal of Crystal Growth. 2002;**240**:152-156. DOI: 10.1016/S0022-0248(02)00843-6

[23] Hirai T, Harada Y, Hasimoto S, Edamatsu K, Itoh T. Lifetimes of excitons in ZnO fine particles dispersed in alkali halide crystals. Journal of Luminescence. 2001;**94-95**:261-265. DOI: 10.1016/S0022-2313(01)00290-3

[24] Valverde-Aguilar G, Manríquez Zepeda JL. Photoluminescence and Photoconductivity studies on amorphous and crystalline ZnO thin films obtained by sol-gel method. Applied Physics A. 2015;**118**(4):1305-1313. DOI: 10.1007/s00339-014-8836-y

[25] Valle GG, Hammer P, Pulcinelli SH, Santilli CV. Transparent and conductive ZnO:Al thin films prepared by sol-gel dip-coating. Journal of the European Ceramic Society. 2004;**24**(6):1009-1013. DOI: 10.1016/S0955-2219(03)00597-1

[26] Wang Y-de, Zhang S, Wu X-hui. Green Light Luminescence from ZnO/Dodecylamine Mesolamellar Nanocomposites synthesized by Self-Assembly. European Journal of Inorganic Chemistry. 2005;**2005**(4):727-731. DOI: 10.1002/ejic.200400450

[27] Pal B, Sharon M. Enhanced photocatalytic activity of highly porous ZnO thin films prepared by sol-gel process. Materials Chemistry and Physics. 2002;**76**(1):82-87. DOI: 10.1016/S0254-0584(01)00514-4

[28] Wang Z, Zhang H, Wang Z, Zhang L, Yuan J. Structure and strong ultraviolet emission characteristics of amorphous ZnO films grown by electrophoretic deposition. Journal of Materials Research. 2003;**18**(1):151-155. DOI: http://dx.doi.org/10.1557/JMR.2003.0021

[29] Tokumoto MS, Briois V, Santilli CV, Pulcinelli SH. Preparation of ZnO Nanoparticles: Structural Study of the Molecular Precursor. Journal of Sol-gel Science and Technology 2003;**26**(1):547-551. DOI: 10.1023/A:1020711702332

[30] Kamalasanan MN, Chandra S. Sol-Gel synthesis of ZnO thin films. Thin Solid Films. 1996;**288**(1-2):112-115. DOI:10.1016/S0040-6090(96)08864-5

[31] Jeong S-H, Kim B-S, Lee B-T. Photoluminescence dependence of ZnO films grown on Si(100) by radio-frequency magnetron sputtering on the growth ambient. Applied Physics Letters. 2003;**82**(16):2625-2627. DOI: 10.1063/1.1568543

[32] Nadarajah K, Chee CY, Tan CY. Influence of annealing on properties of spray deposited ZnO thin films. Journal of Nanomaterials. 2013;**2013**:Article ID 146382, 8 pages. DOI: 10.1155/2013/146382

[33] Chen S, Zhao Z, Tang BZ, Kwok HS. Growth methods, enhanced photoluminescence, high hydrophobicity and light scattering of 4, 4′-bis(1, 2, 2-triphenylvinyl)biphenyl nanowires. Organic Electronics. 2012;**13**(10):1996-2002. DOI:10.1016/j.orgel.2012.06.014

[34] Sun YM. Study on the electronic structure of ZnO and its defects by using FP-LMTO method [thesis]. University of Science and Technology of China; 2000.

[35] Kale RB, Hsu Y-J, Lin Y-F, Lu S-Y. Synthesis of stoichiometric flowerlike ZnO nanorods with hundred percent morphological yield. Solid State Communications. 2007;**142**(5):302-305. DOI: 10.1016/j.ssc.2007.02.022

[36] Zhang DH, Xue ZY, Wang QP. The mechanisms of blue emission from ZnO films deposited on glass substrate by r.f. magnetron sputtering. Journal of Physics D: Applied Physics. 2002;**35**(21): 2837-2840. DOI: 10.1088/0022-3727/35/21/321

[37] Oba F, Nishitani SR, Isotani S, Adachi H, Tanaka I. Energetics of native defects in ZnO. Journal of Applied Physics. 2001;**90**(2):824-828. DOI:10.1063/1.1380994

[38] Kumar V, Kumar, V, Som S, Yousif A, Singh N, Ntwaeaborwa OM, Kapoor A, Swart HC. Effect of annealing on the structur, morphological and photoluminescence properties of ZnO thin films prepared by spin coating. Journal of Colloid and Interface Science. 2014;**428**:8-15. http://dx.doi.org/10.1016/j.jcis.2014.04.035

[39] Lin BX, Fu ZX, Jia YB. Green luminescent center in undoped zinc oxide films deposited on silicon substrates. Applied Physics Letters. 2001;**79**(7):943-945. DOI: 10.1063/1.1394173

[40] Xu PS, Sun YM, Shi CS, Xu FQ, Pan HB. Native Point Defect States in ZnO. Chinese Physics Letters. 2001;**18**(9):1252-1253. DOI: 10.1088/0256-307X/18/9/331

[41] Fu ZX, Guo CX, Lin BX, Liao GH. Cathodoluminescence of ZnO Films. Chinese Physics Letters. 1998;**15**(6):457-459. DOI: 10.1088/0256-307X/15/6/025

Chapter 11

Sol-Gel Materials Used in Phosphoproteomics and Glycoproteomics Applications

Hacı Mehmet Kayili, Mehmet Atakay and Bekir Salih

Additional information is available at the end of the chapter

http://dx.doi.org/10.5772/intechopen.68891

Abstract

Glycosylation and phosphorylation are two of the most commonly seen and important post-translational modifications (PTMs) of proteins. Both play key role in many vital biological processes. Mass spectrometry is the most favored technique used for monitoring the dynamics of PTMs. Mass spectrometric analysis of phosphopeptide and glycopeptide is a crucial step in phosphoproteomics and glycoproteomics to understand the role of these modifications in the cellular pathways. Enrichment and purification of (phospho- and glyco-) peptides and glycans are recommended prior to mass spectrometric analysis because of the lower amount of modified peptides in a peptide mixture. Herein, we present titanium/silica and zirconium/silica sol-gel materials for the enrichment of (phospho- and glyco-) peptides and glycans to enhance MALDI-MS analysis performance. Enrichment of (phospho- and glyco-) peptides was successfully applied using standard proteins including β-casein, fetuin, and IgG as well as some complex medium. The sol-gel materials were compared with commercial metal oxides regarding their modified peptide enrichment performances.

Keywords: sol-gel materials, phosphoproteomics, glycoproteomics, mass spectrometry, enrichment

1. Introduction

The post-translational modification (PTM) of proteins plays a critical role in regulating many cellular pathways [1, 2]. There is a wide range of chemical modifications that have been observed in a protein after protein synthesis and these modifications are vital for protein function, localization, and turnover [3, 4]. The modifications that occur in protein structures after the translation step control many biological activities, so it is very important to examine the roles of these modifications in cell regulation mechanisms. Functional diversity of proteins is increased by PTMs. Also, aberrant posttranslational modifications can cause several cellular diseases. Therefore, analysis of modified proteins is necessary to better understand the relationship between PTMs and functional changes. The detailed analysis of proteins in complex samples and the determination of the locations of modified amino acids are very important in terms of the understanding of cellular signaling mechanisms [5, 6].

Glycosylation and phosphorylation are two of the most commonly seen PTMs of proteins. Both are involved in many biological events, such as immune response, signal transduction, and cell-to-cell interaction [1, 6, 7]. Phosphoproteomics and glycoproteomics are the significant proteomics areas in which phosphorylated and glycosylated proteins are identified and characterized [2, 7, 8]. Many researches have been made in these fields in order to discover new biomarkers and drug targets [9]. Analysis performed for the characterization of (phospho- and glyco-) proteins needed huge efforts and intense operations. Mass spectrometry-based (phospho- and glyco-) proteomics studies using modern mass spectrometers enable fast and sensitive characterization of modified proteins [8, 10, 11]. Bottom-up approach is the most used application for the analysis of (phospho- and glyco-) proteins by mass spectrometry. By this approach, proteins of interest are enzymatically digested into complex peptide mixtures and then the peptide fragments are either separated using a chromatographic technique prior to mass spectrometric analysis or directly analyzed by a mass spectrometer [12]. Several difficulties are observed in the analysis of phosphorylated and glycosylated peptides by mass spectrometry. The abundance of (phospho- and glyco-) peptides in the peptide pool is usually less than ordinary peptides. Because the large amount of ordinary peptides generated by the proteolytic digestion of proteins, the ion signals of (phospho- and glyco-) peptides are suppressed. Moreover, the ionization efficiency of these modified peptides is low in mass spectrometric analysis.

Thus far, numerous enrichment and separation techniques have been developed in order to analyze both (phospho- and glyco-) peptides. Also, various approaches including covalent and noncovalent interaction-based methods can be used for the enrichment of (phospho- and glyco-) peptides. These analytical methods and strategies based on different interaction modes have been reviewed in detail [13–20]. Since these methods are complementary to each other, they can be applied together to improve enrichment and separation efficiency.

Sol-gel process is a chemical synthesis method based on inorganic polymerization reactions [21]. One of the advantages of sol-gel process is that little or no heating is required during reaction. The reaction is carried out in two steps: hydrolysis of metal alkoxides and polycondensation of hydroxyl groups. Consequently, three-dimensional sol-gel network is formed. The resulting sol-gel material is in the form of a solid polymeric network structure with chains having diameters in the micrometer size and chains with average lengths greater than a micrometer [22]. In a general sol-gel method, the precursors with lower molecular weight such as tetramethoxysilane (TMOS) or tetraethoxysilane (TEOS) are hydrolyzed in the presence of water, acid catalyst, and proper solvents. Titanium(IV) isopropoxide, tantalum(V) ethoxide, and zirconium(IV) ethoxide are widely used metal alkoxide precursors for producing metal-based sol-gel materials. Metal alkoxides of titanium, zirconium, and tantalum are more reactive than alkoxysilanes due to their lower electronegativity and higher Lewis acidity [23].

Sol-gel materials are of interest because of their potential to be produced with tailor-made pore size and shape. They have many advantages over organic polymers [24]. For example, the thermal stability of organic polymers is too low to compare with the sol-gel materials. In the sol-gel synthesis method, the control of the thickness, porosity, and surface area is easier [25, 26]. The sol-gel materials can be synthesized very purely in the appropriate temperature range and modified in mild conditions. The compositions of hybrid sol-gel materials can be well controlled. Sol-gel-based materials have been used for various bioapplications, such as enzyme encapsulation [27], immobilization [28], and biosensor [29]. Recently, our group has introduced tantalum-based and amine-functionalized sol-gel materials to enrich phosphopeptides from complex medium [30, 31]. Sol-gel materials that are fully compatible for the entire enrichment studies are not yet available. Methods and materials for enriching phosphopeptides and glycopeptides have to be selected depending on the sample to be analyzed.

In this chapter, we describe specific enrichment methods for the phosphopeptides and glycopeptides using titanium/silica and zirconium/silica sol-gel materials. Several standard (phospho- and glyco-) proteins including beta-casein, fetuin, and IgG were used to test enrichment efficiency of the materials prior to MALDI-MS analysis. Also, 2-AA-labeled human plasma N-glycans were analyzed by MALDI-MS after purification with titanium/silica and zirconium/silica sol-gel materials, allowing it to apply for further proteomics applications.

2. Materials and methods

2.1. Materials

If not otherwise stated, all the reagents, solvents, and proteins were purchased from Sigma-Aldrich (St. Louis, MO, USA) and used without further purification. Sequencing grade trypsin was obtained from Roche Diagnostic GmbH (Mannheim, Germany). Human plasma (lyophilized, P9523) was bought from Sigma-Aldrich (St. Louis, MO, USA). Titanium(IV) oxide with a 21 nm primary particle size [transmission electron microscopy (TEM)] and ≥99.5% purity (trace metals basis) and zirconium(IV) oxide powder, 5 μm, 99% (trace metals basis) were purchased from Sigma-Aldrich (St. Louis, MO, USA). Purified water used in all experiments was obtained using an Expe-Ultrapure Water System (Mirae St. Co., Korea).

2.2. Methods

2.2.1. Synthesis of the titanium/silica and zirconium/silica sol-gel materials

A solution of ethanol and water was mixed (2.0 mL, ethanol/water, 5/1, v/v) in a reaction vessel. Titanium(IV) isopropoxide and TMOS (tetramethyl orthosilicate) were mixed with a molar ratio of 5:1 and added to the reaction vessel. After stirring for 20 min, 20 μL of fuming nitric acid was added drop wise to the sol. The mixture was then stirred overnight to produce a wet gel. The wet gel was then washed with a mixture of water and ethanol to remove residues and dried at 50°C in vacuum prior to use. Some amount of the gel powder was calcined at 500°C for 2 h. The same protocol was also applied as described above to synthesize the zirconium/silica sol-gel material using zirconium(IV) ethoxide instead of titanium(IV) isopropoxide.

2.2.2. Proteolytic digestion of standard proteins

One milligram of fetuin, IgG, bovine serum albumin (BSA), and β-casein were separately dissolved in 78 μL of 25 mM ammonium bicarbonate (ABC) prior to digestion. The reduction of cysteine residues was achieved by the addition of 2 μL of DTT (1,4-dithiothreitol, 200 mM in 25 mM ABC) to these solutions and the samples were incubated at 60°C for 30 min. After adding 6.5 μL of IAA (2-iodoacetamide, 200 mM in 25 mM ABC) to the solutions for alkylation of reduced cysteine residues, the samples were incubated in the dark at room temperature for 20 min. Finally, proteomics grade trypsin was added to these solutions (protein/enzyme, 30/1, w/w) and the samples were incubated overnight at 37 °C. The final solutions of the digested samples were diluted to a proper concentration for the enrichment studies.

2.2.3. N-glycan release

Human plasma (glyco-) proteins (3 mg) were denatured by the addition of 100 μL of 2% SDS and the sample was incubated at 60°C for 10 min. Then, 40 μL of 4% NP-40 and 4 U of PNGase F in 40 μL of 5× phosphate buffered saline (PBS) were added to this solution. The sample was incubated overnight at 37°C in order to release N-glycans.

2.2.4. Labeling of human plasma N-glycans with 2-aminobenzoic acid (2-AA)

The released human plasma N-glycans were labeled by 2-AA as described previously with minor modifications [32]. The released glycan sample (50 μL) was mixed with 25 μL of 2-AA (47 mg/mL) prepared in DMSO/glacial acetic acid (17:3, v/v) and 25 μL of 2-picoline borane (107 mg/mL in DMSO). The incubation was performed at 65°C for 2 h. The reaction mixture was then cooled to room temperature and used for 2-AA human N-glycan analysis by MALDI-TOF-MS after purification with titanium/silica and zirconium/silica sol-gel materials.

2.2.5. Phosphopeptide enrichment by titanium/silica and zirconium/silica sol-gel materials

Five milligram of the materials was washed three times with methanol prior to the enrichment process. The washed materials were then equilibrated three times with 1000 μL of loading solution (ACN/water (1:1, (v/v) containing 1% (v/v) TFA). The mixture of the tryptic digests of BSA and β-casein at a 50:1 mole ratio (BSA:β-casein) was diluted in 50 μL of the loading solution and mixed with the material for 30 min at room temperature. The supernatant was then collected for further processes and the sol-gel particulates were washed with 1000 μL of the loading solution to remove unmodified peptides. The elution of phosphopeptides was achieved using 20 μL of 50 mM ABC solution adjusted to pH 10–11 with the addition of ammonia.

2.2.6. Glycopeptide enrichment and N-glycan purification by titanium/silica and zirconium/silica sol-gel materials

Three milligram of the materials were washed two times with 800 μL of loading buffer, which is composed of ACN/water/TFA, 80:15:5, v/v/v containing 1 M glycolic acid. Subsequently, 5 μL of fetuin and IgG digests (from 1-μg/μL digest samples) were individually diluted in 40 μL of the loading buffer and incubated with the materials for 20 min at room temperature. The supernatant was then collected for further processes and the sol-gel particulates were washed with 800 μL of the loading buffer, washing buffer 1 (ACN/water/TFA, 80:19:1, v/v/v) and washing buffer 2 (ACN/water/TFA, 20:79.9:0.1, v/v/v) to remove interferences. Before elution step, the sol-gel particulates were dried in a speedVac for 5 min. The elution of glycopeptides was achieved using 20 μL of 25 mM ABC solutions adjusted to pH 10–11 with ammonia.

The 2-AA-labeled N-glycans were purified by the materials (3 mg) with a similar method as described above. Briefly, 5 μL of 2-AA-labeled human plasma N-glycan sample was mixed with 40 μL of loading buffer (ACN/water/TFA, 88:11.8:0.2, v/v/v) and incubated with the materials for 10 min at room temperature. After the incubation step, the sol-gel particulates were washed with 800 μL of the loading buffer. Then, the sol-gel particulates were dried with a speedVac for 5 min. Finally, elution was then performed by the addition of 20 μL of 25 mM ABC solutions adjusted to pH 10–11 with ammonia.

2.2.7. MALDI-TOF-MS analysis

A Voyager-DE PRO MALDI-TOF-MS (Applied Biosystems, USA) containing a 337-nm nitrogen laser was used for the analysis. All spectra except for purified 2-AA-labeled N-glycan analysis were obtained in positive ion linear mode (mass range from m/z 1000 to 5000) using an acceleration potential of 20 kV. The instrument was externally calibrated using a standard peptide mixture. A 2,5-dihydroxybenzoicacid (DHB) matrix solution (10 mg/mL) was prepared in ACN:water (1:1, v/v) containing 0.1% orthophosphoric acid). Eluates (1 μL) were spotted onto MALDI target plate and co-crystallized with 1 μL of the DHB MALDI matrix.

3. Results and discussion

3.1. Enrichment of phosphopeptides from the mixture of β-casein/BSA digests

The noncovalent interaction of the phosphate group with metal cations or metal oxides is utilized in the commonly used phosphopeptide enrichment methods [33, 34]. Surfaces of the materials such as titanium dioxide (TiO_2) and zirconium dioxide (ZrO_2) become positively charged at low pH, making them suitable for binding of phosphopeptides. TiO_2 is the most commonly used metal oxide in the phosphopeptide enrichment methods [20, 35]. Oxides of various metals other than titanium are also used in the phosphopeptide enrichment studies [31, 36–41]. The phosphate groups of phosphopeptides form coordination with metal ions present on the metal oxide surface at acidic pH. Herein, the phosphate groups act as the electron pair donor bidentate ligand while metals are the corresponding electron pair acceptor. The metal oxides used in the phosphopeptide enrichment studies are the oxides of the transition metals [42]. However, acidic nonphosphopeptides may also interact with the metal oxide surface at similar conditions. These undesired interactions lower the specificity of the phosphopeptide enrichment method. The interactions of phosphopeptides with metal oxides vary depending on the pH of the medium. By maintaining the pH at an appropriate value, the phosphopeptides can be retained on the surface while the other peptides are removed. The phosphopeptides act as Lewis bases at pH range 2.0–2.5 while the metal oxide surface tends to accept electrons. At this pH range, the carboxylic acid groups of the acidic nonphosphopeptides having pKa (, $-\log_{10}K_{a,}$ K_a is the acid dissociation constant) values higher than phosphopeptides are in neutral state by retaining their protons. In this case, the phosphopeptides adhere to the metal oxide surface, while the surface does not interact with acidic nonphosphopeptides in the neutral state. When the pH is increased, the surface of the metal oxide material becomes neutral this time. The Lewis base effect of the phosphate groups of the phosphopeptides will be greatly reduced due to the increased amount of $^{-}$OH ions in the medium. Thus, phosphopeptides can easily be removed from the surface by breaking the interactions that support the formation of coordination between the species. The pH range is adjusted to 9.0–10.0 at the elution step of phosphopeptide enrichment studies using metal oxides.

The phosphopeptide enrichment performances of sol-gel materials containing titanium and zirconium metals were evaluated by using data obtained from MALDI-MS analysis. The phosphopeptides were attempted to enrich from tryptic digest mixtures of BSA and β-casein at a 50:1 mole ratio (BSA: β-casein). The peptide mixture of the proteins was analyzed using MALDI-MS without an enrichment step. None of the phosphopeptide signals could be observed in MALDI-MS analysis of the non-enriched sample (**Figure 1**).

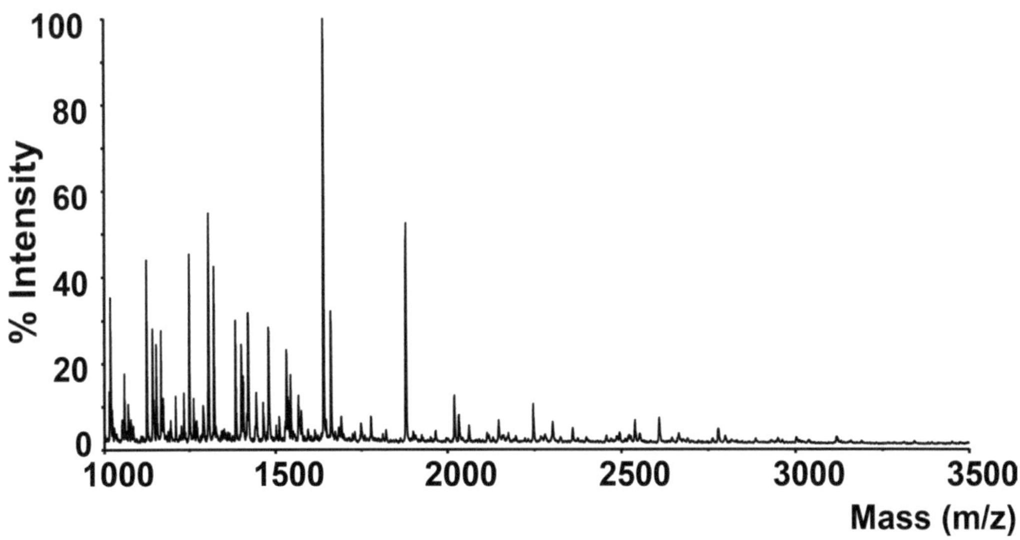

Figure 1. MALDI mass spectrum obtained from the direct analysis of the tryptic digest mixture of BSA and β-casein (in 50:1 mole ratio).

http://dx.doi.org/10.5772/intechopen.68891

The phosphopeptide enrichment method was initially performed using commercial TiO_2 material. In the mass spectrum obtained from the MALDI-MS analysis of the enriched sample using TiO_2, the signals of three phosphopeptides (β1, β3, and β4) could be observed in the presence of intense nonphosphopeptide signals (**Figure 2A**). The list of phosphopeptides obtained from tryptic digestion of β-casein is given in **Table 1**. This result indicates that TiO_2 is not enough selective for phosphorylated peptides.

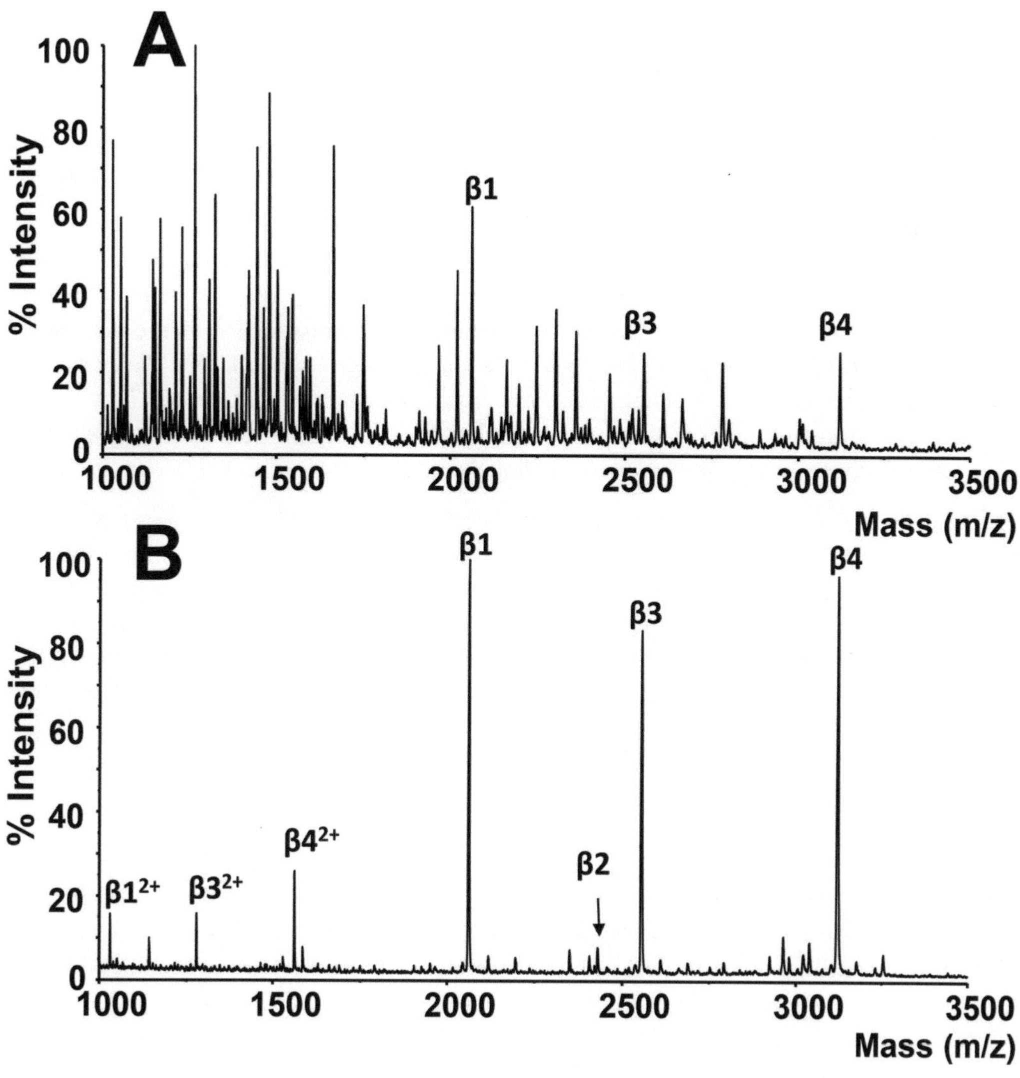

Figure 2. MALDI mass spectra obtained from the tryptic digest mixture of BSA and β-casein after enrichment with (A) commercial TiO_2 and (B) titanium/silica sol-gel material. The signals of the phosphopeptides are indicated by β1, β2, β3, and β4.

Phosphopeptides	Amino acid sequence	$[M + H]^+$
β1	FQ[pS]EEQQQTEDELQDK	2061.81
β2	IEKFQ[pS]EEQQQTEDELQDK	2432.08
β3	FQ[pS]EEQQQTEDELQDKIHPF	2556.03
β4	RELEELNVPGEIVE[pS]L[pS][pS][pS]EESITR	3122.26

Table 1. List of phosphopeptides obtained from tryptic digestion of β-casein.

The intense signals of four phosphopeptides were observed in the mass spectrum obtained from the analysis of the enriched sample using titanium/silica sol-gel material. Signals of doubly charged $[M + 2H]^{2+}$ molecular ions of three phosphopeptides (β1, β3, and β4) were also observed in the mass spectrum (**Figure 2B**). There were no signals of nonphosphopeptides or impurities other than phosphopeptides in the mass spectrum given in **Figure 2B**. The results show that the titanium/silica sol-gel material is far more successful than the commercial TiO_2 in phosphopeptide enrichment performance.

The digest mixture was also used to test the phosphopeptide enrichment efficiency of the zirconium/silica sol-gel material. The commercial ZrO_2 was compared with the zirconium/silica sol-gel in terms of the phosphopeptide enrichment performance. When the mass spectrum obtained from the analysis of the enriched sample using commercial ZrO_2 is evaluated, it is clear that the metal oxide is not selective for phosphopeptides (**Figure 3A**). The signals of three phosphopeptides (β1, β3, and β4) could be observed in the mass spectrum with the great complexity of the nonphosphopeptide signals.

The signals of four phosphopeptides could be clearly observed in the mass spectrum obtained from the analysis of the sample enriched using the zirconium/silica sol-gel material (**Figure 3B**). Signals of doubly charged $[M + 2H]^{2+}$ molecular ions of two phosphopeptides (β3 and β4) were also observed in the mass spectrum with low intensity. The signal at m/z 2351 represents dephosphorylation through neutral loss of HPO_3 (80 Da) from β2 (**Figure 3B**). The results show that the zirconium/silica sol-gel material has a rather high phosphopeptide selectivity compared to the commercial ZrO_2.

3.2. Enrichment of glycopeptides from the fetuin and IgG digests

Metal oxides are commonly used tools for the enrichment of sialylated glycopeptides as in phosphopeptide enrichments. The noncovalent interaction is formed by the way of coordination between the metal ions on the metal oxide and the oxygen atom or the nitrogen atom on the glycan. In addition, the carboxyl groups of sialic acid containing glycopeptides are interacted with positively charged ions on the metal-oxide surface at acidic pH [13]. In order to eliminate the interference effect of phosphopeptides, in the method, specific enzymes such as alkaline phosphatase are usually applied prior to the enrichment of glycopeptides. However, some acidic peptides containing glutamic acid and aspartic acid amino acid residues are also interacted with the surface of metal oxides, thereby decreasing the enrichment selectivity.

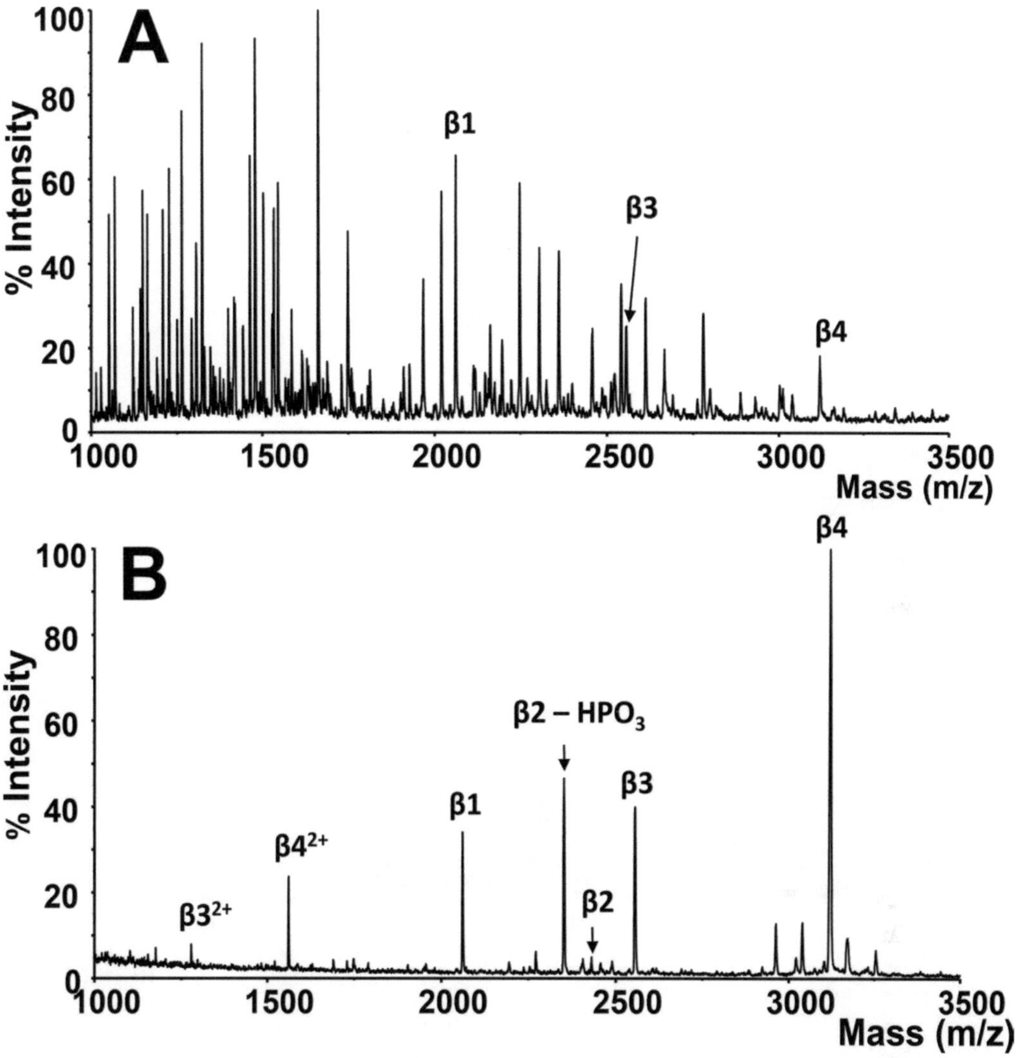

Figure 3. MALDI mass spectra obtained from the tryptic digest mixture of BSA and β-casein after enrichment with (A) commercial ZrO_2 and (B) zirconium/silica sol-gel material. The signals of the phosphopeptides are indicated by β1, β2, β3, and β4.

Elution of glycopeptides from the surface of metal oxides was achieved increasing the pH to 10–11 range.

In the study, fetuin and IgG proteins including sialylated and neutral glycans, respectively, were selected to evaluate glycopeptide enrichment performance of the sol-gel materials. Both proteins were enzymatically digested and the resulting peptide mixtures containing glycopeptides were used to identify their glycosylation sites. To achieve this, 5 μg of fetuin and IgG digests were individually diluted with loading buffer and used for the enrichment studies. Before the enrichment of sialylated glycopeptides of fetuin, direct analysis of tryptic products of fetuin was performed using MALDI-MS and no significant glycopeptide signal was observed in the mass spectrum (**Figure 4A**). When commercial ZrO_2 was used for the

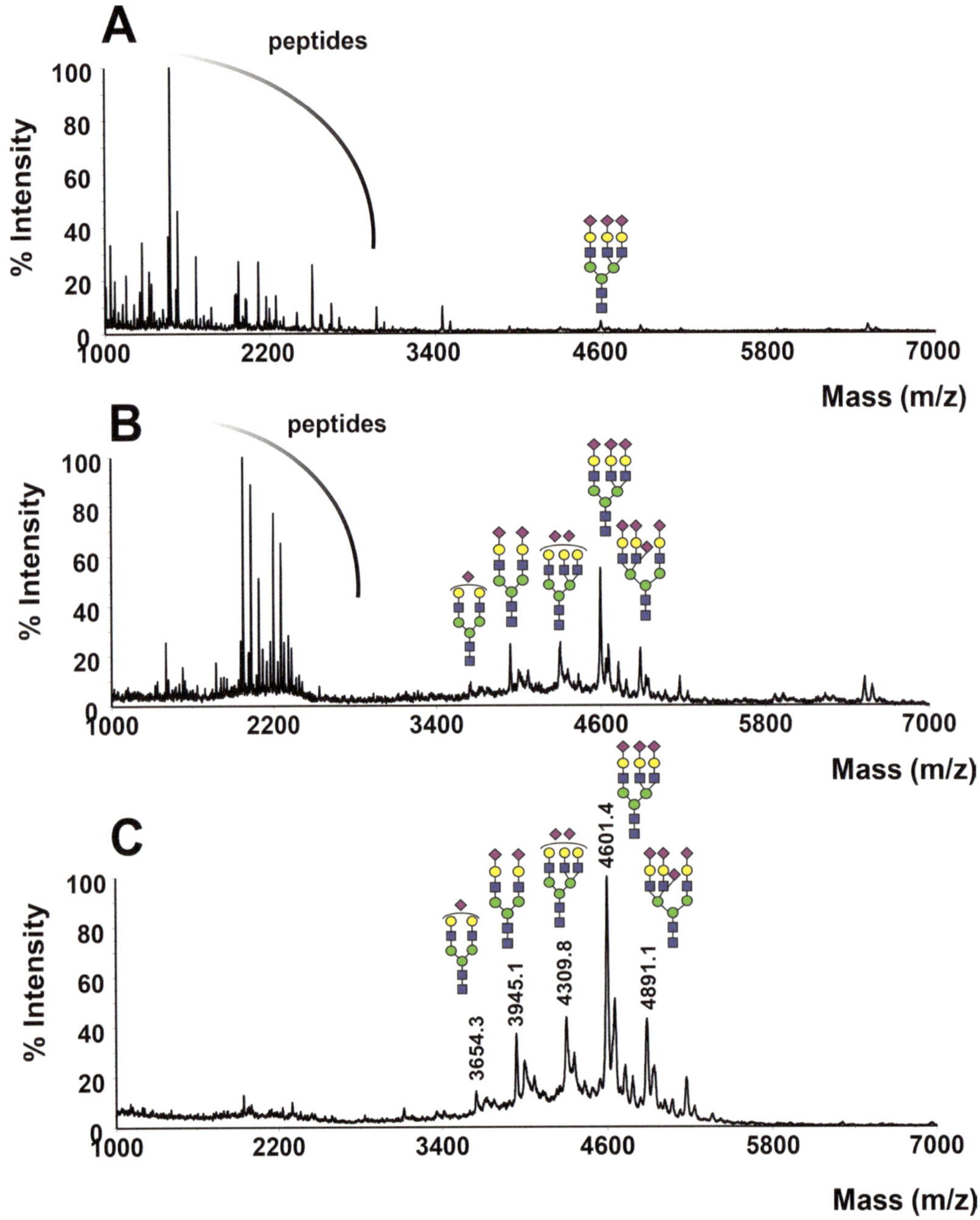

Figure 4. MALDI mass spectra obtained from the tryptic digest of fetuin. (A) Direct analysis, (B) after the enrichment with commercial ZrO_2 and (C) after the enrichment with zirconium/silica sol-gel material.

enrichment of fetuin glycopeptides, the signals of non-glycosylated peptides were observed in the MALDI-MS spectrum together with glycosylated peptide signals (**Figure 4B**). The glycopeptides of fetuin belonging to Asn156 glycosylation site were identified. By the enrichment of glycopeptides using zirconium/silica sol-gel material, glycopeptides of fetuin were detected with improved signal intensities in MALDI-MS analysis (**Figure 4C**). This clearly shows that the zirconium/silica sol-gel material is more selective than commercial ZrO_2 for glycopeptides.

To evaluate the enrichment performance of the metal containing sol-gel materials in the presence of the glycopeptides including neutral glycans, tryptic (glyco-) peptides of IgG were enriched by titanium/silica sol-gel materials. In the direct analysis of tryptic products of IgG with MALDI-MS, a few glycopeptides were identified (**Figure 5**A) and non-glycosylated peptides dominated the spectrum.

In addition, commercial TiO_2 was also used to compare enrichment performance of the material with titanium/silica sol-gel material. When both spectra were obtained from the enrichment of glycopeptides by TiO_2 and titanium/silica sol-gel materials (**Figure 5B** and **C**), the enrichment selectivity and efficiency of titanium/silica sol-gel materials was found to be better than that commercial TiO_2. These results also demonstrate that the glycopeptides containing neutral glycans were efficiently enriched by the sol-gel materials as well as sialylated glycopeptides.

It is stated in the literature that the salt removal is essential for the efficient analysis of glycopeptides by MALDI-MS [43]. Higher salt concentration causes decrease in ionization efficiency and suppresses glycopeptide signals. The developed sol-gel materials successfully remove salts and detergents derived from the reducing and alkylating reagents.

3.3. Purification of 2-AA-labeled N-glycans

N-glycan profiling of human plasma is very significant for clinical glycomics [44]. Glycan biomarker discovery studies focus on glycan analysis using high-throughput techniques [45]. 2-AA labeling of N-glycans is one of the commonly applied sample preparation methods for the analysis of N-glycans by MALDI-MS, HPLC-HILIC-Florescence, and CE-MS [46]. Purification of 2-AA-labeled N-glycans prior to MALDI-MS analysis is required and crucial in order to remove salts (PBS), detergents (SDS, NP-40), and solvents (DMSO).

In the study, N-glycans derived from human plasma were labeled with 2-AA and purified by both titanium/silica and zirconium/silica sol-gel materials, and thus MALDI-MS analysis of N-glycans was successfully achieved. However, the analysis of 2-AA-labeled N-glycans was not performed without a purification step. No N-glycan signals were observed in the mass spectrum (data not shown).

Figure 6A and **B** shows the MALDI-MS spectra of the purified 2-AA-labeled N-glycans after the purification step using zirconium/silica and titanium/silica sol-gel materials, respectively. As shown in **Figure 6A** and **B**, N-glycome profile was observed by the MALDI-MS analysis of purified 2-AA-labeled N-glycans carried out in negative ionization mode. By this approach, not only sialylated glycans but also neutral glycans were purified due to having florescence

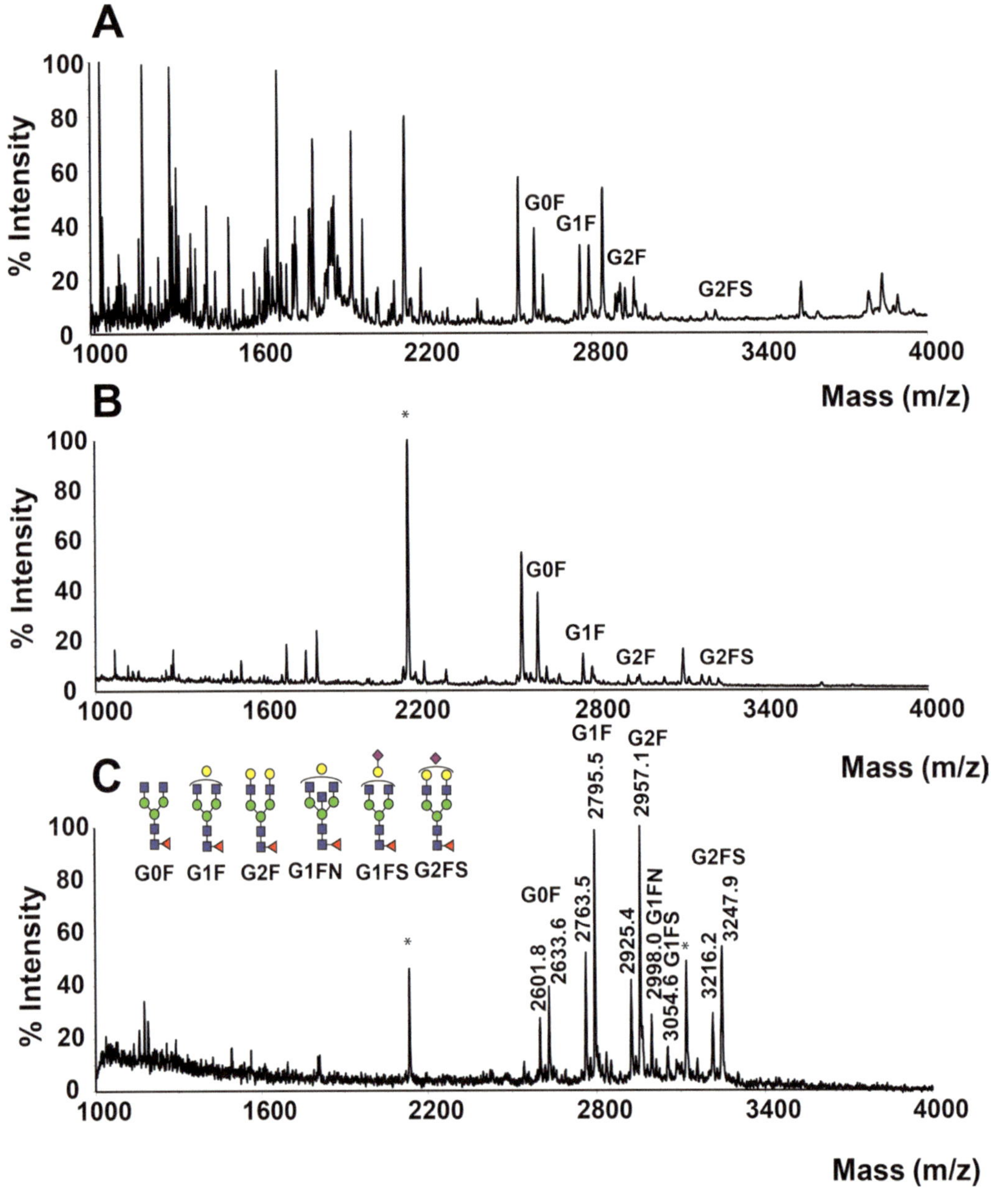

Figure 5. MALDI mass spectra obtained from the tryptic digest of IgG. (A) Direct analysis, (B) after the enrichment with commercial TiO_2, and (C) after the enrichment with titanium/silica sol-gel material.

tag of 2-AA, which has a carboxyl group. The additional negatively charged group on the glycans after 2-AA-labeled makes the interaction between 2-AA-labeled N-glycans and sol-gel particulates stronger.

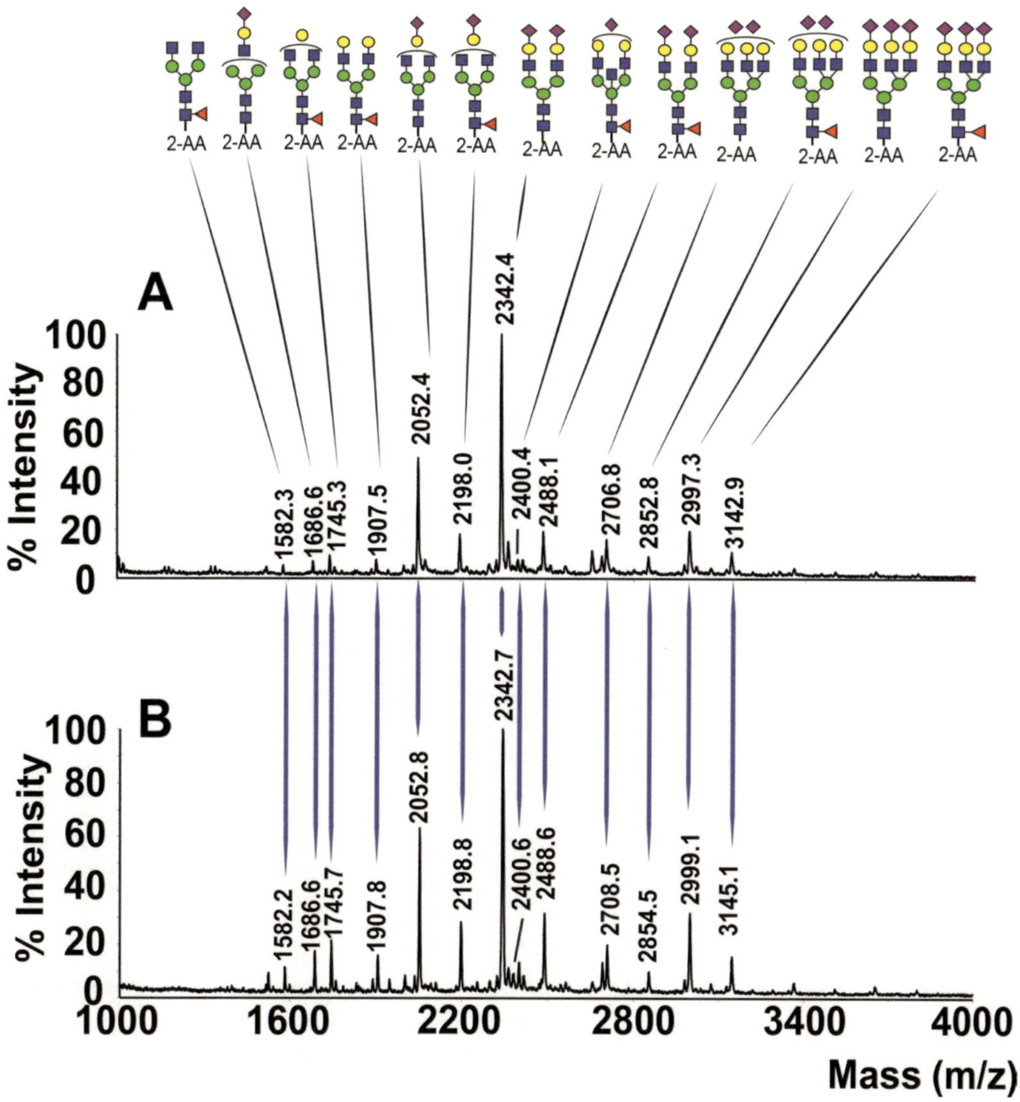

Figure 6. MALDI mass spectra of 2-AA-labeled N-glycans purified by (A) zirconium/silica sol-gel material and (B) titanium/silica sol-gel material.

4. Conclusions and future perspectives

Titanium/silica and zirconium/silica sol-gel materials were synthesized by a facile method and used for the enrichment of (phospho- and glyco-) peptides. Besides that, 2-AA-labeled N-glycans were purified using the sol-gel materials, thereby removing interferences such as salts, detergents, and solvents that are used in the sample preparation methods. This study shows that sol-gel materials promise advantages in phosphoproteomics and glycoproteomics studies regarding sample preparation methods and provide improvements in the (phospho- and glyco-) peptide analysis.

Acknowledgements

This work was funded by the Scientific and Technological Research Council of Turkey (TÜBİTAK) under Contract Nos. 115Z241 and 110T203.

Author details

Hacı Mehmet Kayili[1,2,3†], Mehmet Atakay[1†] and Bekir Salih[1*]

*Address all correspondence to: bekirsal@gmail.com

1 Department of Chemistry, Hacettepe University, Ankara, Turkey

2 Department of Chemistry, Çankırı Karatekin University, Çankırı, Turkey

3 Department of Nutrition and Dietetics, Karabuk University, Karabuk, Turkey

† These authors contributed equally to this chapter

References

[1] Temporini C, Callerli E, Massolini G, Caccialanza G. Integrated analytical strategies for the study of phosphorylation and glycosylation in proteins. Mass Spectrometry Reviews. 2008;**27**(3):207–236. DOI: 10.1002/mas.20164

[2] Blom N, Sicheritz-Ponten T, Gupta R, Gammeltoft S, Brunak S. Prediction of post-translational glycosylation and phosphorylation of proteins from the amino acid sequence. Proteomics. 2004;**4**(6):1633–1649. DOI: 10.1002/pmic.200300771

[3] Karlaftis V, Perera S, Monagle P, Ignjatovic V. Importance of post-translational modifications on the function of key haemostatic proteins. Blood Coagulation & Fibrinolysis. 2016;**27**(1):1–4. DOI: 10.1097/MBC.0000000000000301

[4] Burén S, Ortega-Villasante C, Blanco-Rivero A, Martínez-Bernardini A, Shutova T, Shevela D, et al. Importance of post-translational modifications for functionality of a chloroplast-localized carbonic anhydrase (CAH1) in *Arabidopsis thaliana*. PLoS One. 2011;**6**(6):e21021. DOI: 10.1371/journal.pone.0021021

[5] Reinders J, Sickmann A. State-of-the-art in phosphoproteomics. Proteomics. 2005;**5**(16): 4052–4061. DOI: 10.1002/pmic.200401289

[6] Mann M, Jensen ON. Proteomic analysis of post-translational modifications. Nature Biotechnology. 2003;**21**(3):255–261. DOI: 10.1038/nbt0303-255

[7] Meri S, Baumann M. Proteomics: Posttranslational modifications, immune responses and current analytical tools. Biomolecular Engineering. 2001;**18**(5):213–220. DOI: 10.1016/s1389-0344(01)00106-x

[8] Larsen MR, Roepstorff P. Mass spectrometric identification of proteins and characterization of their post-translational modifications in proteome analysis. Fresenius Journal of Analytical Chemistry. 2000;**366**(6–7):677–690. DOI: 10.1007/s002160051562

[9] Banks RE, Dunn MJ, Hochstrasser DF, Sanchez JC, Blackstock W, Pappin DJ, et al. Proteomics: New perspectives, new biomedical opportunities. Lancet. 2000;**356**(9243):1749–1756. DOI: 10.1016/s0140-6736(00)03214-1

[10] North SJ, Hitchen PG, Haslam SM, Dell A. Mass spectrometry in the analysis of N-linked and O-linked glycans. Current Opinion in Structural Biology. 2009;**19**(5):498–506. DOI: 10.1016/j.sbi.2009.05.005

[11] Seo J, Lee KJ. Post-translational modifications and their biological functions: Proteomic analysis and systematic approaches. Journal of Biochemistry and Molecular Biology. 2004;**37**(1):35–44

[12] Guerrera IC, Kleiner O. Application of mass spectrometry in proteomics. Bioscience Reports. 2005;**25**(1–2):71–93. DOI: 10.1007/s10540-005-2849-x

[13] Chen CC, Su WC, Huang BY, Chen YJ, Tai HC, Obena RP. Interaction modes and approaches to glycopeptide and glycoprotein enrichment. Analyst. 2014;**139**(4):688–704. DOI: 10.1039/c3an01813j

[14] Ongay S, Boichenko A, Govorukhina N, Bischoff R. Glycopeptide enrichment and separation for protein glycosylation analysis. Journal of Separation Science. 2012;**35**(18): 2341–2372. DOI: 10.1002/jssc.201200434

[15] Kolli V, Schumacher KN, Dodds ED. Engaging challenges in glycoproteomics: Recent advances in MS-based glycopeptide analysis. Bioanalysis. 2015;**7**(1):113–131. DOI: 10.4155/bio.14.272

[16] Cao L, Qu Y, Zhang ZR, Wang Z, Prytkova I, Wu S. Intact glycopeptide characterization using mass spectrometry. Expert Review of Proteomics. 2016;**13**(5):513–522. DOI: 10.1586/14789450.2016.1172965

[17] Fila J, Honys D. Enrichment techniques employed in phosphoproteomics. Amino Acids. 2012;**43**(3):1025–1047. DOI: 10.1007/s00726-011-1111-z

[18] Reinders J, Sickmann A. State-of-the-art in phosphoproteomics. Proteomics. 2005;**5**(16): 4052–4061. DOI: 10.1002/pmic.200401289

[19] Thingholm TE, Jensen ON, Larsen MR. Analytical strategies for phosphoproteomics. Proteomics. 2009;**9**(6):1451–1468. DOI: 10.1002/pmic.200800454

[20] Leitner A. Phosphopeptide enrichment using metal oxide affinity chromatography. TrAC – Trends in Analytical Chemistry. 2010;**29**(2):177–185. DOI: 10.1016/j.trac.2009.08.007

[21] Wen JY, Wilkes GL. Organic/inorganic hybrid network materials by the sol-gel approach. Chemistry of Materials. 1996;**8**(8):1667–1681. DOI: 10.1021/cm9601143

[22] Kandimalla VB, Tripathi VS, Ju H. Immobilization of biomolecules in sol-gels: biological and analytical applications. Critical Reviews in Analytical Chemistry. 2006;**36**(2):73–106. DOI: 10.1080/10408340600713652

[23] Díaz-García ME, Laínño RB. Molecular imprinting in sol-gel materials: recent developments and applications. Microchimica Acta. 2005;**149**(1–2):19–36. DOI: 10.1007/s00604-004-0274-7

[24] Wencel D, MacCraith B, McDonagh C. High performance optical ratiometric sol-gel-based pH sensor. Sensors and Actuators B: Chemical. 2009;**139**(1):208–213. DOI: 10.1016/j.snb.2008.12.066

[25] Marx S, Liron Z. Molecular imprinting in thin films of organic-inorganic hybrid sol-gel and acrylic polymers. Chemistry of Materials. 2001;**13**(10):3624–3630. DOI: 10.1021/cm000983z

[26] Lahav M, Kharitonov AB, Katz O, Kunitake T, Willner I. Tailored chemosensors for chloroaromatic acids using molecular imprinted TiO_2 thin films on ion-sensitive field-effect transistors. Analytical Chemistry. 2001;**73**(3):720–723. DOI: 10.1021/ac000751j

[27] Kayili HM, Salih B. Fast and efficient proteolysis by reusable pepsin-encapsulated magnetic sol-gel material for mass spectrometry-based proteomics applications. Talanta. 2016;**155**:78–86. DOI: 10.1016/j.talanta.2016.04.014

[28] Çelikbıçak Ö, Atakay M, Güler Ü, Salih B. A trypsin immobilized sol-gel for protein identification in MALDI-MS applications. Analytical Letters. 2014;**47**(4):707–719. DOI: 10.1080/00032719.2013.831423

[29] Ren J, Wang LH, Han XY, Cheng JF, Lv HL, Wang JY, et al. Organic silicone sol-gel polymer as a noncovalent carrier of receptor proteins for label-free optical biosensor application. ACS Applied Materials & Interfaces. 2013;**5**(2):386–394. DOI: 10.1021/am3024355

[30] Atakay M, Celikbicak O, Salih B. Amine-functionalized sol-gel-based lab-in-a-pipet-tip approach for the fast enrichment and specific purification of phosphopeptides in MALDI-MS applications. Analytical Chemistry. 2012;**84**(6):2713–2720. DOI: 10.1021/ac203008u

[31] Celikbicak O, Atakay M, Guler U, Salih B. A novel tantalum-based sol-gel packed microextraction syringe for highly specific enrichment of phosphopeptides in MALDI-MS applications. Analyst. 2013;**138**(15):4403–4410. DOI: 10.1039/c3an00021d

[32] Ruhaak LR, Huhn C, Waterreus WJ, de Boer AR, Neususs C, Hokke CH, et al. Hydrophilic interaction chromatography-based high-throughput sample preparation method for N-glycan analysis from total human plasma glycoproteins. Analytical Chemistry. 2008;**80**(15):6119–6126. DOI: 10.1021/ac800630x

[33] Zhou H, Ye M, Dong J, Corradini E, Cristobal A, Heck AJ, et al. Robust phosphoproteome enrichment using monodisperse microsphere-based immobilized titanium (IV) ion affinity chromatography. Nature Protocols. 2013;**8**(3):461–480. DOI: 10.1038/nprot.2013.010

[34] Gates MB, Tomer KB, Deterding LJ. Comparison of metal and metal oxide media for phosphopeptide enrichment prior to mass spectrometric analyses. Journal of the American Society for Mass Spectrometry. 2010;**21**(10):1649–1659. DOI: 10.1016/j.jasms.2010.06.005

[35] Han M, Li Z. Mesoporous metal oxide nanoparticles for selective enrichment of phosphopeptides from complex sample matrices. Analytical Methods. 2016;**8**(43):7747–7754. DOI: 10.1039/C6AY02069K

[36] Zhou H, Tian R, Ye M, Xu S, Feng S, Pan C, et al. Highly specific enrichment of phosphopeptides by zirconium dioxide nanoparticles for phosphoproteome analysis. Electrophoresis. 2007;**28**(13):2201–2215. DOI: 10.1002/elps.200600718

[37] Li Y, Liu Y, Tang J, Lin H, Yao N, Shen X, et al. Fe_3O_4@Al_2O_3 magnetic core-shell microspheres for rapid and highly specific capture of phosphopeptides with mass spectrometry analysis. Journal of Chromatography A. 2007;**1172**(1):57–71. DOI: 10.1016/j.chroma.2007.09.062

[38] Lin H-Y, Chen W-Y, Chen Y-C. Iron oxide/niobium oxide core-shell magnetic nanoparticle-based phosphopeptide enrichment from biological samples for MALDI MS analysis. Journal of Biomedical Nanotechnology. 2009;**5**(2):215–223. DOI: 10.1166/jbn.2009.1022

[39] Ma R, Hu J, Cai Z, Ju H. Highly selective enrichment of phosphopeptides with high-index facets exposed octahedral tin dioxide nanoparticles for mass spectrometric analysis. Talanta. 2014;**119**:452–457. DOI: 10.1016/j.talanta.2013.11.049

[40] Huang X, Wang J, Wang J, Liu C, Wang S. Preparation of graphene-hafnium oxide composite for selective enrichment and analysis of phosphopeptides. RSC Advances. 2015;**5**(109):89644–89651. DOI: 10.1039/C5RA17471F

[41] Çelikbıçak Ö, Kaynar G, Atakay M, Güler Ü, Kayili HM, Salih B. Specific enrichment and direct detection of phosphopeptides on insoluble transition metal oxide particles in matrix-assisted laser desorption/ionization mass spectrometry applications. European Journal of Mass Spectrometry. 2013;**19**(3):151–162. DOI: 10.1255/ejms.1228

[42] Nelson CA, Szczech JR, Dooley CJ, Xu Q, Lawrence MJ, Zhu H, et al. Effective enrichment and mass spectrometry analysis of phosphopeptides using mesoporous metal oxide nanomaterials. Analytical Chemistry. 2010;**82**(17):7193–7201. DOI: 10.1021/ac100877a

[43] Alagesan K, Khilji SK, Kolarich D. It is all about the solvent: on the importance of the mobile phase for ZIC-HILIC glycopeptide enrichment. Analytical and Bioanalytical Chemistry. 2017;**409**(2):529–538. DOI: 10.1007/s00216-016-0051-6

[44] Stavenhagen K, Kolarich D, Wuhrer M. Clinical glycomics employing graphitized carbon liquid chromatography-mass spectrometry. Chromatographia. 2015;**78**(5–6):307–320. DOI: 10.1007/s10337-014-2813-7

[45] Shubhakar A, Reiding KR, Gardner RA, Spencer DIR, Fernandes DL, Wuhrer M. High-throughput analysis and automation for glycomics studies. Chromatographia. 2015;**78**(5–6):321–333. DOI: 10.1007/s10337-014-2803-9

[46] Ruhaak LR, Zauner G, Huhn C, Bruggink C, Deelder AM, Wuhrer M. Glycan labeling strategies and their use in identification and quantification. Analytical and Bioanalytical Chemistry. 2010;**397**(8):3457–3481. DOI: 10.1007/s00216-010-3532-z

Chapter 12

Sol-Gel Glass Coating Synthesis for Different Applications: Active Gradient-Index Materials, Microlens Arrays and Biocompatible Channels

Ana Isabel Gómez Varela, María Aymerich,
Daniel Nieto García, Yolanda Castro Martín,
Pieter A.A. de Beule, Ezequiel Álvarez,
Carmen Bao-Varela and María Teresa Flores-Arias

Additional information is available at the end of the chapter

http://dx.doi.org/10.5772/67830

Abstract

The intent of this chapter is to review the use of sol-gel processing of silica and silica-titania optical coatings in recent research by the authors in three different areas: the synthesis of active gradient-index (GRIN) materials by multilayer deposition of erbium- and ytterbium-doped silica-titania films, the improvement of the optical and morphological qualities of microlens arrays fabricated by laser ablation and the functionalization of polydimethylsiloxane (PDMS) channel preclinical devices. Through the use of sol-gel, layers with specific properties can be produced. In this regard, undoped and erbium- and ytterbium-doped SiO_2-TiO_2 films have been produced and characterized using atomic force microscopy (surface topography evaluation) and spectral ellipsometry (determination of optical constants, thickness and porosity of the films). In a second application, a silica sol has been synthesized to coat microlens arrays fabricated by laser ablation. The deposited layer reduces the surface roughness of the microlens array, which yields the improvement of the contrast and the homogeneity of the foci. Finally, PDMS channels fabricated with laser technologies and soft-lithography methods are coated with a sol-gel-derived silica film to avoid the degradation of the material with organic solvents, and their biocompatibility is studied.

Keywords: sol-gel, dip coating, gradient-index media, rare-earth doping, laser writing technique, laser microfabrication, microlens arrays, PDMS devices, cell adhesion

1. Introduction

This chapter reviews the exploration of sol-gel chemical route for different applications. In particular, the synthesis of active gradient-index (GRIN) optical materials and the improvement of the optical qualities of microlens arrays as well as a better functionalization of the polydimethylsiloxane (PDMS) channel preclinical devices are presented. Sol-gel chemistry offers some unique opportunities for the synthesis of optical materials over existing production methods [1–4]. The process allows excellent control of its purity and composition since it starts with pure materials. In addition, one important advantage of the sol-gel process related to the synthesis of these types of materials is that coatings with high optical quality using low sintering temperature can be obtained. The decrease of the sintering temperature is crucial with a view to possible future industrial application.

GRIN materials are characterized by a refractive index distribution that varies spatially in a controlled manner [5]. They are typically manufactured by ion-exchange process [6, 7], although their fabrication has also been accomplished by neutron irradiation [8], chemical vapour deposition (CVD) [9] and sol-gel techniques [10, 11]. Sol-gel allows the incorporation of dopants such as rare-earth elements in an easy way. The ability to induce specific optical properties in a material by simply selecting a deliberate dopant is one of the most important advantages of sol process when it comes to the fabrication of active optical media [12–14]. In this regard, different erbium- and ytterbium-doped films based on SiO_2-TiO_2 systems have been prepared. The fabrication and characterization of monolayers with controlled properties is the first step towards the preparation of active GRIN materials. It is performed by growing successive layers until a multilayer stack described by a complex parabolic-like refractive index profile distribution is obtained [15]. These types of materials are very interesting for laser beam shaping purposes [16].

Sol-gel layers can also be used to optimize the optical and morphological properties of microlens arrays, which can be found in many modern optical devices. There are several important applications of microlens arrays, encompassing fibre couplers in optical communications systems, viewing optics, laser beam shaping elements, charge-coupled device cameras and so on. Microlenses can be fabricated by several methods, such as thermal reflow forming, stereolithography technique, mould insert, pressure difference or decompression method and hot embossing [17–20]. The majority of these methods are appropriated to produce microlenses on photoresist. Nevertheless, in recent years, the fabrication of glass-based microlenses is attracting special attention since it is widely used in photonic systems [21, 22]. Microlenses on glass can be obtained using the laser direct-writing technique, which presents some significant advantages over conventional methods: flexibility in terms of surface shapes, diameter and focal length; the very small dead space between lens lets of any desired shape and profile; and the simplicity of the fabrication setup [23, 24]. But the laser direct-writing technique is in general not able to produce microlenses with comparable imaging and stray light properties to other conventional methods. Hence, in this chapter, a silica sol-gel layer is deposited onto the micro-optical device to reduce the damage created on the glass during the laser ablation process and to enhance the optical properties of the microlens by increasing the quality of the interstices among the microlens of the array. The addition of the sol-gel thin film yields to microlens arrays with comparable imaging and stray light properties to other techniques, preserving at the same time the main advantages of the laser direct-writing technique, including flexibility in terms of design, fast prototyping, low cost, non-contaminant, repeatability, etc.

Furthermore, sol-gel thin film processing can be used to enhance the lifetime of PDMS preclinical devices. PDMS is a polymer widely used for the fabrication and prototyping of fluidic chips. It can be employed for mimicking blood vessels and may be employed in biomedical primarily assays due to: (i) its transparency, which facilitates the observation of contents in channels by direct visual inspection or through a microscope; (ii) its biocompatibility; (iii) it can be coated, which allows the fabrication of PDMS multilayer devices and the integration of micro valves; (iv) its deformability; and (v) its cost-effectiveness [25]. Nonetheless, it also presents one important disadvantage as material degradation is caused when using organic solvents for cleaning, which makes the device non-reusable. This can be overcome by simply coating the PDMS chip with one or more sol-gel layers. This procedure provides the structure with the chemical robustness of the glass and preserves the biocompatibility and transparency properties without significantly altering the geometry of the PDMS device [26, 27]. In this chapter, the fabrication of PDMS devices coated with different sol-gel compositions to overcome the deterioration problem is presented. Additionally, the biocompatibility of each device is determined by evaluating cell adhesion effectiveness of human umbilical vein endothelial cells (HUVECs) cultured onto the coated channels [28].

The remainder of this chapter is organized as follows: in Section 2, the sol-gel synthesis and film deposition method are introduced. Section 3 presents the experimental procedure and methods for the fabrication of sol-gel-derived active GRIN media, microlens arrays by laser ablation and the subsequent functionalization of the micro-optical devices via sol-gel, as well as for the enhancement of the lifetime of PDMS preclinical devices. Characterization techniques and results are also described in this section. Finally, Section 4 draws the main research findings.

2. Sol-gel synthesis and film deposition

2.1. Sol-gel preparation

SiO_2-TiO_2 and SiO_2 thin films were prepared via sol-gel route for different applications. The sol compositions and their corresponding molar ratio concentration are listed in **Table 1**.

SiO_2-TiO_2 thin layers have been produced using methyltriethoxysilane (MTES, $CH_3Si(OCH_2CH_3)_3$, 98%, ABCR GmbH & Co., Karlsruhe, Germany) and titanium isopropoxide (TISP, $Ti[OCH(CH_3)_2]_4$, 97%, ABCR GmbH & Co., Karlsruhe, Germany) as precursors of silica and

Sol	Molar ratio	H_2O/alkoxides	Alkoxides/A_cH	C (g_{oxides}/L)
MTES:TISP	70/30	1.5	1	100
MTES:TISP	80/20	1.5	1	100
MTES:TEOS	60/40	1.75	4	180

Table 1. Composition, molar ratios and concentration of SiO_2-TiO_2 and SiO_2 sols described in this chapter.

titania, respectively (MTES/TISP sol). Both SiO_2 and TiO_2 are widely used in optics and photonics applications mainly due to the wide spectral region of transparency from the visible to the infrared, thermal stability and chemical durability. The MTES/TISP sol preparation is accomplished as follows: in the first stage, MTES is pre-hydrolysed using ethanol as solvent. HCl (0.1N) was then added with stirring for 1 h at room temperature. On the other hand, titanium isopropoxide is complexed by mixing with ethanol and glacial acetic acid. This solution is maintained under agitation process for 1 h. Then, the resulting solution is mixed to the MTES sol and distilled water is added, drop by drop, to complete the hydrolysis process.

On the other hand, hybrid SiO_2 sol was made from MTES and tetraethoxysilane (TEOS, $Si(OCH_2CH_3)_4$, 99%, ABCR GmbH & Co., Karlsruhe, Germany) in the molar rate 60/40 (MTES/TEOS sol). In this case, a two-step sol is prepared at ambient atmosphere by mixing TEOS and MTES with absolute ethanol followed by addition of acidulated water (0.1 M AcH) drop by drop. Next, distilled water is incorporated to the sol, which is then refluxed in a water bath at 40°C for 2 h under continuous magnetic stirring.

2.2. Dip-coating technique

Sol-gel is one of the simplest and most commonly used routes to obtain thin films with a huge variety of inorganic, hybrid and nanocomposite materials. It allows a high degree of control of the critical parameters and enables a flexibility that cannot be obtained using other conventional techniques given that it permits to coat a wide variety of substrates and complex geometries, including substrates with holes or intricate shapes. Several wet thin film coating methods exist, including dip-coating, spin-coating, spray-coating and flow-coating techniques, among others. The stages of dip coating [29] are schematically represented in **Figure 1**. Basically, the process, which is performed under well-controlled temperature and

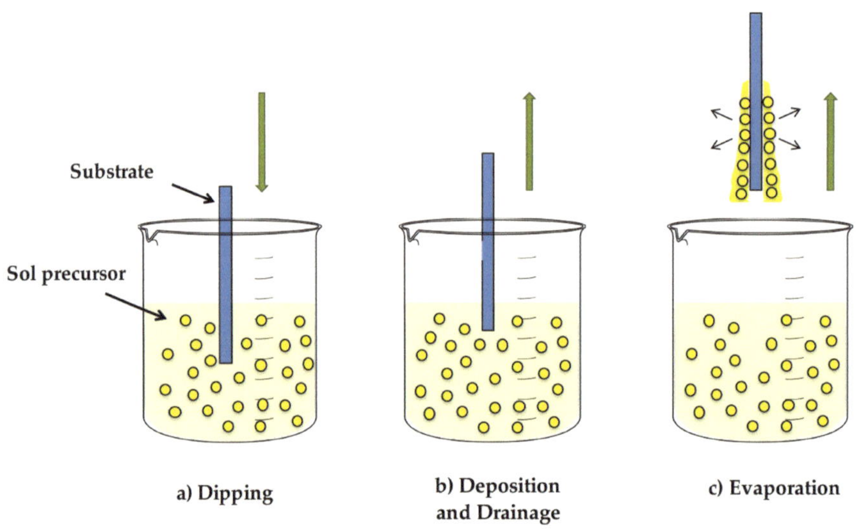

Figure 1. Sequential stages of the dip-coating technique for thin film deposition: (a) the substrate is dipped and immersed in the sol precursor, (b) the substrate is withdrawn at constant speed rate, and (c) solvent evaporation yields to the gelation of the layer.

http://dx.doi.org/10.5772/67830

atmospheric conditions, starts by immersing the substrate to be coated in the initial solution, being subsequently withdrawn at a constant withdrawal speed. Precisely, the control on the withdrawal speed and on the evaporation conditions makes possible a fine-tuning of the film characteristics, including thickness, optical constants and inner structure. The combined effects of viscous drag and capillarity action yield to the solution homogeneously spreading out along the surface of the substrate. In the final stage of the process, evaporation occurs and brings gelation of the film. Usually, a post-heat treatment is applied to the coated substrates, which also influences the characteristics of the films.

3. Sol-gel glass coating synthesis and characterization for different applications

3.1. Active gradient-index optical materials

Active gradient-index optical materials can be used as beam shaping transforming devices with gain or loss. In particular, an active GRIN material, characterized by a parabolic and complex refractive index profile, can turn an input laser beam described by a Gaussian irradiance profile into a beam with a uniform irradiance distribution [16]. **Figure 2** shows the basis of the beam shaping operation by an active GRIN medium. The input Gaussian beam impinges in the active medium and travels through the material up to the beam shaping distance. At this beam shaping distance, which corresponds to the output face of the GRIN medium, the original beam is turned into a nearly uniform beam. If the material has gain, an amplified uniform beam is obtained (a); on the contrary, if the material has loss, the beam suffers from attenuation after propagating in the medium (b).

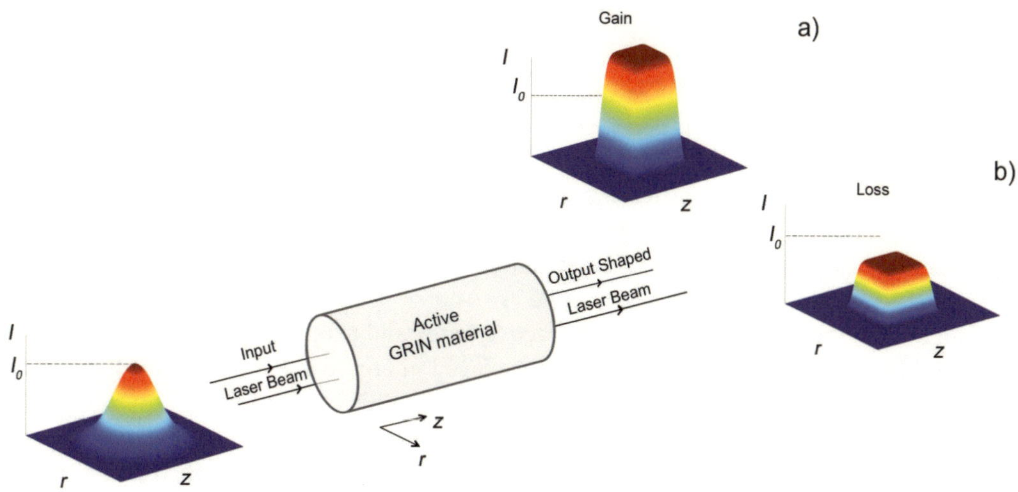

Figure 2. Schematic representation of Gaussian-to-uniform beam shaping operation by an active GRIN material with (a) gain and (b) loss. I_0 is the on-axis irradiance value of the Gaussian beam.

3.1.1. Materials and methods

A series of undoped and Er^{3+}- and Yb^{3+}-doped $70SiO_2$-$30TiO_2$ films with good optical (homogeneous) and mechanical (crack-free) properties were prepared using MTES and TISP (precursors of silica and titania, respectively) in acid conditions. The synthesis was performed in two steps following the indications provided in Section 2.1. In both erbium- and ytterbium-doped sols synthesis, the rare-earth (RE) precursor is added to the mixture after the distilled water and then maintained under vigorous stirring for 1 h. The atomic percentage of erbium nitrate pentahydrate ($Er(NO_3)_3 \cdot 5H_2O$, 99.9%, ABCR GmbH & Co., Karlsruhe, Germany) and ytterbium nitrate pentahydrate ($Yb(NO_3)_3 \cdot 5H_2O$, 99.9%, ABCR) precursors was varied between 0.3 and 2.0 at.% and 0.3 and 1.0 at.%, respectively. Previously to film deposition, commercial glass slides were washed in distilled water, dried and finally washed in ethanol. Films were deposited by dip-coating technique using a withdrawal rate of 10, 20, 30, 40 and 50 cm/min. To obtain the final samples, the sol-gel-derived films were heat treated at 450°C for 30 min using a ramp rate of 10°C/min in air atmosphere.

Selected samples were analysed by using a spectral ellipsometry (SE), an indirect method for thin film characterization, where the change of polarization of an incident light upon reflection with the material is measured. In this work, ellipsometry measurements were performed using a spectral ellipsometer (M-2000UTM, J.A. Co., Woollam) to characterize the films deposited onto glass slides. The spectra were taken in the wavelength range of 400 and 1000 nm at incident angles of 65, 70 and 75 degrees. A Cauchy model was used to fit the data obtained with WVASE32 software. On the other hand, the topographical features of the erbium-doped samples under atomic force microscopy (AFM) for a scan area of 5 μm are shown in **Figure 7**. The instrument used was a Bruker Dimension Icon® atomic force microscope operated under intermittent contact mode in air at a line scan rate of 1–2 Hz. The cantilever tip had a height of 10–15 μm and a radius of 10 nm. Typical cantilever resonance frequency varied between 350 and 367 kHz at force constants between 20 and 80 Nm^{-1}. Tapping mode images were analysed using the Gwyddion software (gwyddion.net).

3.1.2. Results

SE investigation of the films was performed in order to determine the optical constants (refractive index and extinction coefficient) and thickness of the films. As can be seen in **Figure 3**, the film thickness increases when increasing the withdrawal rate for both erbium- and ytterbium-doped samples, as well as for undoped samples. It is also interesting to note that film thickness significantly increases with dopant concentration, an effect observed by Bruynooghe et al. in 1997 [30] and attributed to the increase of the viscosity of the sol with increasing erbium concentration. The spectral dependence of refractive index and extinction coefficient of erbium- and ytterbium-doped silica-titania films are displayed in **Figure 4**. Representative films were obtained using a withdrawal rate of 20 cm/min for SiO_2-TiO_2 coatings and 10 cm/min for Er and Yb-SiO_2-TiO_2 coatings. Figure shows that the refractive index diminishes with dopant concentration (**Figure 4a** and **c**). It shows that positive values of *k* are obtained in the wavelength range from 400 to 1000 nm. These values can be used for the fabrication of active GRIN media with loss. For the erbium-doped samples, the graphs reveal that the 2.0 at.% Er-doped

http://dx.doi.org/10.5772/67830

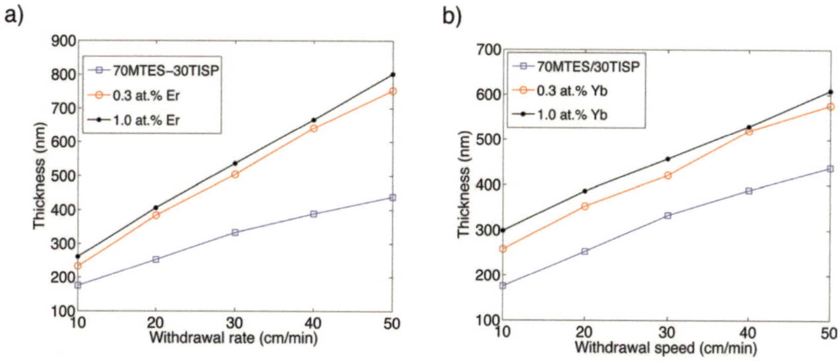

Figure 3. Thickness of the rare-earth-doped 70MTES/30TISP thin films versus withdrawal rate obtained with SE: (a) erbium-doped films and (b) ytterbium-doped films.

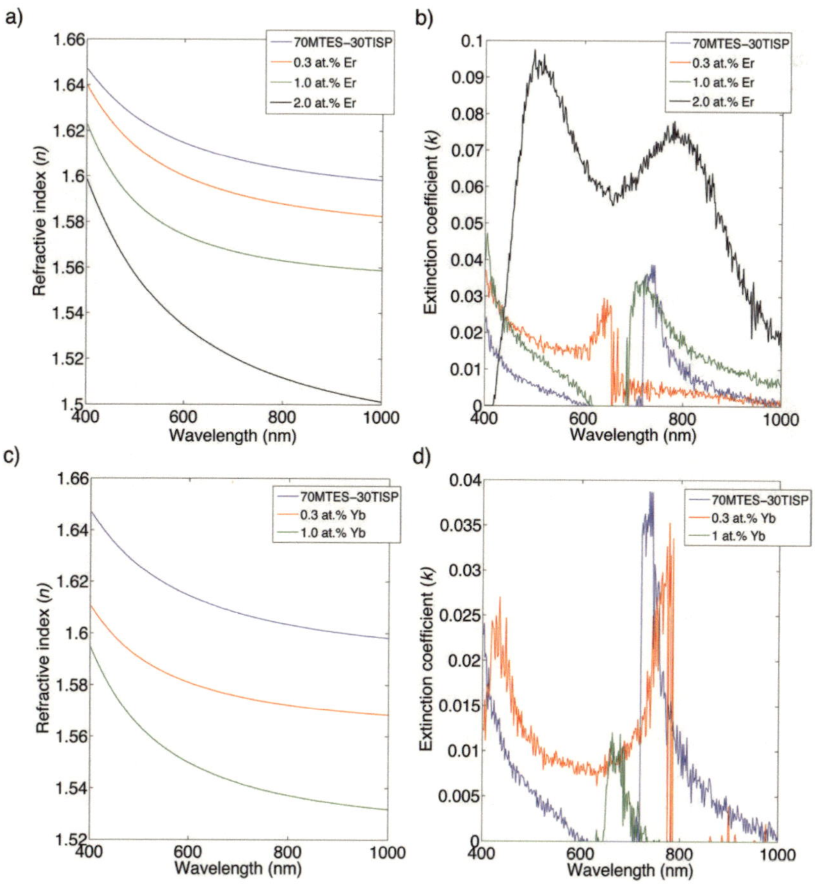

Figure 4. Optical constants of the erbium-doped (top) and ytterbium-doped (bottom) films measured by means of spectral ellipsometry compared to undoped silica-titania film: (a) and (c) refractive index and (b) and (d) extinction coefficient.

film presents the highest value of *k* along the whole spectral range, factor that probably occurs because of the dissimilitude of its surface morphology compared to the other deposited layers. With regard to ytterbium-doped samples, it can be found that for the higher doped film, *k* is close to zero along the whole measuring wavelength interval and achieves its maximum value around 650 nm. The measured values of the extinction coefficient evidence that for the SiO_2-TiO_2 film, *k* reaches its maximum value at ca. 750 nm and it is slightly shifted to the right for 1 at.% Yb-doped silica-titania film, reaching a maximum at approximately 800 nm. For both samples, the extinction coefficient shows a similar behaviour.

For both erbium- and ytterbium-doped samples, the refractive index was found to decrease appreciably when increasing dopant concentration but to remain below the undoped SiO_2-TiO_2 film (see **Figure 4**). This feature suggests that the porosity of the coatings increases with rare-earth concentration. Porosity percentage of the films can be determined using the following expression [31, 32]:

$$\Pi = 1 - \left(\frac{n_p^2 - 1}{n_p^2 + 2} \middle/ \frac{n_b^2 - 1}{n_b^2 + 2} \right) \tag{1}$$

where n_p and n_b represent the refractive index of the undoped silica-titania film (reference film) and the refractive index of the doped film, respectively. This equation assumes that the pores of the coating are filled up solely with air. Hence, the porosity values for the erbium- and ytterbium-doped films were determined from the measured refractive index with spectral ellipsometry at $\lambda = 633$ nm as function of dopant concentration and listed in **Table 2**. As expected, the results demonstrate that film porosity increases with dopant concentration. In addition, higher porosity relative values are obtained for the ytterbium-doped films when the same dopant concentration is employed.

AFM measurements of the surface of the undoped and erbium-doped silica-titania films are compared in **Figure 5**, while the corresponding numerical values for the most typical roughness parameters are gathered in **Table 3**. A visual inspection of AFM images reveals no significant changes on the surface morphology of the 0, 0.3 and 1 at.% Er-doped films. In general, topographical images display irregular-shaped and randomly grown granular surfaces. On the contrary, the sol-gel-derived film prepared with the higher erbium concentration (2 at.%), whose pore size diameter was estimated to 250 nm, presents important structural changes compared to the other films. In this case a honeycomb structured porous surface is found and the size and number of porous has increased, being this feature consistent with the SE results (see **Table 2**), where an increase of the porosity with dopant concentration was evidenced. In **Table 3**, it can be observed from arithmetic roughness (R_a) and root-mean-square roughness (R_q) values that smooth surfaces are obtained and that they increase with erbium concentration. The skewness (R_{sk}) positive values expose that the surface peaks and asperities are predominant over valleys, except for the 2 at.% Er- SiO_2-TiO_2 film where the contrary factor occurs, indicating that the surface is mainly composed of valleys. In addition, kurtosis (R_{ku}) reports the sharpness of profile peaks. The values provided are larger for the undoped SiO_2-TiO_2, implying the existence of a small number of extreme heights. On the contrary, the lower kurtosis in the other samples studied indicates moderate surface height features. A similar behaviour is expected regarding Yb-doped films.

Er at.%	Π (%)	Yb at.%	Π (%)
0.3	2.0	0.3	5
1	5.4	1	10
2	11.25	–	–

Table 2. Relative porosity values of the sol-gel-derived thin films as a function of dopant concentration.

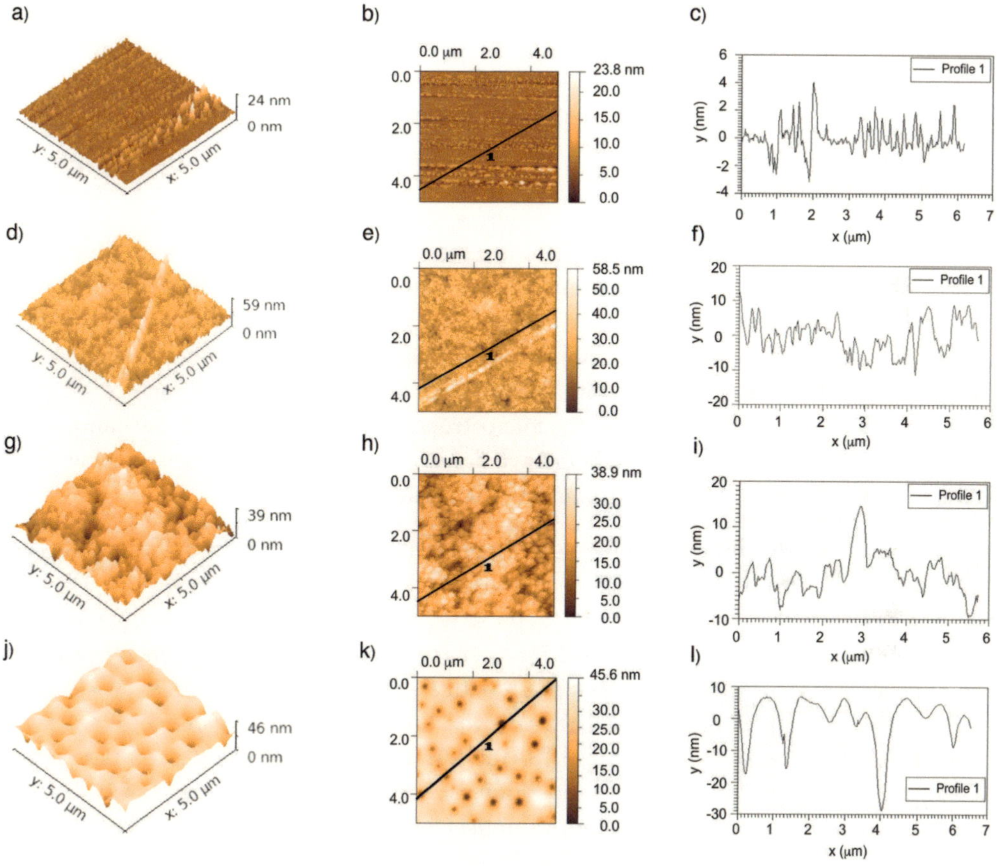

Figure 5. (a), (d), (g) and (j) 3D tapping mode in air AFM images of the thin films prepared onto a commercial glass substrate by sol-gel dip coating with (b), (e), (h) and (k) respective surface morphology and (c), (f), (i) and (l) section profiles recorded along the black line indicated. From top to bottom: 0, 0.3, 1 and 2 at.% Er- SiO_2-TiO_2 film measurements.

3.2. Improvement of the optical and morphological properties of microlens arrays fabricated by laser using a sol-gel coating

In this section, a simple, non-contaminant and repeatable method to fabricate microlens arrays is presented. The micro-optical devices are obtained via laser ablation, and the surface

Er at.%	R_a (nm)	R_q (nm)	R_{sk}	R_{ku}
0	0.82	1.28	2.410	11.70
0.3	3.88	5.25	0.818	2.96
1	4.19	5.31	0.121	0.027
2	4.10	5.74	−1.550	3.93

Table 3. AFM principal surface roughness parameters of the Er-doped SiO_2-TiO_2 films.

roughness of these devices is reduced by depositing a sol-gel-derived silica film, which covers the interstitial spaces between consecutive microlenses. This fact provokes substantial reduction of the stray light at the focal plane of the microlenses, thus increasing their optical quality.

3.2.1. Materials and methods

A commercial float soda-lime window glass is used as substrate, which was provided by a local supplier. The dimensions of the samples were 60 mm × 20 mm × 3 mm. A previous compositional analysis of the soda-lime glass reveals the existence of tin (Sn) only in one side of the samples. This fact is related to the fabrication process of the window glass. As a result, the samples present two different faces with the presence of Sn impurities only in one of them. These impurities will be crucial during the microstructuring of the glass with the nanosecond laser [33]. Nevertheless, some researchers have demonstrated the advantages of using ultra-short femtosecond pulses for microstructuring materials [34].

Experimental setup for fabricating microlens arrays is illustrated in **Figure 6**. The microlenses were obtained on glass using Ti:Sapphire Femtosecond Amplitude Systems S-pulse HP laser operating at 1030 nm wavelength and 500 fs pulse width combined with a galvanometer system (mirror positioning system designed for laser marking) for addressing the output laser beam. The laser is focused on the glass surface with a 100 mm focal length lens. This is a flat-field lens that provides a uniform irradiance distribution over a working area of 80 × 80 mm^2. The laser scan path describes circular trenches for designing each microlens. Laser parameters were a power on the order of 0.8 W, a repetition rate of 10 kHz, a wavelength of 1030 nm, a pulse width of 500 fs and a scan speed of 160 mm/s. Different microlens arrays were fabricated according to the laser setup described in **Figure 6** and varying the number of laser passes from 1 to 5. As can be seen in the figure, each cylindrical structure is obtained by the ablation of a circular trench formed by moving the laser beam, using the galvanometer mirror system, relatively to the glass sample being maintained in the same position. A time of 10 s is taken to expose an area of 2 × 2 mm^2.

Next to the laser fabrication process, MTES/TEOS sol was prepared by sol-gel process according to the synthesis route indicated in Section 2.1. Silica coatings were deposited onto the optical elements or microlens by dip-coating technique using a withdrawal rate of 10 cm/min and heat treated at 450°C for 30 min in air atmosphere.

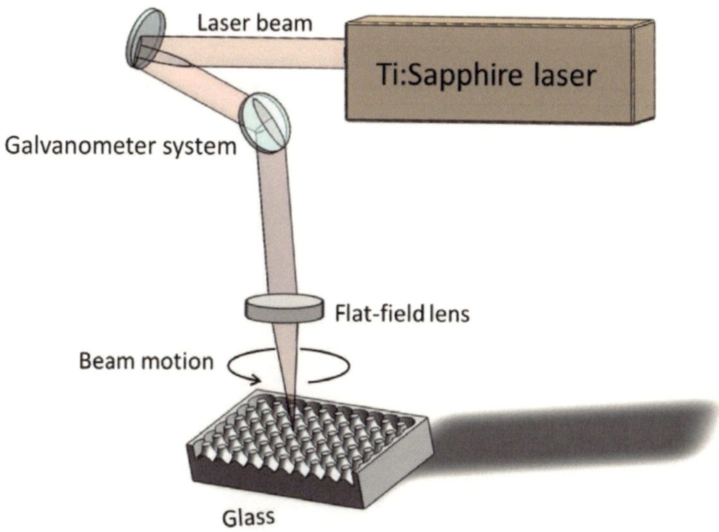

Figure 6. Experimental setup for fabricating microlens arrays.

The optical transmission of glass substrate and sol-gel coating was measured using a PerkinElmer Spectrometer (type Lambda 950 UV/Vis). The thickness and refractive index of the films were determined with the spectral ellipsometer (WVASE32, J.A. Co., Woollam M-2000UTM). Besides, the finished samples were inspected using both optical and confocal microscopes. The optical microscope Nikon MM-400 was used to visualize the sample and predetermine damage at the surface. In addition, a confocal microscope, SENSOFAR PLμ 2300, allowed us to perform topographic measurements on the surface and to obtain 3D images of the generated microlenses. The results presented were acquired using a 20XEPI microscope objective. The irradiance of the focus and the microlens spot size were determined using a beam profiler (BP-109UV).

3.2.2. Results

The topographical data acquired with the confocal microscope were used to measure the diameter, sag and roughness of microlens (see **Table 4**). **Figure 7a** shows a 3D confocal image of the microlens array fabricated by five laser passes and **Figure 7b** shows the same element after being coated with a silica sol-gel film. **Figure 7c** shows the transversal profile of one microlens and the same microlens after being coated with the sol-gel thin film. The thickness and refractive index of the SiO_2 film structure measured by SE are 1384 nm and 1.38 at 699.7 nm, respectively.

The focusing properties of the fabricated microlenses were characterized in terms of its focal length and spot size. The experimental setup for measuring the focal length of these microlenses consists of a He-Ne laser (632.8 nm), a 40X objective lens and a camera. We focused the microscope objective on the surface of the microlenses, this location was recorded, and then

	Number of laser passes		
	1	3	5
Microlenses fabricated by laser without sol-gel coating			
Diameter (μm)	38.17 ± 1.34	41.09 ± 1.46	44.08 ± 0.57
Sag (μm)	2.71 ± 0.14	10.0 ± 0.3	19.2 ± 0.4
Roughness (nm)	450 ± 13	1025 ± 34	1340 ± 25
Focal length (μm)	760 ± 10	150 ± 10	90 ± 10
Spot size (μm)	8.3 ± 0.4	5.3 ± 0.3	4.5 ± 0.2
Microlenses fabricated by laser with sol-gel coating			
Diameter (μm)	41.42 ± 1.49	41.60 ± 0.68	39.50 ± 1.48
Sag (μm)	1.39 ± 0.14	7.5 ± 0.5	17.7 ± 0.7
Roughness (nm)	23 ± 0.58	34 ± 0.26	32 ± 0.22
Focal length (μm)	820 ± 10	180 ± 10	210 ± 10
Spot size (μm)	6.2 ± 0.3	3.2 ± 0.1	2.3 ± 0.1

Table 4. Morphological and optical parameters of microlens arrays.

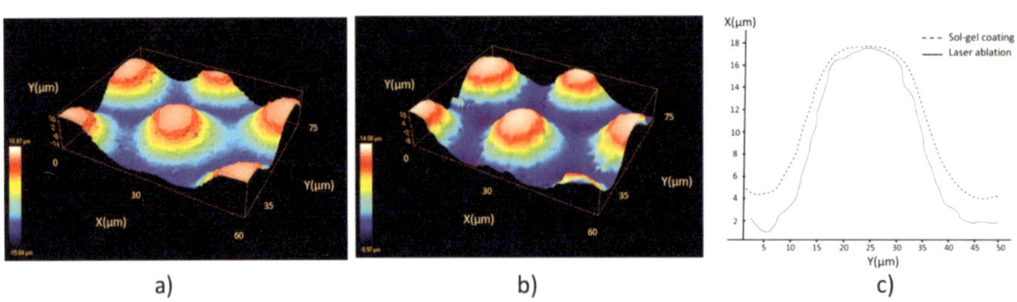

Figure 7. (a) Confocal image of one microlens array fabricated by laser ablation, (b) the same microlens with sol-gel coating and (c) profile of the microlens before (solid line) and after (dashed line) the sol-gel coating deposition.

the microscope was moved until the image of an arbitrary object is observed. The image was approximately located at the focal plane of the microlenses. The difference between these positions is a measure of the focal distance of the tested microlens arrays [35]. The data acquisition accuracy of this setup is ±5 μm. For determining the quality of focuses, the beam profiler BP109-VIS of Thorlabs was used. The focus spot size is determined by calculating the width at $1/e^2$ (see **Figure 8**).

3.3. Sol-gel coatings for PDMS device lifetime enhancement and biocompatibility

The design, fabrication and development of 3D prototypes that mimic blood vessel and lab-on-a-chip systems have aroused a huge interest in the last years due to the amount of

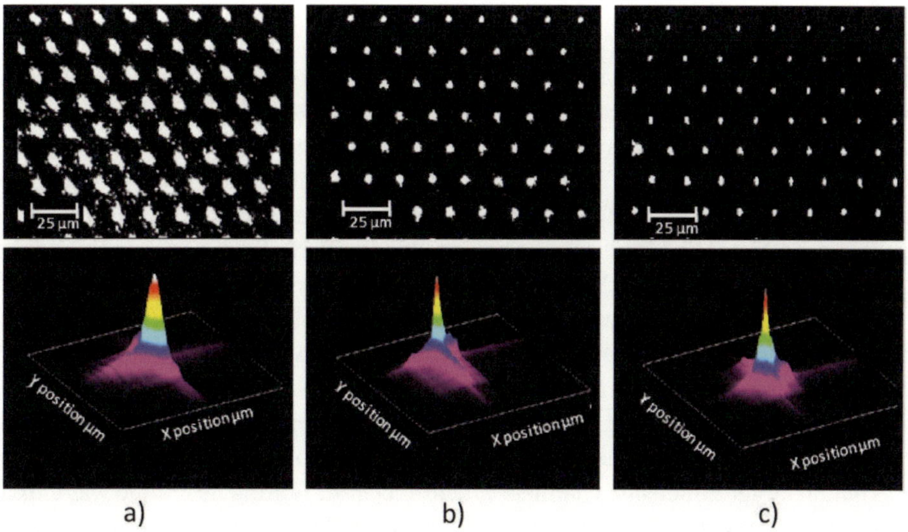

Figure 8. Focal plane image and focal irradiance distribution for the microlenses obtained at different laser passes with and without silica sol-gel coating: (a) one laser pass, (b) three laser passes and (c) five laser passes.

applications they present. By using these devices, the study of different pathologies, such as cardiovascular diseases [36] or tumour cells [37], can be performed. These experiments are carried out in controlled laboratory conditions, and the obtained results are directly linked with a better understanding of these illnesses. The structures that shape these devices can be as complex as a micron-scale laboratory or as simple as a single channel [38, 39]. Several materials have been employed for this purpose, from glass [40] to polymers [41] or silicon [42]. One of the most commonly used materials when manufacturing these devices is PDMS because of its advantages. Its permeability to gases, optical transparency, biocompatibility and replicating ability make this material very suitable for the fabrication of these preclinical chips, where cell cultures under flow conditions are performed. When using PDMS, the most common technique for the replica procedure is the soft-lithography method [25], which provides an accurate replica of the initial structure or master. For the master fabrication, techniques like lithography [43] or cutting plotter [44] have been reported, but the most promising of all is laser direct writing [45]. Its accuracy, versatility, speed and non-contact nature make it a very appropriate technique for the structuring of materials. Laser ablation can be performed in numerous materials, but glass substrates are very suitable for the master role due to its hardness and resistance [35].

As reported previously, PDMS is an optimal material for the fabrication of preclinical devices, but it presents an important disadvantage: it degrades when it is in contact with organic solvents. These solutions are very common in biomedical assays, and when PDMS is exposed to them, the device cannot be reused anymore due to the fact that it will lead to an abnormal cellular growth. Sol-gel coatings present themselves as a solution to overcome this problem by providing the structure with the robustness of the glass against these solvents but maintaining the geometry, transparency and biocompatibility of the chip [26, 27].

3.3.1. Materials and methods

Structures are commonly replicated in PDMS by using soft-lithography methods. By using this technique, a reliable replica of the initial structure or master is easily and quickly achieved. In this work, the master was fabricated over soda-lime glass using indirect laser writing technique. A Nd:YVO_4 laser (Rofin, Plymouth, MI, USA) in Q-switch mode and in the nanosecond regime was employed. This temporal regime is more implemented in industrial lasers than the femtosecond one. The laser beam has a wavelength of 1064 nm and a pulse duration of 20 ns. Laser technology allows the creation of very complex structures, but in this section, a single channel was the chosen design due to the fact that most of the microfluidic chips or preclinical devices are a combination of different channels. The laser setup was combined with a galvanometer system, conformed by mirrors that addressed the laser beam in the desired direction in order to create the structure. A 100 mm focal length flat-field lens was employed to focus the laser beam over the substrate. This lens ensures a working area of 80 × 80 mm^2. In order to obtain larger dimension channels, an indirect laser writing technique was employed. This procedure, known as laser backwriting [46], consists of focusing the laser beam on a metal target placed below the soda-lime glass substrate. The plume generated during the metal ablation process expands until it reaches the glass, and the impact of the expelled particles of the metal foil ablates the glass substrate, creating the desired structure. In this case, channels with millimetre dimensions were fabricated, so several laser passes were applied in order to obtain these structures. Laser parameters were 8 W of average power, 12 kHz of repetition rate and a scan speed of 1000 mm/s.

Due to its ablative nature, during the backwriting process, a high roughness was generated on the surface of the structure. This roughness must be reduced in order to obtain high-quality channels that allow optical microscopy inspections. For this purpose, a combined thermal technique was employed to enhance the quality of the laser-generated structures. The setup has a CO_2 laser (Easy Mark, Tinley Park, IL, USA) and a roller furnace [47]. The thermal procedure was developed as follows: the soda-lime glass master was introduced into the roller furnace and it was gradually heated. The sample was moved at 1000 mm/h and achieved a temperature of 500°C, which was below the transition temperature of the soda-lime glass but enough to avoid any cracks derived from the thermal shock that occurs when the laser beam is focused over the material surface. Once the sample was heated, the CO_2 laser beam interacted with the surface of the channel and melted the material, which was redistributed from the top of the channel to the bottom. Consequently, the roughness value of the device, as well as the depth of the channel, decreased its value. The CO_2 laser has a fundamental wavelength of 10.6 μm and a pulse duration of 10 μs, operating in the Q-switch regime. The laser beam was focused over the substrate by passing through a flat-field lens of 1 m focal length that provided a working area of 120 × 120 mm^2. Laser parameters were 12 kHz repetition rate and 90 cm/s scan speed. Channels had 2 mm width and their depths decreased from 1.215 to 1.005 mm after the thermal treatment. Once the master was fabricated by laser techniques over soda-lime glass, the channel was replicated in PDMS by following soft-lithography methods.

http://dx.doi.org/10.5772/67830

As previously mentioned, PDMS degrades when exposed to organic solvents, which are commonly employed in biomedical applications. One example is ethanol, which is frequently used for the sterilization of laboratory material. Once PDMS comes into contact with this solvent, the device becomes automatically non-reusable. If any cell culture is performed over this material, an abnormal cell growth will be observed. **Figure 9** clearly depicts this situation. In this figure, a cell culture over a PDMS channel, fabricated as it was previously described, is shown. In particular, human umbilical vein endothelial cells (HUVECs) were seeded. These cells were obtained from umbilical cords donated from the mothers after their informed consent, following the protocol approved by the Ethics Committee for Human Studies in Galicia (Spain) according to the Helsinki Declaration of 1975. In **Figure 9a**, we can appreciate the cell growth when cells were cultured over the PDMS channel for the first time. They regularly spread and form a monolayer, with the characteristic cobblestone structure of endothelial cells. **Figure 9b** presents the situation of the cell culture when the channel was reused for the third time and sterilized with ethanol between each use. The cellular behaviour is totally different from the previous one. In this case, we can appreciate how cells do not form a monolayer but instead they attach to each other, gathering in clusters and minimizing the contact surface with the degraded PDMS, leading to an abnormal cell culture and organization. Human endothelial cells were stained with green calcein acetoxy methyl (AM) (Invitrogen, Thermo Fisher Scientific, Waltham, MA, USA) for four min (incubation conditions: 37°C temperature, 95% air, 5% CO_2 atmosphere and more than 80% of humidity). Calcein is a cell-permeant dye that acts as a viability indicator by emitting in green wavelength when it is excited by fluorescence.

As previously mentioned, a possible solution to overcome this problem is the use of sol-gel. By coating the PDMS with sol-gel, we provide the channel with the robustness of the glass and the coated material does not degrade when it is exposed to organic solvents. In

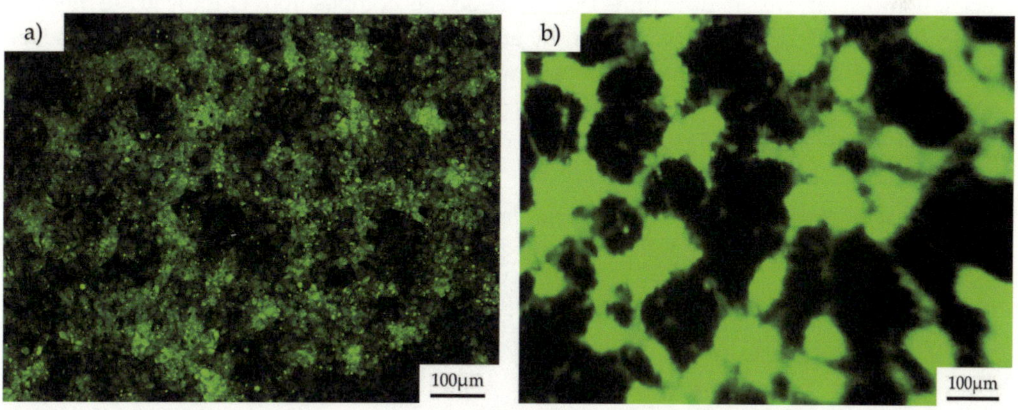

Figure 9. Fluorescence microscopy images of the human umbilical vein endothelial cells (HUVECs) seeded over PDMS channels after the (a) first and (b) third use of the device. The channel was cleaned with ethanol in between usage for sterilization purposes. HUVECs were stained with green calcein AM.

this particular case, different silica and silica-titania sol-gel coatings were applied over the PDMS structures. Hence, 60MTES/40TEOS, 70MTES/30TISP and 80MTES/20TISP compositions were employed in order to determine if they were biocompatible and if cells grew over these different coatings. The study of the differences in the cell behaviour among these compositions was also carried out. For performing the coating, the dip-coating technique was employed (see **Figure 1**). In this case, the dip-coating withdrawal speed was 6 cm/min. Before applying the post-heat treatment to the films, the substrates were allowed to dry several minutes at room temperature and then placed into a static Nanetti furnace for 2 h at a temperature of 150°C. A ramp rate of 5°C/min in air atmosphere was used to gradually heat up the sample. After the manufacturing process, it was observed that the sol-gel layer reduced the device irregularities but the original shape and dimensions of the original structure were maintained.

3.3.2. Results

Once the PDMS channels were coated with the above-mentioned compositions, HUVECs were cultured over them in order to analyse the cell behaviour over these substrates. The HUVECs were isolated and cultured following the protocol described by Rodiño-Janeiro et al. [48]. The PDMS structures were sterilized in autoclave at 120°C for 30 min and immersed in endothelial growth medium (EGM-2; Lonza Basle, Switzerland) for 30 min as pretreatment to enhance cell adhesion to the surfaces. HUVECs were dyed with calcein acetoxy methyl (AM) for determining the biocompatibility of the sol-gel compositions, since this dye only expresses in green fluorescence when cells are alive. After that, cells were washed twice in medium for removing the excess of fluorescent signal that could interfere in the image acquisition. Cells were seeded over the channels at a concentration of 10^6 cells/mL and incubated for a day. After this time, channels were washed with medium to remove non-adherent cells. The samples were observed under fluorescence microscopy and the results are shown in **Figure 10**. As calcein is a viability indicator, we can determine that all the sol-gel compositions employed are biocompatible, since cellular signal was collected by an Axio Vert A.1 Zeiss fluorescence microscope (Oberkochen, Germany). Nevertheless, there were significant differences in the growth and spreading of the HUVECs depending on the composition. The 80MTES/20TISP sol-gel coating is the most suitable environment for cells to grow and spread as we can appreciate a higher

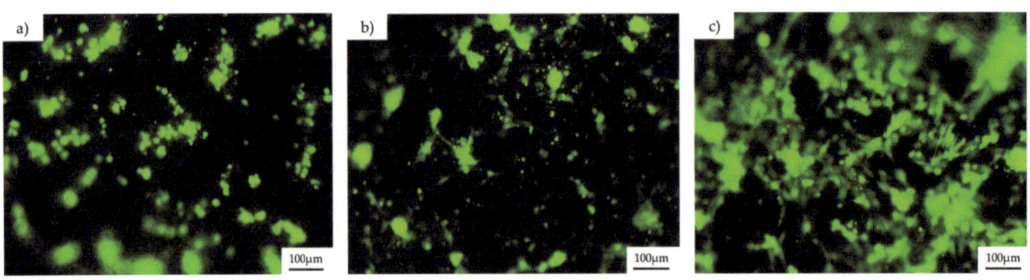

Figure 10. Fluorescence microscopy images of the human umbilical vein endothelial cells stained with calcein AM over the next sol-gel coatings: (a) 60MTES/40TEOS, (b) 70MTES/30Tisp and (c) 80MTES/20TISP after one-day culture.

proliferation in this case than in the others. Cell morphology was also the best; cells were spread and attempt to form a monolayer. On the other hand, the 60MTES/40TEOS composition seems to be the most hostile environment for cell growth. Cellular proliferation was low and cells presented a round shape, which suggests that the anchoring to the surface is week. The 70MTES/30TISP presents an intermediate situation where part of the HUVECs was attached to the surface. In all of the cases, we can appreciate that, even after washing, some cells remain adhered to others instead to the surface, generating an excess of fluorescent signal.

4. Conclusions

Sol-gel chemical route is suitable for the fabrication of active GRIN optical materials, for the functionalization of laser-derived micro-optical arrays as well as for the improvement of PMDS fluidic chips. Sol-gel-derived thin film multilayers may be a good approach to obtain GRIN media due to sol-gel powerfulness and versatility for tailoring the characteristics of the films. Erbium- and ytterbium-doped silica-titania films were prepared by sol-gel, finding that film thickness varied with rare-earth element concentration and increased linearly with increasing withdrawal rate of dip-coating process. The highest thickness values were achieved for larger dopant concentration. Additionally, the refractive index of the films decreased with rare-earth concentration. On account of the fact that this was only a preliminary attempt to obtain active GRIN media by sol-gel process, as to the extinction coefficient, a general trend to increase with doping concentration could be noticed, although the mechanisms of light loss in are worth original investigations. Porosity of the Er- and Yb-doped silica-titania films also increases with RE concentration. Finally, AFM images of the deposited and annealed films revealed smooth surfaces and the formation of porous granular nanostructures where valleys, mountains and island clusters became larger as dopant concentration increases.

Good-quality microlens arrays can be obtained using a combination of laser direct-writing technique and a subsequent sol-gel process. The proposed technique allows us to obtain microlens with a diameter in the range of 38–45 μm and depth in the range of 1–17 μm. The applied sol-gel silica layer reduces the surface roughness and increases the quality of the interstices between the microlenses generated by the ablation process. The obtained sol-gel-coated microlens improves the contrast and the homogeneity of the foci with respect to those uncoated. UV-Vis analysis reveals that the sol-gel layer enhances the transmission of the micro-optical element. Spectral ellipsometry measurements show that the refractive index of the deposited silica layer decreases with the increment of the wavelength, for a coated thickness of 1384 nm.

In this chapter, it has also been shown that sol-gel chemistry offers an interesting route to functionalize preclinical devices. In this regard, a technique that combines laser and soft lithography to fabricate the master and the replica of PDMS channels, respectively, was presented. Since these structures imitate blood vessel geometries, they can be used for preclinical biomedical applications. Different sol-gel coatings were deposited onto the channels in order to provide the structure with the chemical robustness of the glass and preserve the biocompatibility and transparency properties without significantly altering the geometry. Hence, HUVECs were seeded over them to study the biocompatibility of the devices. All the coatings analysed allowed the cells to live. Additionally, the study revealed that the 80MTES/20TISP sol-gel coating was the most appropriate composition when working with HUVECs due to the stretched form of the cells and their attempt to form a monolayer, which indicates that cells were properly attached to the surface.

Acknowledgements

This work has been supported under contracts MAT2010-18519, Ministerio de Ciencia e Innovación, MAT2015-71119-R AEI/FEDER, UE, Ministerio de Economía y Competitividad, Spain, and ISCIII/PI14-01140/ FEDER, Instituto de Salud Carlos III, Spain. Ana I. Gómez-Varela acknowledges the financial support from the FPU Doctoral Fellowship (Ref. number AP2009-3874) FPU, Ministerio de Educación, Cultura y Deporte, Spain. M. Aymerich acknowledges a Pre-Doctoral Fellowship from Xunta de Galicia (Spain) financed by the Secretaría Xeral de Universidades and Fondo Social Europeo. D. Nieto thanks to the Consellería de Cultura,Spain under the Galician Program for Research Innovation and Growth (2011–2015) (I2CPlan).

Author details

Ana Isabel Gómez Varela[1]*, María Aymerich[1], Daniel Nieto García[1], Yolanda Castro Martín[2], Pieter A.A. de Beule[3], Ezequiel Álvarez[4,5], Carmen Bao-Varela[1] and María Teresa Flores-Arias[1]

*Address all correspondence to: anaisabel.gomez@usc.es

1 Photonics4 Life Research Group, Department of Applied Physics, Faculty of Physics, University of Santiago de Compostela, Santiago de Compostela, Spain

2 Institute of Ceramics and Glass (CSIC), Kelsen 5, Campus de Cantoblanco, Madrid, Spain

3 Applied Nano-Optics Laboratory, International Iberian Nanotechnology Laboratory, Braga, Portugal

4 Health Research Institute of Santiago de Compostela (IDIS), University Hospital Complex of Santiago de Compostela (CHUS) SERGAS, Santiago de Compostela, Spain

5 CIBERCV, Madrid, Spain

References

[1] Klein LC. Sol-Gel Optics: Processing and Applications. Massachusetts: Springer Science & Business Media; 1994. DOI: 10.1007/978-1-4615-2750-3.

[2] Aegerter MA, Mennig M. Sol-Gel Technologies for Glass Producers and Users. New York: Springer Science & Business Media; 2004. DOI: 10.1007/978-0-387-88953-5.

[3] Reisfeld R. New materials for nonlinear optics. In: Jørgensen CK, Reisfeld R, editors. Optical and Electronic Phenomena in Sol-Gel Glasses and Modern Application. Structure and Bonding, **85**. Berlin: Springer; 1996. pp. 99–147. DOI:10.1007/BFb0111489.

[4] Rey-García F, Gómez-Reino C, Flores-Arias MT, Fuente GFDL, Durán A, Castro Y. Sol-gel coatings: an alternative route for producing planar optical waveguides. Thin Solid Films. 2011; **519**: 7982–7986. DOI: 10.1016/j.tsf.2011.05.054.

[5] Gómez-Reino C, Perez MV, Bao C. Gradient-Index Optics: Fundamentals and Applications. Berlin Heidelberg: Springer Verlag; 2002.

[6] Ohmi S, Sakai H, Asahara Y, Nakayama S, Yoneda Y, Izumitani T. Gradient-index rod lens made by a double ion-exchange process. Applied Optics. 1988; **27**: 496–499. DOI: 10.1364/AO.27.000496.

[7] Visconti AJ, Bentley JL. Fabrication of large-diameter radial gradient-index lenses by ion exchange of Na^+ for Li^+ in titania silicate glass. Optical Engineering. 2013; **52**: 112103–112103. DOI: 10.1117/1.OE.52.11.112103.

[8] Sinai P. Correction of optical aberrations by neutron irradiation. Applied Optics. 1971; **10**: 99–104. DOI: 10.1364/AO.10.000099.

[9] Pickering MA, Taylor RL, Moore DT. Gradient infrared optical material prepared by a chemical vapor deposition process. Applied Optics. 1986; **25**: 3364–3372. DOI: 10.1364/AO.25.003364.

[10] Yamane M, Yasumori A, Iwasaki M, Hayashi K. Graded index materials by the sol-gel process. In Proceedings of SPIE 1328, Sol-Gel Optics, 1 November 1990. pp. 133–144. DOI: 10.1117/12.22553.

[11] Laczka M, Wegrzynek JND, Wychowaniec M. GRIN glasses prepared by sol-gel method. In: Proceedings of SPIE 2943, Gradient-Index Optics in Science and Engineering, 22 October 1996. pp. 95–104. DOI: 10.1117/12.255522.

[12] Yeatman EM, Ahmad EM. Sol-gel fabrication of rare-earth doped photonic components. Journal of Sol-Gel Science and Technology. 2000; **11**: 231–236. DOI: 10.1023/A:1008792423076.

[13] Gonçalves RR, Messaddeq Y. Erbium-activated silica-zirconia planar waveguides prepared by sol-gel route. Thin Solid Films. 2008; **516**: 3094–3097. DOI: 10.1016/j.tsf.2007.07.183.

[14] Almeida RM, Marques AC. Rare earth-doped photonic crystals via sol-gel. Journal of Materials Science: Materials in Electronics. 2009; **20**: 307–311. DOI: 10.1007/s10854-008-9596-2.

[15] Gómez-Varela AI, Castro Y, Durán A, De Beule PAA, Flores-Arias MT, Bao-Varela C. Synthesis and characterization of erbium-doped SiO_2-TiO_2 thin films prepared by sol-gel and dip-coating techniques onto commercial glass substrates as a route for obtaining active gradient-index materials. Thin Solid Films. 2015; **583**: 115–121. DOI: 10.1016/j.tsf.2015.03.028.

[16] Gómez-Varela AI, Flores-Arias MT, Bao-Varela C, Gómez-Reino C. Focusing, collimation and beam shaping by active GRIN rod lenses: theory and simulation. Optics and Lasers in Engineering. 2012; **50**: 1706–1715. DOI: 10.1016/j.optlaseng.2012.07.011.

[17] Su L, Chen Y, Yi AY, Klocke F, Pongs G. Refractive index variation in compression molding of precision glass optical components. Applied Optics. 2008; **47**: 1662–1667. DOI: 10.1364/AO.47.001662.

[18] Kunnavakkam MV, Houlihan FM, Schlax M, Liddle JA, Kolodner P, Nalamasu O, Rogers JA. Low-cost, low-loss microlens arrays fabricated by soft-lithography replication process. Applied Physics Letters. 2003; **82**: 1152–1154. DOI: 10.1063/1.1555694.

[19] Fu YQ, Kok N, Bryan A. Microfabrication of microlens array by focused ion beam technology. Microelectronic Engineering 2000; **54**: 211–221. DOI: 10.1016/S0167-9317(00)00416-0.

[20] Pan CT, Wu TT, Chen MF, Chang YC, Lee CJ, Huang JC. Hot embossing of microlens array on bulk metallic glass. Sensors and Actuators A: Physical. 2008; **141**: 422–431. DOI: 10.1016/j.sna.2007.10.040.

[21] Wakaki M, Komachi Y, Kanai G. Microlenses and microlens arrays formed on a glass plate by use of a CO_2 laser. Applied Optics. 1998; **37**: 627–631. DOI: 10.1364/AO.37.000627.

[22] Delgado T, Nieto D, Flores-Arias MT. Soda-lime glass microlens arrays fabricated by laser: comparison between a nanosecond and a femtosecond IR pulsed laser. Optics and Lasers in Engineering. 2016; **86**: 29–37. DOI: 10.1016/j.optlaseng.2016.05.001.

[23] Nieto D, Arines J, Gomez-Reino C, O'Connor GM, Flores-Arias MT. Fabrication and characterization of microlens arrays on soda-lime glass using a combination of direct laser write and thermal reflow techniques. Journal of Applied Physics. 2011; **110**: 023108. DOI: 10.1063/1.3609085.

[24] Delgado T, Nieto D, Flores-Arias MT. Fabrication of microlens arrays on soda-lime glass using a laser direct-write technique and a thermal treatment assisted by a CO_2 laser. Optics and Lasers in Engineering. 2015; **76**: 1–6. DOI: 10.1364/AO.49.004979.

[25] McDonald JC, Whitesides GM. Poly (dimethylsiloxane) as a material for fabricating microfluidic devices. Accounts of Chemical Research. 2002; **35**: 491–499. DOI: 10.1021/ar010110q.

[26] Roman GT, Culbertson CT. Surface engineering of poly (dimethylsiloxane) microfluidic devices using transition metal sol-gel chemistry. Langmuir. 2006; **22**: 4445–4451. DOI: 10.1021/la053085w.

[27] Gomez-Sjoberg R, Leyrat AA, Houseman BT, Shokat K, Quake SR. Biocompatibility and reduced drug absorption of sol-gel-treated poly (dimethyl siloxane) for microfluidic cell culture applications. Analytical Chemistry. 2010; **82**: 8954–8960. DOI: 10.1021/ac101870s.

[28] Aymerich M, Gómez-Varela AI, Álvarez E, Flores-Arias MT. Study of different sol-gel coatings to enhance the lifetime of PDMS devices: evaluation of their biocompatibility. Materials. 2016; **9**: 728. DOI: 10.3390/ma9090728.

[29] Brinker CJ, Frye GC, Hurd AJ, Ashley CS. Fundamentals of sol-gel dip coating. Thin Solid Films. 1991; **201**: 97–108. DOI: 10.1016/0040-6090(91)90158-T.

[30] Bruynooghe S, Chabli A, Bertin F, Pierre F, Leflem G. Preparation and characterization of Nd^{3+} and Er^{3+}-doped silica sol-gel coatings by Rutherford backscattering spectroscopy and spectroscopic ellipsometry. Journal of Materials Research. 1997; 12: 2779–2783. DOI: 10.1557/JMR.1997.0370.

[31] Baklanov MR, Mogilnikov KP, Polovinkin VG, Dultsev FN. Determination of pore size distribution in thin films by ellipsometric porosimetry. Journal of Vacuum Science and Technology B. 2000; **18**: 1385–1391. DOI: 10.1116/1.591390.

[32] Skoczek E, Jaglarz J, Karasinsk P. Ellipsometric and spectrophotometric investigations of porous silica thin films produced by sol-gel method. Acta Physica Polonica A. 2011; **120**: 732–735. DOI: 10.12693/APhysPolA.120.732.

[33] Pilkington LAB. The float glass process. Proceeding of the Royal Society A. 1969; **314**: pp. 1–25. DOI: 10.1098/rspa.1969.0212.

[34] Nieto D, Arines J, Flores-Arias MT. Fluence ablation threshold dependence on tin impurities in commercial soda-lime glass. Applied Optics. 2014; **53**: 5416–5420. DOI: 10.1364/AO.53.005416.

[35] Nieto D, Flores-Arias MT, O'Connor GM, Gomez-Reino C. Laser direct-write technique for fabricating microlens arrays on soda-lime glass with a Nd:YVO_4 laser. Applied Optics. 2010; **49**: 4979–4983. DOI: 10.1364/AO.49.004979.

[36] Fiddes LK, Raz N, Srigunapalan S, Tumarkan E, Simmons CA, Wheeler AR, Kumacheva E. A circular cross-section PDMS microfluidics system for replication of cardiovascular flow conditions. Biomaterials. 2010; **31**: 3459–3464. DOI: 10.1016/j.biomaterials.2010.01.082.

[37] Ziober BL, Mauk MG, Falls EM, Chen Z, Ziober AF, Bau HH. Lab-on-a-chip for oral cancer screening and diagnosis. Head Neck. 2008; **30**: 111–121. DOI: 10.1002/hed.20680.

[38] Hongbin Y, Guangya Z, Siong CF, Shouhua W, Feiwen L. Novel polydimethylsiloxane (PDMS) based microchannel fabrication method for lab-on-a-chip application. Sensors and Actuators B: Chemical. 2009; **137**: 754–761. DOI: 10.1016/j.snb.2008.11.035.

[39] Xu BB, Zhang YL, Xia H, Dong WF, Ding H, Sun HB. Fabrication and multifunction integration of microfluidic chips by femtosecond laser direct writing. Lab on a Chip. 2013; **13**: 1677–1690. DOI: 10.1039/c3lc50160d.

[40] Stjernström M, Roeraade J. Method for fabrication of microfluidic systems in glass. Journal of Micromechanics and Microengineering. 1998; **8**: 33. DOI: 10.1088/0960-1317/8/1/006.

[41] Tsao CW, DeVoe DL. Bonding of thermoplastic polymer microfluidics. Microfluidics and Nanofluidics. 2009; **6**: 1–16. DOI: 10.1007/s10404-008-0361-x.

[42] Carlen ET, Bomer JG, van Nieuwkasteele JW, van den Berg A. Silicon and glass Micromachining. In: Herold KE, Rasooly A, editors. Lab-on-a-Chip Technology for Biomedical and Biological Applications. Volume 1: Fabrication and Microfluidics. Caister Academic Press, Norfolk, UK; 2009. pp. 83–114. ISBN: 978-1-904455-46-2.

[43] Li CW, Cheung CN, Yang J, Tzang CH, Yang M. PDMS-based microfluidic device with multi-height structures fabricated by single-step photolithography using printed circuit board as masters. Analyst. 2003; **128**: 1137–1142. DOI: 10.1039/B304354A.

[44] Bartholomeusz DA, Boutté RW, Andrade JD. Xurography: rapid prototyping of microstructures using a cutting plotter. Journal of Microelectromechanical Systems. 2005; **14**: 1364–1374. DOI: 10.1109/JMEMS.2005.859087.

[45] Etsion I. State of the art in laser surface texturing. Journal of Tribology. 2005; **127**: 248–253. DOI: 10.1115/1.1828070.

[46] Castelo A, Nieto D, Bao C, Flores-Arias MT, Pérez MV, Gómez-Reino C, López-Gascón C, De La Fuente GF. Laser backwriting process on glass via ablation of metal targets. Optics Communications. 2007; **273**: 193–199. DOI: 10.1016/j.optcom.2006.12.005.

[47] Estepa LC, De la Fuente GF. Continuous Furnace with Coupled Laser for the Surface Treatment of Materials. U.S. Patent 20090230105 A1, 7 March 2006.

[48] Rodiño-Janeiro BK, González-Peteiro M, Ucieda-Somoza R, González-Juanatey JR, Álvarez E. Glycated albumin, a precursor of advanced glycation end-products, up-regulates NADPH oxidase and enhances oxidative stress in human endothelial cells: Molecular correlate of diabetic vasculopathy. Diabetes-Metabolism Research and Reviews. 2010; **26**: 550–558. DOI: 10.1002/dmrr.1117.

Chapter 13

Sol-Gel Applications in Textile Finishing Processes

Nurhan Onar Camlibel and Buket Arik

Additional information is available at the end of the chapter

http://dx.doi.org/10.5772/67686

Abstract

In the chapter, sol-gel applications in textile finishing process such as flame retardancy, water, oil repellency, ultraviolet (UV) protection, self-cleaning, and antibacterial and anti-wrinkle processes are reviewed. Sol-gel technology is well known in materials, metallurgy, ceramic, and glass industry since 1960 and has been researched in textile industry in the last decade. Sol-gel technology has some advantages when compared to conventional textile finishing process. Sol-gel technology, which is a method applied to the inorganic metal alk-oxide or metal salts to organic textile materials, could impart the high, durable activity and multifunctional properties to different textile materials in the same bath at one step using low concentration of precursors. In addition, sol-gel technology presents alternatively economical, ecological, and environmental friendly process due to one-step application, using low concentration of chemicals, nonhalogenated chemicals, and nonformaldehyde release when compared to conventional processes.

Keywords: sol-gel, textile, cotton, polyester, flame retardancy, water, oil repellency, antibacterial, antiwrinkle, UV protection, self-cleaning, photocatalytic, multifunctional, silica, titania, silver, trialkoxy silanes

1. Introduction

Functional properties can be given to textile materials by textile finishing processes. The use of sol-gel technology in textile finishing processes has taken very much attention in the last decades. Sol-gel technology has advantages such as less chemical usage, low-temperature treatment, easy application and treatment facilities in textile mills, and no requirement for special equipment [1, 2].

Moreover, sol-gel technology makes enabling multifunctional properties in one step which is not possible by conventional textile finishing methods because of incompatibility of chemical materials. In conventional textile finishing methods, the treatments have separately been applied in several steps to obtain functional properties and these multi-step processes increase the costs of the textile mills. Therefore, enabling multifunctional properties in one step is a challenging process to overcome in terms of decreasing water and energy consumption and costs. On the other hand, besides the various advantages, sol-gel technology has a disadvantage of high costs of precursor materials used in sol-gel technology. In sol-gel technology, metal alkoxides and metallic salts have been used as precursors and by the catalytic effect of an acid or alkali, hydrolysis and condensation reactions occur [1, 2]. The reactions of precursor materials with textile surfaces are given in **Figure 1**. Nanosols can be applied to fabric by pad-dry-cure process. Application steps of nanosols to textiles are illustrated in **Figure 2**.

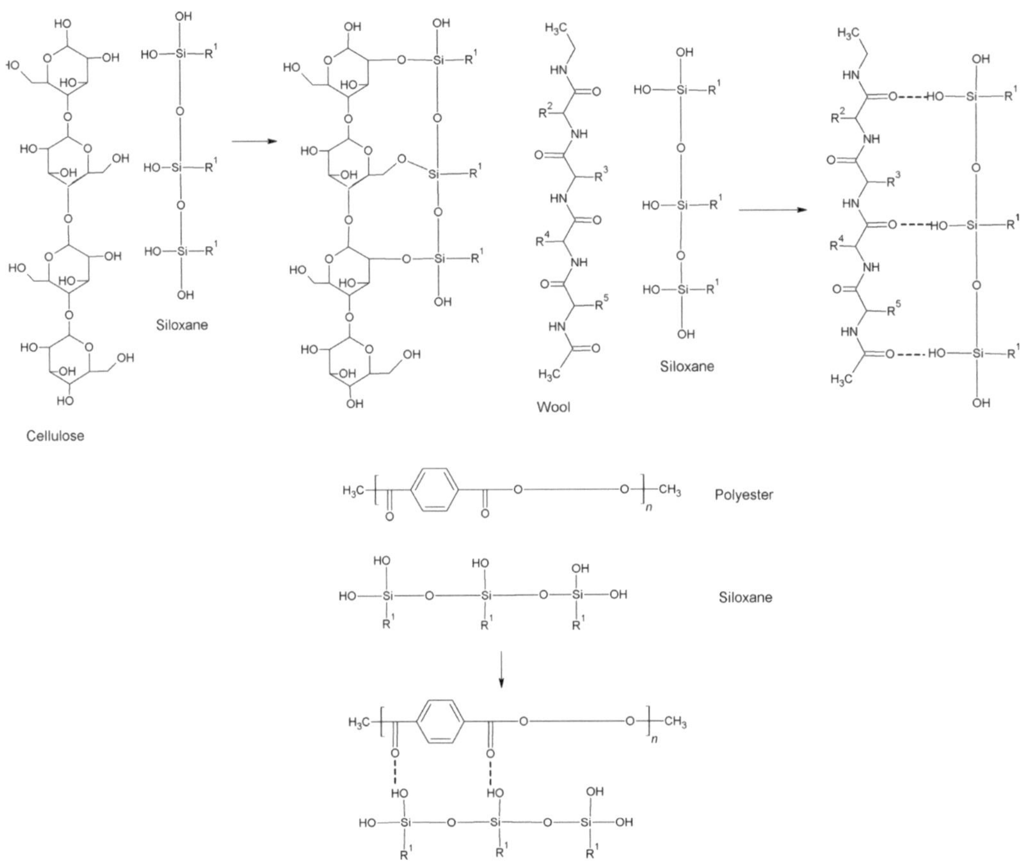

Figure 1. The reactions of precursor materials with some textile surfaces.

By sol-gel technology, various functional properties can be given to textile materials or some properties that already exist can be developed in one or more steps. The use of sol-gel technology in textile industry for flame retardancy, antimosquito [3], water or oil repellency [4], antibacterial activity [5], antiwrinkle, anticrease, durable press or easy-care effect [6], ultraviolet (UV) protection, self-cleaning or soil repellency, photocatalytic activity [7], development in fastness behavior, abrasion resistance, tensile and thermal properties, and in microencapsulation process have been studied in detail. In **Figure 3**, the application areas of sol-gel technology in textile finishing processes have been demonstrated. The aim of this review is to introduce the usage possibilities of sol-gel technology in textile applications.

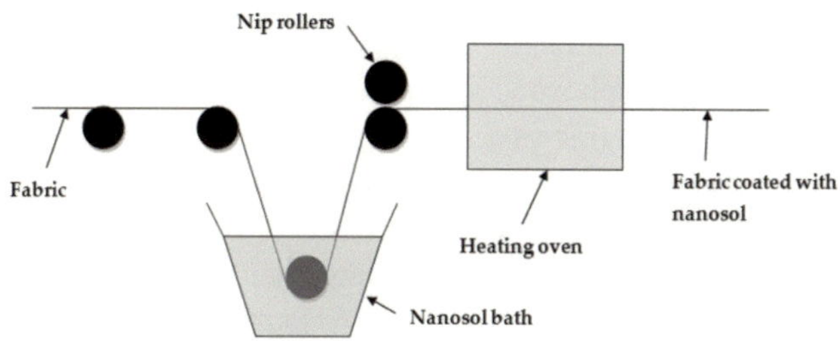

Figure 2. Application steps of nanosols to textiles.

Figure 3. Application areas of sol-gel technology in textile finishing process.

2. Textile finishing applications by sol-gel method

2.1. Flame retardancy finishing

There has been a significant importance to improve flame retardancy properties of textile materials for hindering the fire damages because they are primary combustion sources in a fire. Chloride-, bromide-, phosphorus-, antimony-, and boron-based materials have been used for textile materials to give flame-retardant activity. In addition, the combination of these chemical materials (e.g., antimony-halogen or phosphorus-nitrogen) can also be used due to their synergistic effect. By sol-gel method, high chemical concentration (nearly 300–500 g/l) used in conventional methods can be significantly decreased or by the use of chemical materials without halogen, ecological, and economical flame-retardant activity can be obtained. Alongi and Malucelli [8] reported the detailed review of flame retardancy treatment of cotton fabrics by sol-gel process until 2015.

Grancaric et al. [9] studied the flame retardancy properties of cotton fabric by using diethylphosphatoethyltriethoxysilane (DPTES) as a sol-gel precursor together with monoethanolamine (MEA) as a neutralizer for acidic conditions of DPTES with or without the addition of tetraethoxysilane (TEOS) and 3-glycidoxypropyltriethoxysilane (GPTES) to increase the durability of finishing process. Their results showed that the cotton samples were in self-extinguished classification and had the LOI (limited oxygen index) value of 29. It was observed that the amount of residue of treated fabrics at T_{max} compared with that of untreated fabric increased from 37.5 to 60% and the total heat release (THR) of treated fabrics with respect to untreated fabric decreased from 12 to 4.2 kJ/g. Thermal shielding effect of silica phases and char-forming by synergistic effect of phosphorous and nitrogen were attributed to be the causes of flame retardancy and thermal behavior of cotton samples. However, the LOI values of fabric samples reduced from 29 to 21 after one washing process [9].

Boukhriss et al. [11] investigated the flame retardancy and water repellency properties of cotton fabric treated with sols containing 1-methylimidazolium chloride propyltriethoxysilane (MCPTS) and 1-pyridinium chloride propyltriethoxysilane (PCPTS) salts synthesized. The results indicated that their flame retardancy and water repellency properties improved. Water uptake values with dip test with regard to untreated fabric decreased from 340 to 50%. The weight of residue of treated fabric samples regarding untreated fabric increased from 1 to 93% which is higher than the test results (48%) reported by Alongi et al. [10]. The fabric samples lost their flame retardancy properties after three washing process and entirely burnt [11].

Kappes et al. [12] coated the cotton, PET fabric, and polyester/cotton blends with (3-trimethoxysilylpropyl)diethylenetriamine (TRIAMO) by sol-gel process. Then, coated and uncoated fabrics were treated with phenylphosphonic acid. Hole formation, flame spread to upper or side edges, and burning time after flame application of samples were evaluated by surface and edge-ignition tests according to EN ISO 15025. The two-step treatment of cotton and blend fabrics improved their flame retardancy properties and burning and hole formation after flame treatment of the samples were not observed while cotton fabrics treated with only sol-gel and only phenylphosphonic acid treatment showed burning and hole formation. Polyester fabric samples displayed hole formation and burning after flame treatment, but flame did not reach the upper or side edges for the fabric samples treated with two-step process and only phenylphosphonic acid due to recede of fabric from flame. Sol-gel treatment gave rise to prolong burning time after flame treatment and to prevent fabrics receding from flame that is known to be undesired. Peak heat release rate (pHRR) of all fabric samples treated with two- and one-step process decreased compared with untreated fabric. The total heat evolved (THE) of polyester fabric slightly increased, while these of cotton and blend fabrics slightly decreased [12].

Zhang et al. [13] treated silk fabric to give flame retardancy by using TEOS and boric acid by means of sol-gel process. They used 1,2,3,4-butanetetracarboxylic acid (BTCA) as crosslinking agent to improve durability of flame retardancy properties of the fabric samples. The results showed that LOI values of silk fabric containing P, N, or S with inherent difficult flammability increased from 25.3 to 32.3% by treatment with BTCA solution and then nanosol-containing boric acid as a flame-retardant agent. They found that the flame retardancy of silk fabric was durable to even 30 washing cycles. They demonstrated that the pHRR, temperature at pHRR, the total heat release (THR), weight loss, and the density of released smoke of the fabric samples compared to control fabric decreased as indicating low flammability [13].

Šehić et al. [14] investigated the synergistic effect of the presence of Si- and P- on polyamide 6 fabric surface treated by sol-gel process using 9,10-dihydro-9-oxa-10-phosphaphenanthrene-10-oxide (DOPO)-modified vinyltriethoxysilane (VTS) and tetraethoxysilane (TEOS) as precursors. The increase in the percentage of residue at T_{max1} and the decrease in T_{onset} were attributed to promoting char formation. The heat release capacity (HRC), pHRR, and THR of fabric samples were found to be reduced. Ignition of cotton under burning samples, after-flame, and after-glow times were analyzed by vertical flame-spread test, DIN 53906. The combination of Si and P and the presence of only Si contributed to no dripping, no ignition of cotton substrate under burning sample, longer after-flame time, and no after-glow. Synergistic effectiveness values of samples proved the synergistic activity of Si and P in condensed phase. **Figure 4** shows the combustion cycle of polymer [15].

Deh et al. [16] studied the synergistic effect of N and Si added to phosphorous compounds on flame retardancy of cotton fabric treated by phosphorylation using phosphoric acid and urea and then by sol-gel process using TEOS as a precursor. They found that the cotton fabrics treated by phosphorylation and sol-gel process had high LOI values (>25%) with durability to 10 washing processes. The presence of Si together with N and P on cotton fabric contributed to developing the formation of char acting as a thermal-insulating barrier and to decrease the number of byproducts which are not promoting the dehydration process, is requested flame retardant activity. Dehydration of cotton at low temperature was observed to support char formation and synergistic effect reduced the formation of levoglucosan whose decomposition generates flammable gases and product after pyrolysis. **Figure 5** illustrates the thermal degradation of cotton with heating [16].

Liu et al. [17] synthesized di-(triethoxysilylpropyl) phenylphosphamide (PPD-PTES) by reaction of 3-aminopropyl triethoxysilane (APTES) and phenylphosphonic dichloride (PPDC) and then treated cotton fabric with PPD-PTES for 5, 10, 20, and 30 min of immersion times by sol-gel process. As a result of vertical burning test, the fabric treated with PPD-PTES containing P-, Si- and N-elements self-extinguished as ignition source was found to be withdrawn and minimum char area left was observed on the fabric treated with 30 min of immersion time. As long as immersion times were increased, char residue increased while T_{max}, pHRR, and THR decreased. Nevertheless, they found that the absorbance intensity of CO_2, which is one of the non-flammable gases, did not significantly change in treated fabrics compared to that of untreated fabric while their intensity of C=O and C-O-C groups, which are found

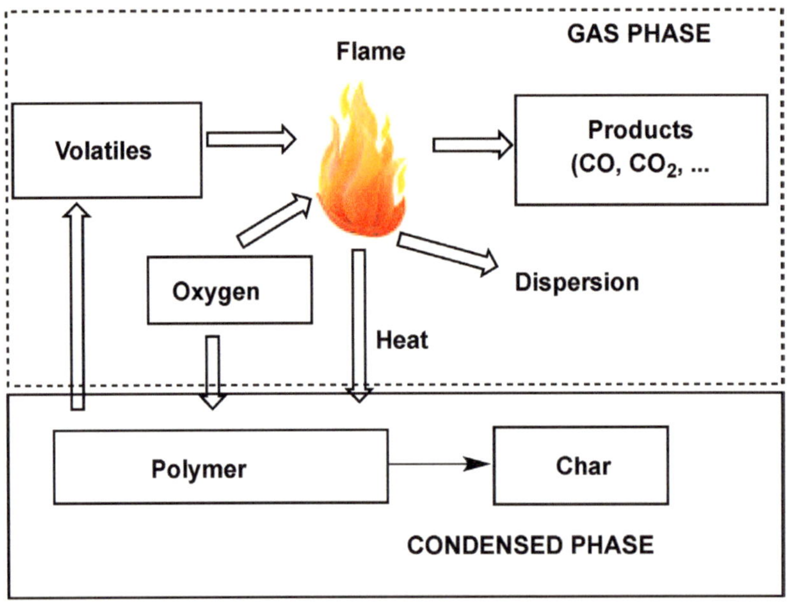

Figure 4. Schematic illustration of combustion cycle of polymer [15].

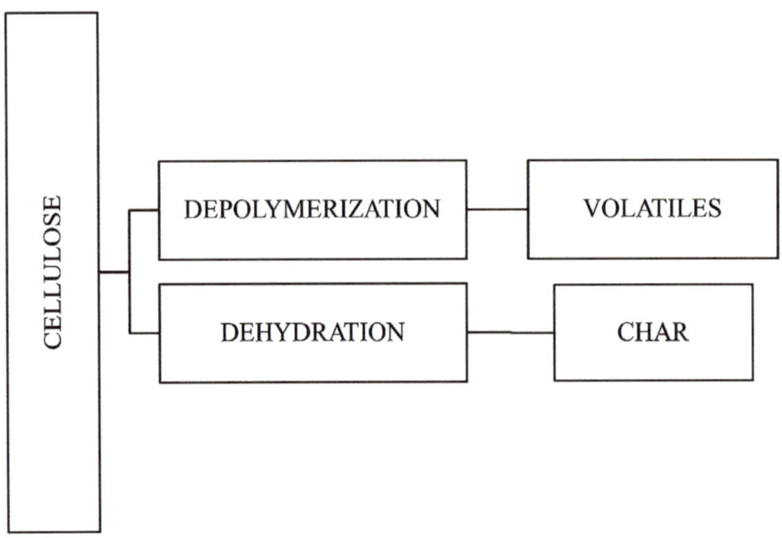

Figure 5. Schematic illustration of the thermal degradation of cotton with heating. Copyright 2015, Royal Society of Chemistry [8].

in flammable gases and compounds, decreased by virtue of Fourier transform infrared spectroscopy(FTIR) analysis [17].

Ren et al. [18] investigated the flame retardancy properties of polyacrylonitrile (PAN) fabric treated by one-step sol-gel process using TEOS and polyphosphoric acid (PPA). The increase at their LOI and char residue to, respectively, 30.1 and 69.48% and the decrease at their THR, pHRR, the fire growth rate index (FIGRA), average mass loss rate (aMLR), the peak of smoke production rate (pSPR), and total smoke production (TSP) as a result of cone calorimetry test [18] were determined.

Aksit et al. [19] proved the synergistic effect of Si, P, and N on flame-retardant cotton fabrics treated with (3-aminopropyl) triethoxysilane as N- and Si-source, (3-glycidyloxypropyl) trimethoxysilane as Si-source, and guanidine phosphate as P-source by sol-gel technique. They determined that the fabric samples had 45.7% LOI values [19].

2.2. Water or oil repellency finish (hydrophobic effect)

Hydrophobic effect can be given to textile materials by decreasing the surface tension of textile material against the liquid [1]. In conventional methods, fluorocarbon components have been used for this purpose. However, sol-gel technology as an alternative method has also been studied because of the harmful effects of fluorocarbon components to the environment. Many researches have been conducted about the use of modified silane precursors by sol-gel technology in order to enable hydrophobic effect on textile materials. It was aimed to gain surface roughness together with creating low-surface energy on substrate surface based on the theory of Wenzel or Cassie and Baxter [20–22] in order to produce superhydrophobic materials in many researches. Moreover, the durability to abrasion, washing and UV exposure, and so on of their superhydrophobic properties for especially outdoor materials were maintained as a significant challenge. Schematic illustration of hydrophilicity-hydrophobicity properties of substrate as contact angles is given in **Figure 6**.

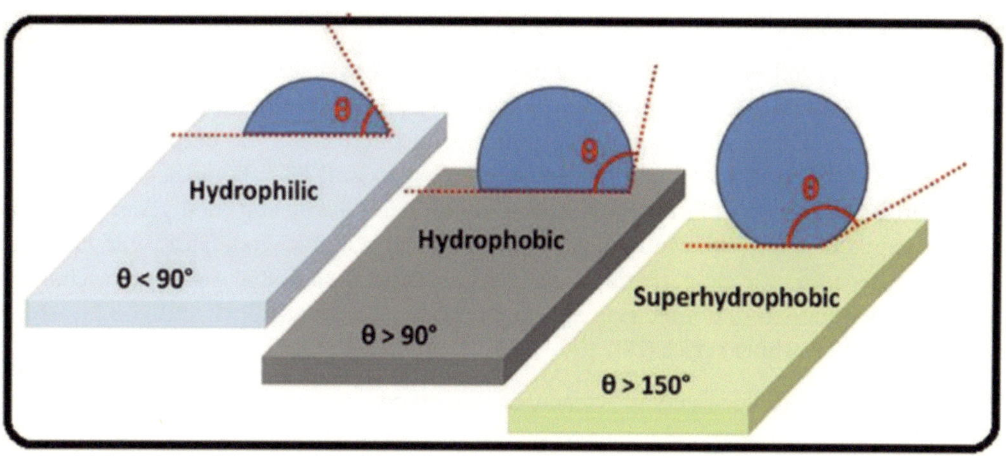

Figure 6. Schematic illustration of hydrophilicity-hydrophobicity properties of substrate as contact angles [23].

Zhao et al. [24] treated the cotton, cotton/polyester, and polyester fabric by using nanosol containing silica nanoparticles, (3-aminopropyl) triethoxysilane (APTES, 99%), and hexadecyltrimethoxysilane (HDTMS) by spray process. They achieved the production of fabrics with contact angle (CA) >150° and hysteresis <10° required to gain superhydrophobic properties to fabric. The increasing amount of HDMS or silica nanoparticles led to higher contact angle and lower-surface energy than that of uncoated fabrics. Surface roughness values measured by atomic force microscopy (AFM) were found to increase for cotton and cotton/polyester fabric compared to uncoated fabrics. The cotton and polyester fabric did not lose their superhydrophobic properties after five washing processes. In addition, the fabrics kept their hydrophobic properties even after 600 abrasion cycles [24].

Xue et al. [25] coated polyester fabric with polymer paste containing silica nanoparticles and polydimethylsiloxane (PDMS) by spread coating method and then posttreated by nanosols containing TEOS and cetyltrimethoxysilane (CTMS). They determined the hydrophobic properties (CA >90°) of fabrics coated with PDMS-silica nanoparticles and superhydrophobic properties (162.5 of CA, 5.4 of sliding angle (SA)) of fabric posttreated by TEOS and CTMS. From the study, it was observed that water pressure resistance properties (38.6 kPa) of fabrics developed, their CAs a little decreased, and their SAs slightly increased as the number of coatings increased. Their contact angles and water resistance properties did not demonstrate significant change after basic and acidic treatment with different pH values or after UV-light exposure. However, water pressure resistance of fabrics decreased to 10 kPa after abrasion test for even 10 cycles because of the insufficient binding force between coating layer and fabric [25].

Hao et al. [26] investigated polymer nanocomposites (PNs) synthesized by *in situ* sol-gel process to achieve multifunctional properties of cotton fabrics. Crosslinked polysiloxane (CLPS) with aminopropyltriethoxysilane (APTES) and then obtained APTES-CLPS with silica sols (CLPS-SiO_2) were treated by *in situ* sol-gel process. The cotton fabric was immersed in APTES-CLPS or CLPS-SiO_2 solutions in ethyl acetate and padded, dried, and cured. They concluded that the fabric samples treated with CLPS-SiO_2 had better thermal stability with lower weight loss in TGA analysis, higher the root-mean-square roughness (4.528 nm) than that treated with APTES-CLPS due to the presence of silica nanoparticles. The contact angles of the fabric reached 158° and they did not lose their superhydrophobic properties until 10 washing cycles [26].

Zhang et al. [27] produced an oligomer of hexadecyltriethoxysilane (HD-oligomer) (HDTES) and HDTES-modified silica nanoparticles (HD-silica) by Stöber method based on sol-gel process. In the second step, HD-silica/HD-oligomer nanocomposites were prepared by adding HDTES, TEOS, ethanol, water, and ammonia to HD-silica and HD-oligomer. Subsequently, wool fabrics were immersed in nanocomposite solution in toluene and triethylamine as catalyst and exposed to ultrasonication to produce superhydrophobic wool fabric with excellent abrasion resistance, excellent chemical and environmental stability, excellent mechanical resistance for adhesion of double side tape and scratching with a scalpel, and good washing durability for 10 washing cycles [27].

Vasiljević et al. [28] studied the improvement of water repellency properties of the cotton fabric by first treating with oxygen plasma and then with 1H,1H,2H,2H-perfluorooctyltriethoxysilane (SiF) in ethanol by sol-gel process. The fabrics pretreated by plasma for 20 s at an operating current of 0.3 A resulted in nearly 150° of contact angle and 24° of sliding angle [28].

Przybylak et al. [29] previously impregnated cotton fabric with octaanion solution synthesized by using TEOS, tetramethylammonium hydroxide, methanol, and water and squeezed, dried, and subsequently impregnated with bifunctional silsesquioxanes containing octafluoropentyloxypropyl groups and trimethoxysilylethyl synthesized. Squeezed, dried, and cured fabric samples were found to have superhydrophobic properties with 151° of contact angle and durability against 10 washing cycles [29].

Xue et al. observed that the fabrics treated with nanosols containing tetrabutyl titanate and then with 1H, 1H, 2H, 2H-perfluorodecyltrichlorosilane in toluene and/or stearic acid in acetone and cured at 110°C for 1 h had superhydrophobic properties and good UV-protection properties [30].

Onar and Mete [31] investigated water repellency properties of the cotton fabric treated with two-step process as surface roughening with ZnO, Al_2O_3, TiO_2, and ZrO_2 nanoparticles and hydrophobic modification with long-chain alkyl silanes (hexadecyltrimethoxy silane). From the study, it was observed that the cotton fabric treated with Al_2O_3 nanoparticles and HDMS had superhydrophobic properties and all fabric samples treated with nanoparticles and HDMS maintained their hydrophobic properties after washing process [31].

Onar et al. [32] purposed to achieve the production of water-repellent cotton fabric by sol-gel process using tridecafluorooctyltriethoxysilane (FTES), hexadecyltrimethoxysilane (HDTMS), glycidyloxypropyltriethoxysilane (GLYE), phenyltriethoxysilane (FES), vinyltrimethoxysilane (VTEO), 3-aminopropyl trimethoxysilane (AMMO), titanium (IV) isopropoxide (TIPT), and zirconium (IV) acetylacetonate (ZrAc) as precursors. They found that the treatment with HDTMS under acidic and alkaline conditions or FTES under alkaline conditions led to provide superhydrophobicity on cotton fabric. The fabric samples treated with FES, VTEO, TIPT, and GLYE precursors lost their hydrophobic properties after washing process, while the fabrics treated with HDMS, FTES, AMMO, and ZrAc precursors preserved the hydrophobic properties after washing process [32].

Colleoni et al. [33] treated the cotton and polyester fabric with octyltriethoxysilane (OTES) together with melamine-based crosslinking agent. They revealed that the cotton and polyester fabrics had good hydrophobic properties with, respectively, 130° and 150° of contact angle values [33].

Gao et al. [34] researched the hydrophobic properties of the cotton and polyester fabric treated with TEOS and then HDTMS by sol-gel process (**Figure 7**). The process imparted highly hydrophobic properties to the cotton (155° CA) and polyester fabric (143° CA) with durability till 30 washing processes and a slight reduction of their tensile strength (<10%) [34].

Teli and Annaldewar [35] studied the water repellency and UV-protection properties of cotton fabric dip coated with silica nanoparticles prepared by sol-gel process and subsequently impregnated with zinc nitrate hexahydrate in hexamine and then hydrophobically modified with ethanolic sodium stearate solution. As a result of that, the fabric samples imparted excellent UV protection with +50 UPF and superhydrophobic properties. The fabric samples had still excellent UV protection and hydrophobic properties after 10 washing cycles [35].

de Ferri et al. [36] coated the silk fabric with nanosols composed of 3-glycidoxypropyltrimethoxysilane (GLYMO) and DynF8815 with acid catalyst by sol-gel process. The treatment caused the high abrasion resistance to 8500 cycle, high contact angle (nearly 123° CA) to four oils, and hydrophobic properties with 148° of contact angle. The washing durability of oil and water repellency of the fabric samples was also analyzed. It was observed that the fabric had still hydrophobic (130° CA) and oleophobic properties (100° of CA) after, respectively, 15 washing and 5 washing cycles [36].

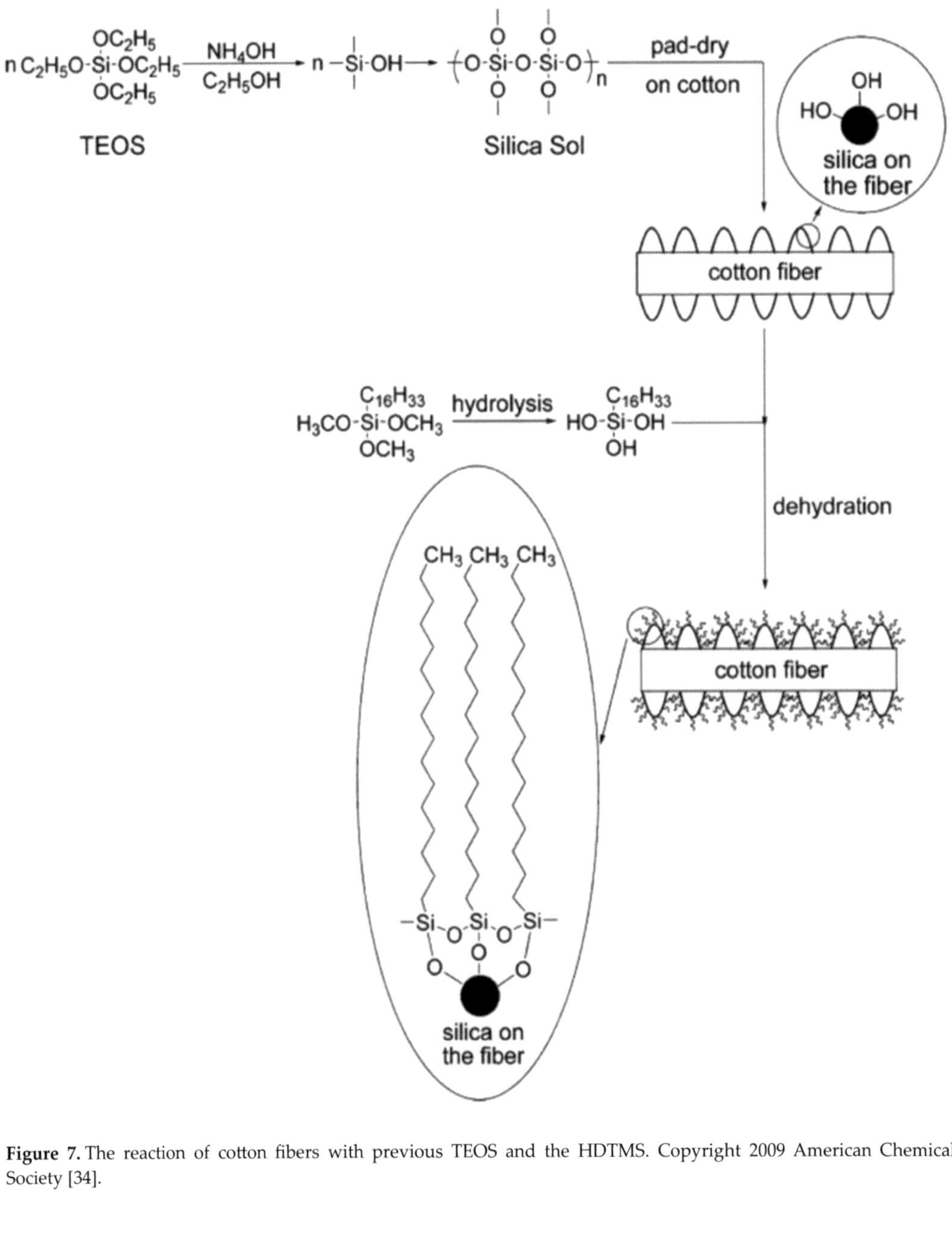

Figure 7. The reaction of cotton fibers with previous TEOS and the HDTMS. Copyright 2009 American Chemical Society [34].

Vasiljević et al. [37] aimed to produce highly oleophobic and superhydrophobic cotton fabric by three-step sol-gel process. The first step was the treatment of cotton fabric with silica nanoparticles with different average particle sizes (50 ± 15, 230 ± 20, and 780 ± 30 nm) prepared according to Stöber method based on sol-gel process; the second step was the application of the *in situ* growth of polysiloxane layer on cotton fabric in basic TEOS solution, and the third step was the padding with a fluoroalkyl functional water-borne oligosiloxane (FAS, Dynasylan F 8815). They obtained fabric samples with low sliding angle (2°), high oleophobicity (6), and superhydrophobicity durable to 10 washing cycles [37].

2.3. UV protection, self-cleaning, and soil repellency (photocatalytic effect)

Long-time exposure to UV light may result in early skin aging, allergy, sun burns, or even skin cancer in especially fair-skinned human. Various chemical agents such as benzophenone derivatives, ortho-hydroxy phenyl, and diphenyl triazine derivatives as organic UV absorbers and TiO_2 and ZnO particles as inorganic UV absorbers have been used [38]. Alternatively, UV protection can be obtained by sol-gel method with generally TiO_2 and ZnO nanoparticles incorporated to textile materials. TiO_2 nanoparticles provide self-cleaning as well as UV protection by means of photocatalytic reaction. Anatase phase of TiO_2 nanoparticles, which is one of three phases of TiO_2 (anatase, brookite, and rutile), has photocatalytic activity. The formation of anatase phase at low temperature on textile material coated with titania nanosols is a big challenge because textile materials cannot endure high temperature [39]. Researchers are investigating some solutions such as steaming process [40, 41], hydrothermal process [42], or solvothermal process [43] to produce anatase crystals at low temperature so far.

Photocatalytic reaction leads to the decomposition of organic and inorganic pollutants under the ultraviolet light [43]. In addition, organic components such as microorganisms in textile surfaces and dyestuff residues in textile wastewater can also be decomposed by photocatalytic effect. By this way, photocatalytic effect ensures not only UV protection and self-cleaning but also antibacterial effect and ecological and economical wastewater treatment [43, 44]. **Figure 8** illustrates the self-cleaning properties of cotton fabric treated with titania nanosols.

Liu et al. [46] investigated the self-cleaning and UV-protection properties of raw wool fabric and Kroy-process wool fabric (removed scales with chlorination process) treated with polycarboxylic acid such as 1,2,3,4-butanetetracarboxylic acid (BTCA), citric acid (CA) as crosslinking agent and then nanosols prepared by using titanium (IV) butoxide (TBT) and TEOS as precursors. The self-cleaning properties of the fabric samples were evaluated by photo-induced decomposition test of methylene blue. Self-cleaning properties of raw wool fabric were found to be better than that of Kroy-process wool fabric, while UPF values of raw material (356) were lower than that of Kroy-process wool fabric (992) [46].

Dhineshbabu et al. [47] studied the UV-protection, antibacterial, and flame retardancy properties of cotton fabric treated with colloidal sols containing TEOS as precursor and polyethylene glycol (PEG) as stabilizer and TiO_2 nanoparticles produced by sol-gel process. They found that UPF values of the fabric samples were greater than 50 UPF corresponding to excellent UV protection after even 10 washing cycles. According to ASTM D1230 standard, their results showed that the total burning time and residual weight after burning of the fabric samples were higher than that of untreated fabric according to the flame test at 45° angle. Thus, their burning performance improved after the treatment and nearly 70% of this performance was preserved after 10 washing cycles. The study revealed that the fabric samples had good antibacterial activity against *Staphylococcus aureus* and *Escherichia coli* with, respectively, 18 and 15 mm diameter of inhibition zone because of the reactive oxygen species (ROS) generation mechanism of TiO_2 nanoparticles [47].

Rilda et al. [48] previously produced TiO_2-SiO_2 powder by using titanium isopropoxide and TEOS as precursors with different molar ratios, diethanol amine (DEA) as stabilizer, isopropanol as solvent, and chitosan in acetic acid as dispersing agent by sol-gel process. The fabric samples coated with acrylic acid binder were immersed to TiO_2-SiO_2 suspension containing the powder, surfactant, and isopropanol and then coated with the suspension to ensure homogeneous coating by spin-coating machine. Falling at absorbance of methylene blue under UV irradiation for 120 min was attributed to self-cleaning properties of fabrics treated with TiO_2-SiO_2 powders with especially 1:2 molar ratios [48]. The discoloration of wine stain on cotton fabric treated with TiO_2-SiO_2 colloidal solution increased exposure times under solar light simulator for 0, 4, 8, and 24 h [49].

Wang et al. [39] treated the cotton fabric with TiO_2 acidic hydrosol prepared with titanium (IV) butoxide as precursor by dip-pad-steam process. Photocatalytic activity of TiO_2 prepared by low-temperature steaming process was higher than that of TiO_2 dried at 150°C. They stated that the fabric samples had excellent durable self-cleaning and UV-protection properties. The crystallinity and photocatalytic activity of TiO_2 improved with the increase of the water content in TiO_2 hydrosol [39].

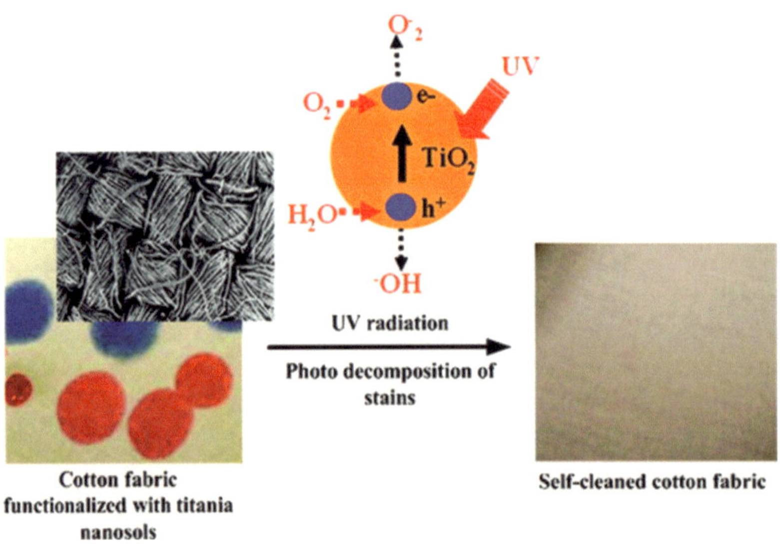

Figure 8. Self-cleaning properties of cotton fabric treated with titania nanosols. Copyright 2009 American Chemical Society [45].

Shaban et al. [50] investigated the self-cleaning properties of cotton fabrics coated with ZnO nanoparticles or ZnO solution containing zinc acetate dihydrate as precursor, 2-methoxy-ethanol as solvent, and monoethanolamine as stabilizer by sol-gel spin-coating process. Their results indicated that ZnO nanoparticles had photocatalytic activity especially under UV irradiation due to their high band-gap energy and the methyl orange dye on cotton fabrics coated with ZnO nanoparticles or ZnO solution degraded under sunlight and 200-W lamp illumination [50].

Noorian et al. [51] prepared precursor solutions based on copper sulfate and/or zinc chlorite containing folic acid, NaOH, and water by in situ synthesis. They immersed the cotton fabric in the solutions, dried and washed. They stated that the combination of ZnO and Cu_2O and the addition of folic acid improved the UV protection and anticrease properties of the fabric samples with, respectively, 87.31% enhanced UVP and 100.75° of crease recovery angle [51].

2.4. Antibacterial and antimicrobial activity

Antibacterial and antimicrobial activities are important for sport, outdoor, home, automotive, and medical textiles and for garments like hosiery and underwear. For antibacterial and antimicrobial textile finishing, various biocides like silver, triclosan, chitosan, quaternary ammonium salts, N-halamines, biguanide derivatives, synthetic dyes, and peroxyacids have been applied to textile surfaces [52–58]. Sol-gel technology as an alternative method can bring advantages like ecological treatment, less chemical use, low-temperature processing, low toxicity to human health, protection of inherent properties of textile material, and excellent washing and usage durability in this finishing process [52, 55, 59–61].

Different types of sol-gel systems are known to have antibacterial or antimicrobial effect. These systems are photoactive titania coatings with anatase modification and sol-gel coatings with embedded colloidal metal or metal compounds like silver, silver salts, copper compounds, zinc, or biocidal quaternary ammonium salts [5, 62–65]. On the other hand, these systems have several disadvantages like high operation temperatures to produce highly photoactive thin films and the use of strong acids to keep aqueous sols in the peptized state that cause destruction on textiles. In addition, titania coatings require UV radiation to perform antimicrobial or antibacterial activity [66–69]. Because of harmful effects of strong acids and high temperatures on textiles, in recent methods, organic solvents, for example, alcohol-based solutions that can be applied at low temperatures, were developed [5, 59, 60]. Simoncic et al. compiled the antibacterial activity of silver nanoparticles prepared by *in situ* method and colloidal solutions [70].

Poli et al. prepared zinc-based coatings by sol-gel method in neutral hydro-alcoholic medium and applied on cotton fabric to obtain antibacterial activity. They reported that this method was simple, cheap, reproducible, and less harmful when compared to conventional methods [5]. Other than titania-based sol-gel coatings, silica-based coatings are also under examination for antibacterial or antimicrobial effectiveness. To enable the antibacterial or antimicrobial effect, polycationic components are incorporated within the silica layer matrix and by this means positively charged polycationic components interact with negatively charged microbial cell membranes and damage their cellular functions [5, 61, 62, 63]. The use of these systems was researched in detail [5, 59–63, 67].

Mahltig et al. reported that the sol-gel processing exhibited the antimicrobial activity based on controlled release, contact action, or photocatalytic activity [71].

Rivero and Goicoechea presented the detailed review about improving antibacterial properties of textiles by sol-gel process [72].

Zhang et al. classified the application methods on cotton methods for colloidal suspensions of metal nanoparticles and precursor solutions of metal ions, respectively, titled as sol and solution to gain antimicrobial properties [52].

Mohamed et al. [73] prepared the colloidal solutions of silver nanoparticles (10–25 nm) synthesized by chemical reduction process using dextran as stabilizing and reducing agent and modified TEOS using ascorbic acid as scavenging agent and (3-aminopropyl)triethoxysilane (APTEOS) by using BTCA or vinyltriethoxysilane (VTEOS) and applied it to cotton fabrics by sol-gel process by dip coating. Silver nanoparticles modified with TEOS and VTEOS were cured with UV irradiation by means of a photoinitiator. They proved that the modified fabrics had antibacterial activity against *S. aureus* and *E. coli* with nearly 90% of bacterial reduction even after 20 washing cycles and thermo-regulating properties [73].

The combination of antibacterial biopolymers with titania and silica matrix is a different method and this method is observed to give ecological advantage and possessed antibacterial activity both in the presence or absence of light. Arik and Seventekin [55] studied chitosan/titania and chitosan/silica hybrid coatings in terms of antibacterial activity. Sol-gel method was used to prepare coating solutions and then solutions were applied to cotton fabric. From the study, hybrid coatings were found to have better antibacterial activity and washing durability than only chitosan, only titania, or only silica coatings [55].

2.5. Antiwrinkle, anticrease, durable press or easy-care effect

Antiwrinkle, anticrease, durable press, or easy-care effect has been enabled by the fixation of the crosslinking agents to textile materials. Since dimethyloldihydroxyethylene urea (DMDHEU), which was used commonly in conventional antiwrinkle finishing treatments, releases free formaldehyde that is harmful to human health and environment and causes negative effects due to the acidic catalysts on tensile properties and whiteness index of textile materials, new alternative crosslinking agents and methods have been recently researched. By means of sol-gel technology, it is possible to enable antiwrinkle effect herewith good tensile and whiteness properties by the use of silane coupling agents [74, 75] like glycidoxy propyltrimethoxysilane as epoxy silanes, vinyltrimethoxysilane as vinyl silanes [76, 77], and tetraethoxysilane that do not release formaldehyde [78–80].

The main cause of the tensile strength loss and yellowing in conventional antiwrinkle finishing is the disruptive effect of the acidic catalyst on fibers at high temperatures. But in sol-gel method, metallic alkoxides and organic solvents are used in mild conditions that provide tensile strength and color retention [78–80].

In antiwrinkle sol-gel applications, mixed sols with conventional agents can also be applied and different combinations like DMEU/SiO_2 (dimethylolethyleneurea/silica), TEOS-TTB/

DMDHEU (tetra ethoxysilane-titanium (IV) n-butoxide/dimethylol dihydroxyethylene urea), BTCA-MAH/SiO_2 (1,2,3,4-butanetetracarboxylic acid-maleic anhydride/silica), BTCA-SHP/TEOS-GPTMS (1,2,3,4-butan tetracarboxylic acid-sodium hypophosphite/tetraethoxysilane-glycidylpropyloxytrimethoxysilane), DMDHEU/TEOS-GPTMS were evaluated for antiwrinkle finishing and their effects on physical properties of cotton in previous studies [75, 78–81]. These combinations were also found to have advantages in terms of abrasion resistance [75], heat resistance [78], and UV-light resistance [79] as well as antiwrinkle effect.

Schramm and Rinderer treated the cotton fabric with BTCA and/or nanosol containing GPTMS together with metal alkoxides such as aluminum isopropoxide (AIP), titanium tetra isopropoxide, and zircon tetrabutoxide (ZTB) or hydrophobic trialkoxysilanes such as methyltriethoxysilane (MTEOS), octyltriethoxysilane (OTEOS), or Dynasylan F8815 (fluoroalkyl functional oligosiloxane) to impart antiwrinkle properties. Treatment only with metal alkoxides led to no improvement of their anticrease properties, while treatment with GPTMS developed their anticrease properties. Moreover, the addition of AIP and OTEOS or MTEOS to GPTMS solution caused the increase of dry-crease recovery angle (DCRA) of the fabrics treated with the solution, respectively, from 263° to 289°, 290°, and 294°. Tensile properties of fabric treated with GPTMS solution decreased 10% when compared to that of untreated fabric. The addition of AIP solution to GPTMS solution brought on further 10% loss of their tensile strength, while the addition of OTEOS and MTEOS solutions to solution containing GPTMS and AIP induced to increasing 10% of the tensile strength with regard to tensile strength of fabrics treated with solution containing GPTMS and AIP. The treatment with GPTMS solution containing AIP or AIP together with F8815, MTEOS, and OTEOS solutions ensured the enhancement of contact angle values of the fabrics, respectively, from hydrophilic to 112°, 147°, 145°, and 136° corresponds to excellent hydrophobic properties. They also found that BTCA pretreatment improved the anticrease properties of the fabrics treated with GPTMS solution until 299° of DCRA and led to further loss of tensile strength of the fabric samples treated with GPTMS solution. BTCA pretreatment also did not cause significant change on contact angle values of the fabric samples-treated GPTMS together with metal alkoxides and/or hydrophobic trialkoxysilanes [22].

2.6. Multifunctional finishing

It is possible to give various functional properties to textile materials in one step by the combination of compatible inorganic precursors, thanks to sol-gel technology and the finishing like this is called multifunctional finishing. Multifunctional finishing has drawn much attention in the last decades with many studies on this subject. **Table 1** summarized some researches about various functionalities on textile materials treated with *in situ* sol-gel method and sols with nanoparticles [82–101].

Memon et al. treated polyester fabrics with titania-doped silica nanosols. It was observed that the antiwrinkle and UV-protection properties developed, while air permeability and whiteness index values decreased only slightly [102].

Onar and Mete developed water-oil–repellent and flame-retardant cotton fabrics by sol-gel method or knife-over-roll coating using a one-step treatment. The nanosol containing TEOS as a precursor, hexadecyltrimethoxysilane as a precursor with water-repellent properties, and

Sol preparation	Functionality	Textile	Precursor or NPs	Solvent	Catalyst	NP binding	Other chemicals/ function	Ref. no.
NP sols	Anti-UV, rubbing fastness	Co	TEOS, GPTMS	Ethanol, water	NH_4OH	Polyacrylate/ adhesive, NM-5263/ thickening agent	Dodecanol/ nonsurfactant template	[82]
NP sols	UPF, hydrophobicity, antibacterial, mechanical strength	Co/Pes (65/35, 20/80)	TiO_2NP, ZnONP	Toluene	–	Acrylate-based binder	–	[83]
NP sols	Electroconductive, antibacterial	Co, CV	$AgNO_3$	Ethanol	–	–	PVP/stabilizer, EG/ reducing agent, NaCl/mediating agent	[84]
In situ sol-gel	Antibacterial, self-cleaning	Co	$AgNO_3$	Water	–	Acrylic acid/ plasma polymerization monomer	$NaBH_4$/reducing agent	[85]
In situ sol-gel	UV-resistance, thermal properties	UHMWPE	Titanium butoxide	Ethanol, water	Acetic acid	–	MPTMS/graft polymerization monomer	[86]
In situ sol-gel	Antibacterial, antifungal, abrasion, pilling resistance	Co/PES	GPTMS, aluminum isopropoxide	Ethanol, water	HCl	PVA	Ag/CuNP/ bioactive additive	[87]

Table 1. (continued).

Sol preparation	Functionality	Textile	Precursor or NPs	Solvent	Catalyst	Other chemicals/ function	Ref. No
In situ sol-gel	UV-protection, antibacterial, dyeing	Co	TIP	Ethanol, water	Acetic acid	CHPTMAC/cationization agent, dyes	[88]
In situ sol-gel	Surface bonding, carrier for chemicals / drugs / therapeutic	PES, Co	TEOS, APTES, AEAPTMS	Ethanol, water	-	Sodium silicate solution/ nucleation agent	[89]
In situ sol-gel	Flame retardancy	Silk	TEOS	Ethanol, water	–	Dimethyl phosphonate	[90]
In situ sol-gel	Thermal resistance, water-repellency	Co	PVS (with VTES)	Ethanol, water, toluene	HCl	–	[91]
In situ sol-gel	Antibacterial, self-cleaning	Co	Zinc nitrate	–	NaOH	*Tragacanth gum*/reducing, stabilizing, binding agent	[92]
NP sols	Superhydrophobicity, oil-water separation, stain repellency, self-cleaning	Co, Bamboo, lignocellulosic	TiO_2NP, SiO_2NP, VTES	Ethanol	NaOH	–	[93]
In situ sol-gel	Antibacterial, abrasion resistance	Co	TEOS, zinc acetate	Ethanol, water	HF	–	[94]
In situ sol-gel	Photocatalytic activity, UV-protection	Co	$Ti(OBu)_4$	*tert*-butanol	Acetic acid	–	[95]
In situ sol-gel	Thermal resistance	Co	GPTMS	Water	HCl	–	[96]
In situ sol-gel	Photocatalytic activity	PES, PLA	TIP	Ethanol	Acetic acid, HCl	–	[97, 98]
In situ sol-gel	Sweat pH sensing	Co	GPTMS	Water	$BF_3.OEt_2$	Halochromic dye/litmus	[99]
In situ sol-gel	Flame retardancy	Wo	TEOS	Ethanol, water	HCl	H_3BO_3, $Zn_3B_2O_6$, $NH_4HB_4O_7$/ flame retardants	[100]
In situ sol-gel	Hydrophobic, conductive	Co	MTMS	methanol	HCl	MWCNT	[101]

NP: nanoparticles, NH_4OH: ammonium hydroxide, UPF: UV-protection factor, CV: viscose, PVP: polyvinylpyrrolidone, EG: ethylene glycol, UHMWPE: ultra-high-molecular-weight polyethylene, PVA: poly(vinyl alcohol), MPTMS: γ-methacryloxy propyl trimethoxy silane.

TIP: Titanium(IV) isopropoxide, CHPTMAC: 3- chloro-2-hydroxypropyl trimethylammonium chloride, AEAPTMS: N-[3-(trimethoxysilyl)propyl]-ethylene-diamina PVS: polyvinylsilsesquioxane, VTES: Vinyltriethoxysilane, PLA: poly(lactic acid), HF: Hydrofluoric acid, $BF_3.OEt_2$: Boron trifluoride diethyl etherate, MTMS: methyltrimethoxysilane, MWCNT: Multiwall carbon nanotube.

Table 1. Some researches about various functionalities on textile materials treated with *in situ* sol-gel method and sols with nanoparticles.

guanidine phosphate as a flame-retardant agent were prepared for using at pad-dry-cure process. On the other hand, the combination of the nanosols and various polymers such as polyacrylate, polyvinyl acetate (VA), and polyurethane dispersions and fluorocarbon polymer was applied to cotton fabric by knife-over-roll process. They found that the fabric coated with nanosols containing VA polymer relatively has good durable water-oil repellency and flame retardancy properties [103].

Chakrabarti and Banerjee prepared multifunctional cotton fabric that shows enhanced mechanical strength, water and stain repellence, antimicrobial properties, UV-blocking capacity, and self-cleaning characteristics by zinc oxide nanoparticles [104].

Ibahim et al. enhanced both the functionality and pigment-printability of linen/cotton-blended fabric in a one-step process by the incorporation of inorganic nanomaterials such as silver (Ag-NPs), zinc oxide (ZnO-NPs), zirconium oxide (ZrO_2-NPs), or titanium dioxide (TiO_2-NPs) [105].

Mahltig et al. obtained photobactericidal and photochromic textile materials with UV-protective properties by sol-gel technology and stated that especially epoxysilane-modified sols led to promising results [106].

Pan et al. prepared superhydrophobic (water contact angle of 146.27°) and UV-blocking (UPF value of 164.06) cotton fabric by nano-alumina sol via sol-gel method [107].

Simoncic et al. compared the sol-gel application procedures as one step or two steps in terms of multifunctional water-oil–repellent and antimicrobial properties of cotton and found that one-step procedure was more effective in the production of coatings with water- and oil-repellent properties, while two-step procedure provided higher antibacterial activity, as well as better washing fastness [108].

Mahltig and Fischer applied sol-gel coatings to viscose and polyamide textiles in order to give functional properties like water repellency and antimicrobial activity at the same time and found that textile comfort properties were also preserved [109].

Textor and Mahltig described a sol-gel–based surface treatment for the preparation of water-repellent antistatic textiles and applied it to two types of textiles (100% polyester fabric and 65% polyester and 35% cotton-blended fabric). They obtained sufficient water repellency and antistatic properties simultaneously and suggested that this concept can be adapted to the alternative multifunctional surface coatings other than combining water repellence with antistatic properties for textile materials [110].

El-Shafei et al. purposed to develop flame-retardant and antibacterial cotton fabric by treatment with BTCA solution containing sodium hypophosphite (SHP) as a catalyst and chitosan phosphate synthesized and then TiO_2 nanosols. Increase of TiO_2, SHP and chitosan-phosphate concentration led to increase of LOI values of the fabric treated with them. The combination of BTCA and chitosan phosphate causes lower crease recovery angle than that of treatment with only BTCA. The antibacterial activity of the fabrics against *S. aureus, E. coli, Candida albicans* (fungus), and *Aspergillus flavus* (fungus) developed with the increase of TiO_2 and chitosan-phosphate concentration [111].

http://dx.doi.org/10.5772/67686

Rana et al. investigated the water repellency, UV protection, and antibacterial properties of cotton fabric treated with silica nanosols containing dimethyl dimethoxysilane (DMDMS) and TEOS as precursors and AgBr-TiO_2 nanoparticles, which were synthesized by using silver nitrate, potassium bromide, and titanium tetrachloride as precursors, by spray coating process. The treatment with silica nanosols with and without AgBr-TiO_2 particles gave rise to, respectively, 145.8°, 149.1° and 136°, 144.3° of contact angle values of the fabrics before and after washing process. Nevertheless, the fabrics treated with silica nanosols along with AgBr-TiO_2 particles have excellent UV protection with 41.9 of UPF value and good antibacterial activity with 99.88% bacterial reduction against *E. coli* [112].

Vasiljević et al. prepared six different solutions containing 1H,1H,2H,2H-perfluorooctyl-triethoxysilane (SiF), 3-(trimethoxysilyl)-propyldimethyloctadecyl ammonium chloride (SiQ), and P,P-diphenyl- N-(3-(trimethoxysilyl)propyl) phosphinic amide (SiP) together with TEOS or organocyclotetrasiloxane 2,4,6,8-tetrakis (2-(diethoxy(methyl)silyl)ethyl)-2,4,6,8-tetramethyl-cyclotetrasiloxane. Optimization of water, oil repellency, flame-retardant, and antibacterial properties of the fabric treated were carried out. They revealed that the fabrics treated at optimized conditions have superhydrophobic properties with 161° of contact angle and 4° of sliding angle and good antibacterial properties with 100 and 81.6% bacterial reduction against, respectively, *S. aureus* and *E. coli* but not good oil repellency properties (3) [113].

Gu et al. proved the flame retardancy and water repellency properties of cotton fabric treated with phosphorus-doped silica hydrosol containing hexadecyltrimethoxysilane. They found that the LOI value, residue amount at 600°C, and contact angle of the fabrics increased from, respectively, 18.5, 13.3% and hydrophilic to 29.4, 29.7, and 134.6° when compared to that of untreated fabric. Their onset temperature and pHRR values decreased from, respectively, 320 and 235°C to 217 and 81.46°C [114].

El-Naggar et al. stated that the cotton fabrics treated with titania nanosols in situ synthesized and then urea nitrate solution by two-step process possessed excellent antibacterial properties with, respectively, 99.4 and 99.37% bacterial reduction against to *S. aureus* and *E. coli* and excellent UV-protection properties with 42.11% UPF value with good durability to washing [115].

Sivakumar et al. studied the antibacterial, UV-protection, soil release, and self-cleaning properties of the cotton/polyester fabric subjected to TiO_2 nanoparticles, which were synthesized using titanium isopropoxide as precursor and then modified with 3-aminopropyltriethoxysilane and silicon oil, by pad-dry-cure process. The maximum photocatalytic and antibacterial activities were obtained for the fabrics treated with modified TiO_2, while the maximum UPF values were obtained for the fabric treated with unmodified TiO_2 [116].

Behzadnia et al. developed the photosonochemical synthesis of N-Ag/TiO_2 on wool fabric using titanium isopropoxide and silver nitrate as precursors and ammonia as nitrogen doping. The effect of pH, sonication, and precursor concentration on antibacterial and photocatalytic properties of wool fabrics treated were researched. Ag doping to N-TiO_2 nanocomposites on wool improved self-cleaning, photocatalytic, and antibacterial properties of N-doping TiO_2 nanocomposites on wool, while sono-treated samples containing N-Ag/TiO_2 possess higher self-cleaning and photocatalytic properties when compared to stirred fabrics containing N-Ag/TiO_2 [117].

3. Conclusion

Since 1960s, sol-gel-coating method on the substrates like metal, glass, and ceramic has been studied extensively. In the last decades, researches about sol-gel technology have focused on enabling functional properties to textile materials as an alternative to conventional textile finishing treatments. In textile finishing, flame retardancy, water or oil repellency, UV protection, antimicrobial activity, and antiwrinkle effect were already studied by sol-gel method. In addition, multifunctionality means that two or more of these properties are evaluated in combination and investigated in terms of washing durability, textile comfort, and physical properties. It is an important fact to ensure multifunctional properties in one step, and by nanotechnological sol-gel method, this problem can be achieved and moreover, eco-friendly treatments can be applied in textile finishing due to less harmful chemical use. Therefore, sol-gel technology is expected to be used in plenty of textile finishing processes and promising studies are the starting points for further developments in future.

Author details

Nurhan Onar Camlibel* and Buket Arik

*Address all correspondence to: nonar@pau.edu.tr

Textile Engineering Department, Engineering Faculty, Pamukkale University, Denizli, Turkey

References

[1] Ismail WNW. Sol–gel technology for innovative fabric finishing—A review. Journal of Sol–Gel Science and Technology. 2016;**78(3)**:698–707. DOI: 10.1007/s10971-016-4027-y

[2] Gashti MP, Pakdel E & Alimohammadi F. Nanotechnology-based coating techniques for smart textiles. In Active Coatings for Smart Textiles. UK: Woodhead Publishing; 2016, 243–268 p.

[3] Ardanuy M, Faccini M, Amantia D, Aubouy L, Borja G. Preparation of durable insecticide cotton fabrics through sol–gel treatment with permethrin. Surface and Coatings Technology. 2014;**239**:132–137. DOI: 10.1016/j.surfcoat.2013.11.031

[4] Colleoni C, Guido E, Migani V, Rosace G. Hydrophobic behaviour of non-fluorinated sol–gel based cotton and polyester fabric coatings. Journal of Industrial Textiles. 2015;**44(6)**:815–834. DOI: 10.1177/1528083713516664

[5] Poli R, Colleoni C, Calvimontes A, Polášková H, Dutschk V, Rosace G. Innovative sol–gel route in neutral hydroalcoholic condition to obtain antibacterial cotton finishing by zinc precursor. Journal of Sol–Gel Science and Technology. 2015;**74(1)**:151–160. DOI: 10.1007/s10971-014-3589-9

[6] Huang KS, Nien YH, Hsiao KC, Chang YS. Application of DMEU/SiO_2 gel solution in the antiwrinkle finishing of cotton fabrics. Journal of Applied Polymer Science. 2006;**102(5)**:4136–4143. DOI: 10.1002/app.24246

[7] Colleoni C, Massafra MR, Rosace G. Photocatalytic properties and optical characterization of cotton fabric coated via sol–gel with non-crystalline TiO_2 modified with poly (ethylene glycol). Surface and Coatings Technology. 2012;**207**:79–88. DOI: 10.1016/j.surfcoat.2012.06.003

[8] Alongi J, Malucelli G. Cotton flame retardancy: state of the art and future perspectives. RSC Advances. 2015;**5(31)**:24239–24263. DOI: 10.1039/C5RA01176K

[9] Grancaric AM, Colleoni C, Guido E, Botteri L, Rosace G. Thermal behaviour and flame retardancy of monoethanolamine-doped sol–gel coatings of cotton fabric. Progress in Organic Coatings. 2017;**103**:174–181. DOI: 10.1016/j.porgcoat.2016.10.035

[10] Alongi J, Ciobanu M, Malucelli G. Sol–gel treatments on cotton fabrics for improving thermal and flame stability: effect of the structure of the alkoxysilane precursor. Carbohydrate Polymers. 2012;**87(1)**:627–635. DOI: 10.1016/j.carbpol.2011.08.036

[11] Boukhriss A, Gmouh S, Hannach H, Roblin JP, Cherkaoui O, Boyer D. Treatment of cotton fabrics by ionic liquid with PF6- anion for enhancing their flame retardancy and water repellency. Cellulose. 2016;**23(5)**:3355–3364. DOI: 10.1007/s10570-016-1016-9

[12] Kappes RS, Urbainczyk T, Artz U, Textor T, Gutmann JS. Flame retardants based on amino silanes and phenylphosphonic acid. Polymer Degradation and Stability. 2016;**129**:168–179. DOI: 10.1016/j.polymdegradstab.2016.04.012

[13] Zhang QH, Gu J, Chen GQ, Xing TL. Durable flame retardant finish for silk fabric using boron hybrid silica sol. Applied Surface Science. 2016;**387**:446–453. DOI: 10.1016/j.apsusc.2016.06.119

[14] Šehić A, Tomšič B, Jerman I, Vasiljević J, Medved J, Simončič B. Synergistic inhibitory action of P-and Si-containing precursors in sol–gel coatings on the thermal degradation of polyamide 6. Polymer Degradation and Stability. 2016; **128**:245–252. DOI: 10.1016/j.polymdegradstab.2016.03.026

[15] Malucelli G. Surface-engineered fire protective coatings for fabrics through sol–gel and layer-by-layer methods: an overview. Coatings. 2016;**6(3)**:33. DOI: 10.3390/coatings6030033

[16] Deh S, Gähr F, Buchmeiser MR. Synergistic effects in the pyrolysis of phosphorus-based flame-retardants: the role of Si-and N-based compounds. Polymer Degradation and Stability. 2016;**130**: 155–164. DOI: 10.1016/j.polymdegradstab.2016.06.009

[17] Liu Y, Pan YT, Wang X, Acuña P, Zhu P, Wagenknecht U, Heinrich G, Zhang XQ, Wang R, Wang DY. Effect of phosphorus-containing inorganic–organic hybrid coating on the flammability of cotton fabrics: synthesis, characterization and flammability. Chemical Engineering Journal, 2016;**294**:167–175. DOI: 10.1016/j.cej.2016.02.080

[18] Ren Y, Zhang Y, Zhao J, Wang X, Zeng Q, Gu Y. Phosphorus-doped organic–inorganic hybrid silicon coating for improving fire retardancy of polyacrylonitrile fabric. Journal of Sol–Gel Science and Technology. 2016:1–9. DOI: 10.1007/s10971-016-4273-z

[19] Aksit A, Onar N, Kutlu B, Sergin E, Yakin I. Synergistic effect of phosphorus, nitrogen and silicon on flame retardancy properties of cotton fabric treated by sol–gel process. International Journal of Clothing Science and Technology. 2016;**28(3)**:319–327. DOI: 10.1108/IJCST-03-2016-0029

[20] Wenzel RN. Resistance of solid surfaces to wetting by water. Industrial & Engineering Chemistry, 1936;**28(8)**:988–994. DOI:10.1021/ie50320 a024

[21] Cassie ABD, Baxter S. Wettability of porous surfaces. Transactions of the Faraday Society. 1944; **40**:546–551. DOI:10.1039/TF9444000546

[22] Schramm C, Rinderer B. Non-formaldehyde, crease-resistant modification of cellulosic material by means of an organotrialkoxysilane and metal alkoxides. Cellulose. 2015;**22(4)**:2811–2824. DOI: 10.1007/s10570-015-0664-5

[23] Liu H, Gao SW, Cai JS, He CL, Mao JJ, Zhu TX, Chen Z, Huang JY, Meng K, Zhang KQ, Al-Deyab SS, Lai YK. Recent progress in fabrication and applications of superhydrophobic coating on cellulose-based substrates. Materials. 2016;**9(3)**:124. DOI: 10.3390/ma9030124

[24] Zhao Q, Wu LYL, Huang H, Liu Y. Ambient-curable superhydrophobic fabric coating prepared by water-based non-fluorinated formulation. Materials & Design. 2016;**92**:541–545. DOI: 10.1016/j.matdes.2015.12.054

[25] Xue CH, Li M, Guo XJ, Li X, An QF, Jia ST. Fabrication of superhydrophobic textiles with high water pressure resistance. Surface and Coatings Technology. 2017;**310**:134–142. DOI: 10.1016/j.surfcoat.2016.12.049

[26] Hao L, Gao T, Xu W, Wang X, Yang S, Liu X. Preparation of crosslinked polysiloxane/SiO_2 nanocomposite via in-situ condensation and its surface modification on cotton fabrics. Applied Surface Science. 2016;**371**:281–288. DOI: 10.1016/j.apsusc.2016.02.204

[27] Zhang J, Li B, Wu L, Wang A. Facile preparation of durable and robust superhydrophobic textiles by dip coating in nanocomposite solution of organosilanes. Chemical Communications. 2013;**49(98)**:11509–11511. DOI: 10.1039/C3CC43238F

[28] Vasiljević J, Gorjanc M, Jerman I, Tomšič B, Modic M, Mozetič M, Orel B, Simončič B. Influence of oxygen plasma pre-treatment on the water repellency of cotton fibers coated with perfluoroalkyl-functionalized polysilsesquioxane. Fibers and Polymers. 2016;**17(5)**:695–704. DOI: 10.1007/s12221-016-5652-3

[29] Przybylak M, Maciejewski H, Dutkiewicz A. Preparation of highly hydrophobic cotton fabrics by modification with bifunctional silsesquioxanes in the sol–gel process. Applied Surface Science. 2016;**387**:163–174. DOI: 10.1016/j.apsusc.2016.06.094

[30] Xue CH, Jia ST, Chen HZ, Wang M. Superhydrophobic cotton fabrics prepared by sol–gel coating of TiO_2 and surface hydrophobization. Science and Technology of Advanced Materials. 2008;**9(3)**:035001. DOI: 10.1088/1468-6996/9/3/035001

[31] Onar N, Mete G. Development of water repellent cotton fabric with application of ZnO, Al_2O_3, TiO_2 and ZrO_2 nanoparticles modified with ormosils. Tekstil ve Konfeksiyon. 2016;**26(3)**;295–302.

[32] Onar N, Mete G, Aksit A, Kutlu B, Çelik E. Water- and oil-repellency properties of cotton fabric treated with silane, Zr, Ti based nanosols. International Journal of Textile Science. 2015;**4(4)**:84–96. DOI: 10.5923/j.textile.20150404.03

[33] Colleoni C, Guido E, Migani V, Rosace G. Hydrophobic behaviour of non-fluorinated sol–gel based cotton and polyester fabric coatings. Journal of Industrial Textiles. 2015;**44(6)**:815–834. DOI: 10.1177/1528083713516664

[34] Gao Q, Zhu Q, Guo Y, Yang CQ. Formation of highly hydrophobic surfaces on cotton and polyester fabrics using silica sol nanoparticles and nonfluorinated alkylsilane. Industrial & Engineering Chemistry Research. 2009;**48(22)**:9797–9803. DOI: 10.1021/ie9005518

[35] Teli MD, Annaldewar BN. Superhydrophobic and ultraviolet protective nylon fabrics by modified nano silica coating. The Journal of the Textile Institute. 2016:1–8. DOI: 10.1080/00405000.2016.1171028

[36] de Ferri L, Lorenzi A, Carcano E, Draghi L. Silk fabrics modification by sol–gel method. Textile Research Journal. 2016:0040517516676061. DOI: 10.1177/0040517516676061

[37] Vasiljević J, Zorko M, Tomšič B, Jerman I, Simončič B. Fabrication of the hierarchically roughened bumpy-surface topography for the long-lasting highly oleophobic "lotus effect" on cotton fibres. Cellulose. 2016;**23(5)**:3301–3318. DOI: 10.1007/s10570-016-1007-x

[38] Saravanan D. UV protection textile materials. AUTEX Research Journal. 2007;**7(1)**: 53–62.

[39] Wang L, Shen Y, Xu L, Cai Z, Zhang H. Thermal crystallization of low-temperature prepared anatase nano-TiO_2 and multifunctional finishing of cotton fabrics. The Journal of the Textile Institute. 2016;**107(5)**:651–662. DOI: 10.1080/00405000.2015.1054227

[40] Imai H, Hirashima H. Preparation of porous anatase coating from sol–gel derived titanium dioxide and titanium dioxide-silica by water-vapor exposure. Journal of the American Ceramic Society. 1998;**82**:2301–2304. DOI: 10.1111/j.1151-2916.1999.tb02083.x

[41] Liau LK, Chang H, Yang TK, Huang CL. Kinetic investigation of photocatalytic TiO_2 films prepared by the sol–gel method and low temperature treatments. Journal of Chemical Engineering of Japan. 2005;**38**:813–817. DOI: 10.1252/jcej.38.813

[42] Daoud WA, Xin JH. Nucleation and growth of anatase crystallites on cotton fabrics at low temperatures. Journal of the American Ceramic Society. 2004;87(5):953–955. DOI: 10.1111/j.1551-2916.2004.00953.x

[43] Onar N, Aksit AC, Sen Y, Mutlu M. Antimicrobial, UV-protective and self-cleaning properties of cotton fabrics coated by dip-coating and solvothermal coating methods. Fibers and Polymers. 2011;**12(4)**:461–470. DOI: 10.1007/s12221-011-0461-1

[44] Fulekar MH. Environmental Biotechnology. USA: CRC Press; 2010. 597 p.

[45] Abidi N, Cabrales L, Hequet E. Functionalization of a cotton fabric surface with titania nanosols: applications for self-cleaning and UV-protection properties. ACS Applied Materials & Interfaces. 2009;**1(10)**:2141–2146. DOI: 10.1021/am900315t

[46] Liu S, Zhang Q, Xu Z, Yang S, Liu H. Surface modification of TiO_2/SiO_2 composite hydrosol stabilized with polycarboxylic acid on Kroy-process wool fabric. Journal of Adhesion Science and Technology. 2016:1–20. DOI: 10.1080/01694243.2016.1249687

[47] Dhineshbabu NR, Arunmetha S, Manivasakan P, Karunakaran G, Rajendran V. Enhanced functional properties of cotton fabrics using TiO_2/SiO_2 nanocomposites. Journal of Industrial Textiles. 2016;**45(5)**:674–692. DOI: 10.1177/1528083714538684

[48] Rilda Y, Fadhli F, Syukri S, Alif A, Aziz H, Chandren S, Nur H. Self-cleaning TiO_2-SiO_2 clusters on cotton textile prepared by dip-spin coating process. Jurnal Teknologi. 2016;**78(7)**:113–120. DOI: 10.11113/jt.v78.9165

[49] Yuranova T, Mosteo R, Bandara J, Laub D, Kiwi J. Self-cleaning cotton textiles surfaces modified by photoactive SiO_2/TiO_2 coating. Journal of Molecular Catalysis A: Chemical. 2006;**244(1)**:160–167. DOI: 10.1016/j.molcata.2005.08.059

[50] Shaban M, Abdallah S, Khalek AA. Characterization and photocatalytic properties of cotton fibers modified with ZnO nanoparticles using sol–gel spin coating technique. Beni-Suef University Journal of Basic and Applied Sciences 2016;**5(3)**:277–283. DOI: 10.1016/j.bjbas.2016.08.003

[51] Noorian SA, Hemmatinejad N, Bashari A. One-pot synthesis of Cu_2O/ZnO nanoparticles at present of folic acid to improve UV-protective effect of cotton fabrics. Photochemistry and Photobiology. 2015;**91(3)**:510–517. DOI: 10.1111/php.12420

[52] Zhang Y, Xu Q, Fu F, Liu X. Durable antimicrobial cotton textiles modified with inorganic nanoparticles. Cellulose. 2016;23(5):2791–2808. DOI: 10.1007/s10570-016-1012-0

[53] Demir A, Arık B, Ozdogan E, Seventekin N. A new application method of chitosan for improved antimicrobial activity on wool fabrics pretreated by different ways. Fibers and Polymers. 2010;**11(3)**:351–356. DOI: 10.1007/s12221-010-0351-y

[54] Demir A, Arık B, Ozdogan E, Seventekin N. The comparison of the effect of enzyme, peroxide, plasma and chitosan processes on wool fabrics and evaluation for antimicrobial activity. Fibers and Polymers. 2010;**11(7)**:989–995. DOI: 10.1007/s12221-010-0989-5

[55] Arık B, Seventekin N. Evaluation of antibacterial and structural properties of cotton fabric coated by chitosan/titania and chitosan/silica hybrid sol–gel coatings. Tekstil ve Konfeksiyon. 2011;**2**:107–115.

[56] Arik B, Demir A, Ozdogan E, Gulumser T. Effects of novel antibacterial chemicals on low temperature plasma functionalized cotton surface. Tekstil ve Konfeksiyon. 2011:21(4).

[57] Bozaci E, Akar E, Ozdogan E, Demir A, Altinisik A, Seki Y. Application of carboxymethylcellulose hydrogel based silvernanocomposites on cotton fabrics for antibacterial property. Carbohydrate Polymers. 2015;**134**:128–135. DOI: 10.1016/j.carbpol.2015.07.036

[58] Liu Y, Ren X, Liang J. Antibacterial modification of cellulosic materials. Bioresources. 2015;**10(1)**:1964–1985.

[59] Mahltig B, Fiedler D, Fischer A, Simon P. Antimicrobial coatings on textiles–modification of sol–gel layers with organic and inorganic biocides, Journal of Sol–Gel Science and Technology. 2010;**55(3)**:269–277. DOI: 10.1007/s10971-010-2245-2

[60] Mahltig B, Textor T. Silver containing sol–gel coatings on polyamide fabrics as antimicrobial finish-description of a technical application process for wash permanent antimicrobial effect. Fibers and Polymers. 2010;**11(8)**;1152–1158. DOI: 10.1007/s12221-010-1152-z

[61] Mahltig B, Haufe H, Böttcher H. Functionalisation of textiles by inorganic sol–gel coatings. Journal of Materials Chemistry. 2005;**15(41)**:4385–4398. DOI: 10.1039/B505177K

[62] Tarimala S, Kothari N, Abidi N, Hequet E, Fralick J, Dai LL. New approach to antibacterial treatment of cotton fabric with silver nanoparticle–doped silica using sol–gel process. Journal of Applied Polymer Science. 2006;**101**:2938–2943.

[63] Mahltig B, Fiedler D, Böttcher H. Antimicrobial sol–gel coatings. Journal of Sol–Gel Science and Technology. 2004;**32(1–3)**:219–222. DOI: 10.1007/s10971-004-5791-7

[64] Xu P, Liu X, Wang W, Chen S. Improving the antibacterial and UV-resistant properties of cotton by the titanium hydrosol treatment. Journal of Applied Polymer Science. 2006;**102**:1478–1482. DOI: 10.1002/app.24340

[65] Mahltig B, Gutmann E, Meyer DC, Reibold M, Dresler B, Gunther K, Fabler D, Bottcher H. Solvothermal preparation of metallized titania sols for photocatalytic and antimicrobial coatings. Journal of Material Chemistry. 2007;**17**:2367–2374. DOI: 10.1039/B702519J

[66] Mahltig B, Fischer A. Inorganic/organic polymer coatings for textiles to realize water repellent and antimicrobial properties—a study with respect to textile comfort. Journal of Polymer Science: Part B: Polymer Physics. 2010;**48**:1562–1568. DOI: 10.1002/polb.22051

[67] Daoud WA, Xin, JH, Zhang YH. Surface functionalization of cellulose fibers with titanium dioxide nanoparticles and their combined bactericidal activities. Surface Science. 2005; **599**:69–75. DOI: 10.1016/j.susc.2005.09.038

[68] Wei SW, Peng B, Chai LY, Liu YC, Li ZY. Preparation of doping titania antibacterial powder by ultrasonic spray pyrolysis. Transactions of Nonferrous Metals Society of China. 2008;**18(5)**:1145–1150. DOI: 10.1016/S1003-6326(08)60196-X

[69] Daoud WA, Xin JH. Low temperature sol–gel processed photocatalytic titania coating. Journal of Sol–Gel Science and Technology. 2004;**29**:25–29. DOI: 10.1023/B:JSST.0000016134.19752.b4

[70] Simončič B, Klemenčič D. Preparation and performance of silver as an antimicrobial agent for textiles: a review. Textile Research Journal. 2016;**86(2)**:210–223. DOI: 10.1177/0040517515586157

[71] Mahltig B, Grethe T, Haase H. Handbook of Sol-Gel Science and Technology. Antimicrobial coatings obtained by sol–gel method. 2016:1–5. DOI: 10.1007/978-3-319-19454-7_102-1

[72] Rivero PJ, Goicoechea J. Sol–gel technology for antimicrobial textiles. Antimicrobial Textiles. 2016:47–70.

[73] Mohamed AL, El-Naggar ME, Shaheen TI, Hassabo AG. Laminating of chemically modified silan based nanosols for advanced functionalization of cotton textiles. International Journal of Biological Macromolecules. 2017; 95:429–437. DOI: 10.1016/j.ijbiomac.2016.10.082

[74] Xie Y, Hill CA, Xiao Z, Militz H, Mai C. Silane coupling agents used for natural fiber/polymer composites: a review. Composites Part A: Applied Science and Manufacturing. 2010; **41(7)**:806–819. DOI: 10.1016/j.compositesa.2010.03.005

[75] Schramm C, Binder WH, Tessadri R. Durable press finishing of cotton fabric with 1,2,3,4-butanetetracarboxylic acid and TEOS/GPTMS. Journal of Sol–Gel Science and Technology. 2004;**29**:155–165. DOI: 10.1023/B:JSST.0000023850.97771.7d

[76] Jang KO, Yeh K. Effects of silicone softeners and silane coupling agents on the performance properties of cotton fabrics. Textile Research Journal. 1993;**63(10)**:557–565. DOI: 10.1177/004051759306301001

[77] Abidi N, Hequet E, Tarimala S. Functionalization of cotton fabric with vinyltrimethoxysilane. Textile Research Journal. 2007; **77(9)**:668–674. DOI: 10.1177/0040517507080621

[78] Huang KS, Nien YH, Hsiao KC, Chang YS. Application of DMEU/SiO_2 gel solution in the antiwrinkle finishing of cotton fabrics. Journal of Applied Polymer Science. 2006;**102**:4136–4143. DOI: 10.1002/app.24246

[79] Huang K-S, Yang K-L, Lin S-J, Lian W-T. Antiwrinkle treatment of cotton fabric with a mixed sol of TEOS-TTB/DMDHEU. Journal of Applied Polymer Science. 2007;**106**:2559–2564. DOI: 10.1002/app.25281

[80] Huang K-S, Hwang M-C, Chen J-S, Lin S-J, Wang S-P. Application of mixed gel solution in the anti-wrinkle finishing of cotton fabrics. Journal of the Textile Institute. 2007;**98(2)**:169–176. DOI: 10.1533/joti.2005.0300

[81] Schramm C, Rinderer B, Binder WH, Tessadri R, Duelli H. Treatment of 1,3-dimethylol-4,5-dihydroxyimidazolidine-2-one finished cellulosic material with tetraethoxysilane or glycidyloxypropyl-trimethoxysilane solutions. Journal of Materials Science. 2005;**40**:1883 – 1891. DOI: 10.1007/s10853-005-1207-5

[82] Yin Y, Huang R, Xu Y, Wang C. Preparation and characterization of highly dispersed silica nanoparticles via nonsurfactant template for fabric coating. The Journal of the Textile Institute. 2017:1–7. DOI: 10.1080/00405000.2017.1278999

[83] Attia NF, Moussa M, Sheta AM, Taha R, Gamal H. Effect of different nanoparticles based coating on the performance of textile properties. Progress in Organic Coatings. 2017;**104**:72–80. DOI: 10.1016/j.porgcoat.2016.12.007

[84] Giesz P, Mackiewicz E, Nejman A, Celichowski G, Cieślak M. Investigation on functionalization of cotton and viscose fabrics with AgNWs. Cellulose. 2017;**24(1)**:409–422. DOI: 10.1007/s10570-016-1107-7

[85] Wang CX, Ren Y, Lv JC, Zhou QQ, Ma ZP, Qi ZM, Chen JY, Liu GL, Gao DW, Lu ZQ, Zhang W, Jin LM. In situ synthesis of silver nanoparticles on the cotton fabrics modified by plasma induced vapor phase graft polymerization of acrylic acid for durable multifunction. Applied Surface Science. 2017;**396**:1840–1848. DOI: 10.1016/j.apsusc.2016.11.173

[86] Hu J, Gao Q, Xu L, Zhang M, Xing Z, Guo X, Zhang K, Wu G. Significant improvement in thermal and UV resistances of UHMWPE fabric through in situ formation of polysiloxane–TiO_2 hybrid layers. ACS Applied Materials & Interfaces. 2016;**8(35)**:23311–23320. DOI: 10.1021/acsami.6b04914

[87] Kowalczyk D, Brzeziński S, Kamińska I. Multifunctional bioactive and improving the performance durability nanocoatings for finishing PET/CO woven fabrics by the sol–gel method. Journal of Alloys and Compounds. 2015;**649**:387–393. DOI: 10.1016/j.jallcom.2015.06.236

[88] Alebeid OK, Zhao T. Simultaneous dyeing and functional finishing of cotton fabric using reactive dyes doped with TiO_2 nano-sol. The Journal of the Textile Institute. 2016;107(5):625–635.

[89] Matos JC, Avelar I, Martins MBF, Gonçalves MC. Greensilica® vectors for smart textiles. Carbohydrate Polymers. 2017;**156**:268–275. DOI: 10.1016/j.carbpol.2016.08.015

[90] Zhang Q-H, Zhang W, Chen G-Q, Xing TL. Flame retardant finish of silk fabric with dimethyl phosphonate doped silica sol. 2nd Annual International Conference on Advanced Material Engineering; 15–17 April 2016; Wuhan, China, Book Series: AER-Advances in Engineering Research; 2016. p. 877–883

[91] Chen D, Chen F, Zhang H, Yin X, Zhou Y. Preparation and characterization of novel hydrophobic cellulose fabrics with polyvinylsilsesquioxane functional coatings. Cellulose. 2016;**23(1)**:941–953. DOI: 10.1007/s10570-015-0820-y

[92] Ghayempour S, Montazer M. Ultrasound irradiation based in-situ synthesis of star-like Tragacanth gum/zinc oxide nanoparticles on cotton fabric. Ultrasonics Sonochemistry. 2017;**34**:458–465.

[93] Wu Y, Jia S, Qing Y, Luo S, Liu M. A versatile and efficient method to fabricate durable superhydrophobic surfaces on wood, lignocellulosic fiber, glass, and metal substrates. Journal of Materials Chemistry A. 2016;**4(37)**:14111–14121. DOI: 10.1039/C6TA05259B

[94] Belakova D, Vihodceva S, Kukle S, Ziemniewska M. The sol–gel technology application of cotton textile processing. The Journal of the Textile Institute. 2016;**107(11)**:1381–1385. DOI: 10.1080/00405000.2015.1114767

[95] Abid M, Bouattour S, Conceição DS, Ferraria AM, Ferreira LFV, do Rego AMB, Vilar MR, Boufi S. Hybrid cotton–anatase prepared under mild conditions with high photocatalytic activity under sunlight. RSC Advances. 2016;**6(64)**:58957–58969. DOI: 10.1039/C6RA10806G

[96] Rosace G, Guido E, Colleoni C, Barigozzi G. Influence of textile structure and silica based finishing on thermal insulation properties of cotton fabrics. International Journal of Polymer Science. 2016. DOI: 10.1155/2016/1726475

[97] Zgura I, Frunza S, Frunza L, Enculescu M, Florica C, Ganea CP, Negrilla CC, Diamandescu L. Titanium dioxide layer deposited at low temperature upon polyester fabrics. Journal of Optoelectronics and Advanced Materials. 2015;**17(7–8)**:1055–1063.

[98] Zgura I, Frunza S, Frunza L, Enculescu M, Florica C, Cotorobai VF, Ganea CP. Polyester fabrics covered with amorphous titanium dioxide layers: combining wettability measurements and photoinduced hydrophilicity to assess their surface properties. Romanian Reports in Physics. 2016;**68**:259–269.

[99] Caldara M, Colleoni C, Guido E, Re V, Rosace G. Optical monitoring of sweat pH by a textile fabric wearable sensor based on covalently bonded litmus-3-glycidoxypropyltrimethoxysilane coating. Sensors and Actuators B: Chemical. 2016;**222**:213–220. DOI: 10.1016/j.snb.2015.08.073

[100] Zhang Q, Zhang W, Huang J, Lai Y, Xing T, Chen G, Jin W, Liu H, Sun B. Flame retardance and thermal stability of wool fabric treated by boron containing silica sols. Materials & Design. 2015;**85**:796–799. DOI: 10.1016/j.matdes.2015.07.163

[101] Nasirizadeh N, Dehghani M, Yazdanshenas ME. Preparation of hydrophobic and conductive cotton fabrics using multi-wall carbon nanotubes by the sol–gel method. Journal of Sol–Gel Science and Technology. 2015;**73(1)**:14–21. DOI: 10.1007/s10971-014-3488-0

[102] Memon H, Yasin S, Khoso NA, Memon S. Study of wrinkle resistant, breathable, anti-UV nanocoated woven polyester fabric. Surface Review and Letters. 2016;**23(03)**:1650003. DOI: 10.1142/S0218625X16500037

[103] Onar N, Mete G. Development of water-, oil-repellent and flame-retardant cotton fabrics by organic–inorganic hybrid materials. The Journal of the Textile Institute. 2016;**107(11)**:1463–1477. DOI: 10.1080/00405000.2015.1128208

[104] Chakrabarti S, Banerjee P. Preparation and characterization of multifunctional cotton fabric by coating with sonochemically synthesized zinc oxide nanoparticle-flakes and a novel approach to monitor its self-cleaning property. Journal of the Textile Institute. 2015;**106(9)**:963–969. DOI: 10.1080/00405000.2014.955962

[105] Ibahim NA, Eid BM, El-Aziz EA, Elmaaty TM. Functionalization of linen/cotton pigment prints using inorganic nanostructure materials. Carbohydrate Polymers. 2013;**97**:537–545. DOI: 10.1016/j.carbpol.2013.04.084

[106] Mahltig B, Textor T, Kumbasar PA. Photobactericidal and photochromic textile materials realized by embedding of advantageous dye using sol–gel technology. Celal Bayar University Journal of Science. 2015;**11(3)**;306–315.

[107] Pan C, Shen L, Shang S, Xing Y. Preparation of superhydrophobic and UV blocking cotton fabric via sol–gel method and self-assembly. Applied Surface Science. 2012;**259**:110–117. DOI: 10.1016/j.apsusc.2012.07.001

[108] Simoncic B, Tomsic B, Cerne L, Orel B, Jerman I, Kovac J, Zerjav M, Simoncic A. Multifunctional water and oil repellent and antimicrobial properties of finished cotton: influence of sol–gel finishing procedure. Journal of Sol–Gel Science and Technology. 2012;**61**:340–354. DOI: 10.1007/s10971-011-2633-2

[109] Mahltig B, Fischer A. Inorganic/organic polymer coatings for textiles to realize water repellent and antimicrobial properties—a study with respect to textile comfort. Journal of Polymer Science Part B: Polymer Physics. 2010;**48(14)**:1562–1568. DOI: 10.1002/polb.22051

[110] Textor T, Mahltig B. A sol–gel based surface treatment for preparation of water repellent antistatic textiles. Applied Surface Science. 2010;**256**:1668–1674. DOI: 10.1016/j.apsusc.2009.09.091

[111] El-Shafei A, ElShemy M, Abou-Okeil A. Eco-friendly finishing agent for cotton fabrics to improve flame retardant and antibacterial properties. Carbohydrate Polymers. 2015;**118**:83–90

[112] Rana M, Hao B, Mu L, Chen L, Ma PC. Development of multi-functional cotton fabrics with Ag/AgBr–TiO_2 nanocomposite coating. Composites Science and Technology 2016;**122**:104–112. DOI: 10.1016/j.compscitech.2015.11.016

[113] Vasiljević J, Zorko M, Štular D, Tomšič B, Jerman I, Orel B, Medved J, Kovač J, Simončič B. Structural optimisation of a multifunctional water-and oil-repellent, antibacterial, and flame-retardant sol–gel coating on cellulose fibres. Cellulose. 2017;1–18. DOI: 10.1007/s10570-016-1187-4

[114] Gu JL, Zhang QH, Chen YB, Chen GQ, Xing TL. Thermal properties and water repellency of cotton fabric prepared through sol–gel method. Thermal Science. 2016;**20(3)**:863–866. DOI: 10.2298/TSCI1603863G

[115] El-Naggar ME, Shaheen TI, Zaghloul S, El-Rafie MH, Hebeish A. Antibacterial activities and UV protection of the in situ synthesized titanium oxide nanoparticles on cotton fabrics. Industrial & Engineering Chemistry Research. 2016;**55(10)**:2661–2668. DOI: 10.1021/acs.iecr.5b04315

[116] Sivakumar A, Murugan R, Periyasamy S. Evaluation of multifunctional properties of polyester/cotton blend treated with unmodified and modified nano-TiO_2 particles. Materials Technology. 2016;**31(5)**:286–298. DOI: 10.1179/1753555715Y.0000000055

[117] Behzadnia A, Montazer M, Mahmoudi Rad M. In situ photo sonosynthesis of organic/inorganic nanocomposites on wool fabric introducing multifunctional properties. Photochemistry and Photobiology. 2016;**92(1)**:76–86. DOI: 10.1111/php.12546

Chapter 14

Sol-Gel Process for Surface Modification of Leather

Franco Ferrero, Monica Periolatto and
Giuseppe Gozzelino

Additional information is available at the end of the chapter

http://dx.doi.org/10.5772/67837

Abstract

Applications in the textile field of sol-gel processes were widely investigated since coatings of fabrics by materials of nanometric size obtained by sol-gel methods represent a functional surface modification attracting even more attention. However, few experiences of the use of nanostructured coatings are reported for leather finishing. In the experiments reported in the present chapter, a nabuk leather was finished by a sol-gel process to confer hydro and oil-repellency. The silica component could act as a protective coating, improving the rubbing performance of the substrate and conferring a certain grade of hydrorepellency, while the oil repellency was due to a fluorocarbon component. The coatings were applied at low add-on of finishing agent with the aim to keep possibly unvaried the esthetic and hand characteristics of the original leather. Contact angle and sorption time measurements of water and oil drops were carried out on the treated samples and compared with the untreated one. A similar comparison was made by testing color fastness to rubbing and change of esthetic appearance. Finally, chemical surface characterization was carried out by Fourier Transform Infrared in Attenuated Total Reflectance (FTIR-ATR) analysis.

Keywords: leather, sol-gel, TEOS, FOS, Fluorolink S10, hydrorepellency, oil repellency, contact angle, FTIR-ATR

1. Introduction

Applications in the textile field of sol-gel processes were widely investigated [1–28]. In fact, coatings of fabrics by materials of nanometric size obtained by sol-gel methods represent a functional surface modification attracting even more attention. They have exceptional physical properties and resistances: justification to this behavior comes from the huge area of contact generated when the fillers reach the nanometric size, in combination with a polymer layer of interphase with properties much higher than the matrix ones.

Materials with repellent properties against water or oil-based soils are of high commercial interest for textiles and leather-containing materials with high demand efforts for washings and cleaning. The finishing agents are often perfluorinated compounds [25–28], while sol-gel systems or functional silanes are used as nonfluorinated chemical alternative, but the oil-repellency can be achieved only by fluorine compounds.

Manufacture of water-resistant leather has attracted dramatic attention in recent years as the increasing demand for waterproof upper, upholstery and garment leathers. In general, a water-resistant leather is prepared by using chemicals that enable to form hydrophobic coatings with low surface free energy on leather surface. These coatings can be polymeric thin films containing silicones, fluorinated, or long-chain hydrocarbons. However, the permeability of air and water vapor of treated leather could be compromised, and the development of new methods for the production of waterproof leather maintaining its natural characteristics is still a challenge [29]. As an alternative, plasma treatments [29–31] or hybrid nanostructured coatings [32–34] have been proposed.

Generally, pure laboratory-grade alkoxysilanes are used as sol-gel precursors. A hybrid framework composite can be obtained by mixing alkoxysilanes with organic precursors, and the final properties of the coating can be tailored on purpose by choosing proper precursors. Tetraethoxysilane (TEOS) is the most widely used precursor in the sol-gel technology [13], nevertheless, it can be partially or totally replaced by fluorinated alkoxysilanes to confer together hydro and oil repellency typical of these compounds [7, 9, 10, 17, 19, 35]. Therefore, the finishing of cotton fabrics was satisfactory experimented by TEOS nanosol modified with 1H,1H,2H,2H-fluorooctyl-triethoxy-silane (FOS) [19] or with Fluorolink S10 (Solvay Solexis), cheaper commercial perfluoropolyether already used with TEOS to prepare hybrid coatings on glass [20, 36].

In the case of leather, a silica-based fluorinated hybrid coating can act as surface protection, improving the rubbing performance of the substrate and conferring a certain grade of hydro-repellency, while the oil repellency can be due to the fluorocarbon component. Moreover, nano- or microstructured coatings can be applied by the sol-gel process, reducing the amount of applied finishing agent to keep possibly unvaried the esthetic and hand characteristics of the original leather. Therefore, in the experiments reported in the present chapter, TEOS, FOS, and Fluorolink S10 were experimented as sol-gel precursors for leather finishing by analogy with the process experimented on cotton fabrics.

Leather is a material resulting from tanning, i.e., the production from animal skin of a chemically and biologically stable material by a process that leaves the original fiber structure more or less intact. A section of the animal skin is reported in **Figure 1** [37].

The most important part of an animal skin is the corium, consisting mainly of collagen, a fibrous protein that is strongly affected by tanning. The basic collagen structure consists of twined triple left-handed helices of peptide chains of different length. The peptide chains within the triple helices are held together by hydrogen bonding as a consequence of the groups belonging to the typical structure of polypeptide, as shown in **Figure 2** [38]. The amino acid composition of collagen is atypical for proteins, particularly with respect to its high hydroxyproline content. The most common motifs in the amino acid sequence of collagen are glycine-proline-X

http://dx.doi.org/10.5772/67837

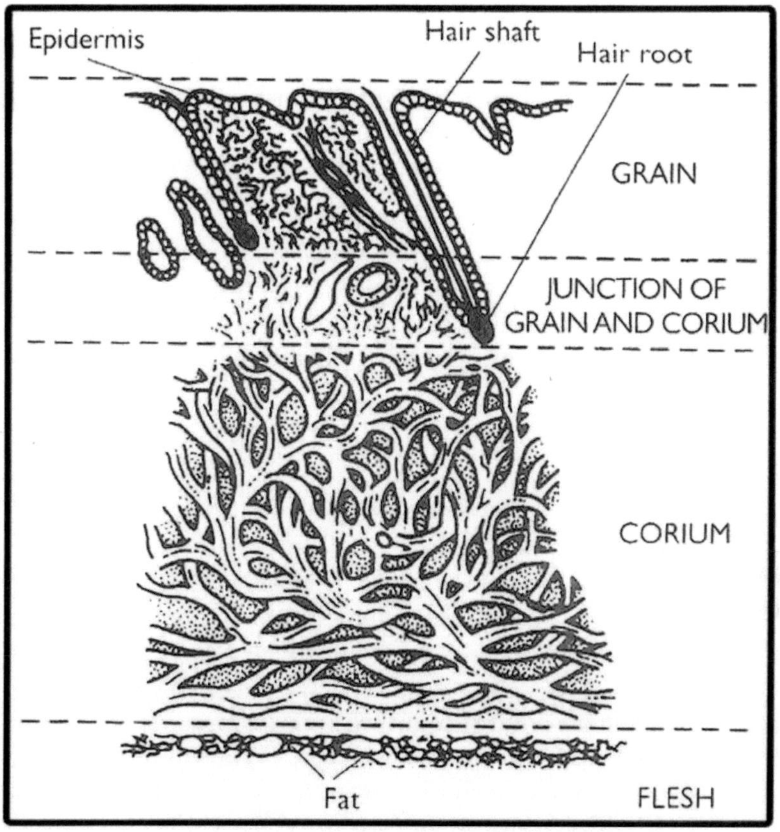

Figure 1. Section of the animal skin.

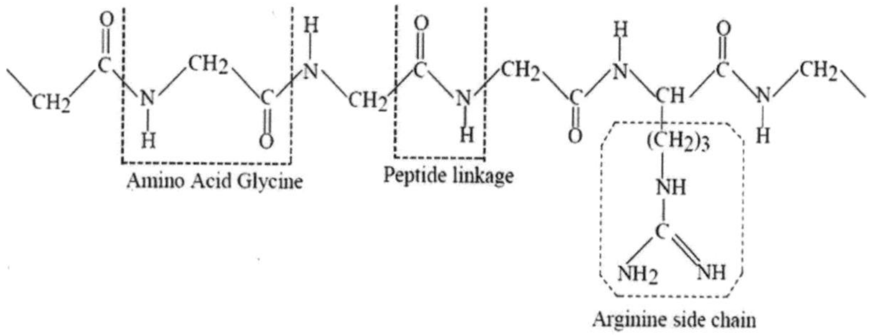

Figure 2. Typical molecular structure of polypeptide.

and glycine-X-hydroxyproline, where X is any amino acid other than glycine, proline, or hydroxyproline. The amino acid composition of a leather collagen is about glycine 26.8% of total nitrogen, alanine 8.0%, arginine 15.3%, proline 9.0%, and hydroxyproline 8.0% [38].

The main chemical processes carried out in sequence to transform animal skin into leather are the following:

- unhairing and liming, by an action of alkali and sodium sulfide to eliminate hair by hydrolysis of the keratin and in the meantime, to induce the modification of collagen yielding hydrolysis of amide groups, arginine residues, and amide linkages, together with swelling and removal of unwanted material;
- deliming and bating, with salts and enzymes to open further the structure;
- pickling, with acids to neutralize alkali;
- tanning with vegetal, synthetic, and chromium(III) salts to impart structure stability;
- neutralizing, dyeing, and fat liquoring with oils to confer desired color and softness;
- drying;
- finishing to improve the quality of the final leather.

The chemical modifications induced by these treatments on collagen make many polar groups, such as –OH, –NH–, =CO, available for adsorption through hydrogen bonding, in the case of sol-gel finishing, with the silanol groups yielded by acid-catalyzed hydrolysis of silane groups of the precursors. Moreover, the condensation reactions occurring in the following thermal step can promote grafting of silanols with polar groups of collagen, improving the adhesion of silica nanoparticles on leather. When TEOS is introduced as a coprecursor, $-O-Si(OH)_3$ groups can be bonded on the leather surface and are able to condense with other $\equiv Si-OH$ or $\equiv Si-O-CH_2-CH_3$ groups of the fluorinated alkoxysilane precursor.

Fluorolink has a fairly high molecular weight compared with FOS and two silane grafting end groups. These characteristics have been recognized to increase the hydrophobic/oleophobic character of coatings and resistance to mechanical removal from the matrix [39].

2. Experimental

In the experiments, a nabuk leather provided by Foglizzo Leather, Torino, Italy, was finished by a sol-gel process to confer hydro and oil-repellency. Besides the standard preparation, it was dyed with anionic dyes in the presence of auxiliaries to improve dye penetration. Samples of 15×4 cm^2 were subjected to the treatment.

Tetraethoxysilane (TEOS) and 1H,1H,2H,2H-Fluorooctyltriethoxysilane (FOS), both pure laboratory-grade reagent purchased from Sigma-Aldrich (Italy), and Fluorolink® S10, commercial product from Solvay Solexis (Italy), were used as precursors. In **Figure 3**, the molecular structures are reported; in particular that of Fluorolink was taken from the literature [39], and the molecular weight of the reagent used was in the range of 1750–1950 g/mol.

Finishing nanosols were synthesized via sol-gel method with acid catalysts in ethanol/water solution. TEOS and FOS, TEOS and Fluorolink S10, or the three products alone were used as

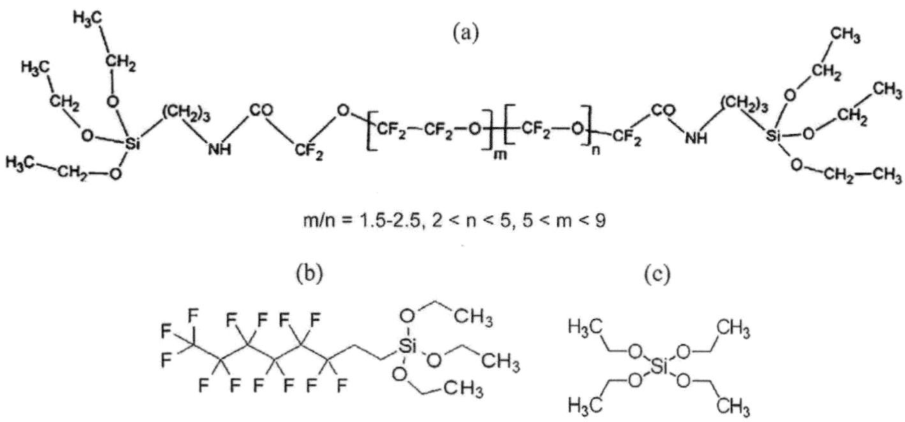

Figure 3. Molecular structure of (a) Fluorolink S10, (b) FOS, (c) TEOS.

precursors obtaining different nanosols by magnetic stirring at 25°C for 24 h. Five nanosols were prepared as detailed in **Table 1**.

The nanosol solutions were applied with the spraying technique, widely used at industrial level in dry finishing, which enables an uniform product distribution, but cannot be utilized in the presence of sediments or precipitates. Hence, before the application, the homogeneity of the solutions was tested by a visual check.

The solution amount to be sprayed on each sample was calculated on the basis of the content of dry resin in each solution providing 2, 5, or 10% add-ons. Higher add-ons were not applied to avoid the overhead of the substrate and maintain the nabuk properties unvaried. Since small solution volumes were needed, these were diluted with ethanol to assure a more homogeneous deposition.

In **Table 2**, the composition of the various samples is reported.

After spraying on both the leather surfaces, an impregnation time of 24 h was adopted, and the samples were maintained in sealed test tubes at room temperature. The following step was a thermal treatment in an oven at 120°C for 1 h to promote the solvent evaporation and at the same time the xerogel formation and the grafting reactions on leather.

Component	Nanosol-1	Nanosol-2	Nanosol-3	Nanosol-4	Nanosol-5
TEOS	0.17	0.21	0.17	–	–
FOS	0.19	–	–	–	0.49
Fluorolink S10	–	–	0.19	0.50	–
Ethanol 96% v/v	0.58	0.74	0.58	0.46	0.46
0.01 M HCl	0.06	0.05	0.06	0.04	0.05

Table 1. Composition, in percent weight, of the prepared nanosols.

Sample	Nanosol type	Add-on (% weight)	
		F	R
1	Untreated	–	–
2	Nanosol-1	10	10
3	Nanosol-1	0.5	0.5
4	Nanosol-2	10	10
5	Nanosol-2	0.5	0.5
6	Nanosol-2	0.2	0.2
7	Nanosol-3	10	10
8	Nanosol-3	0.5	0.5
9	Nanosol-4	10	–
10	Nanosol-4	0.5	–
11	Nanosol-5	10	10
12	Nanosol-5	0.5	0.5

Table 2. Composition of the samples (F: front; R: reverse).

Contact angle and sorption time measurements were carried out on the treated samples and compared with the untreated leather, as well as the color fastness to rubbing, esthetic, and hand properties. The color fastness to rubbing was tested by a Taber crockmeter mod. 418 according to ISO 11640.

Water and oil repellency were tested by static contact angle and sorption time measurements using HPLC-grade water and paraffin oil (Sigma–Aldrich) with a surface tension of 72 and 31.5 mN/m, respectively. The instrument was a DSA20E Easydrop standard tensiometer (Krüss, Germany) equipped with DSA software. Measuring liquid drops were deposited from a glass syringe on the sample surface by means of the software controlled dosing. The contact angles were the average of three measurements for each sample with a standard deviation of about 2%.

Finally, chemical surface characterization was carried out by FTIR-ATR analysis performed on a Nicolet 5700 instrument equipped with a Smart Orbit ATR single bounce accessory mounting a diamond crystal. A 128 scans at 4 cm^{-1} resolution in the 4000–500 cm^{-1} wavelength range were averaged.

3. Results and discussion

3.1. Contact angles and absorption time measurements

The results of static contact angles with water and paraffin oil as well as the absorption time measurements are reported and compared in **Table 3**.

Sample	Contact angle (°)		Absorption time (min)	
	Water	Paraffin oil	Water	Paraffin oil
1F	133	0	30	0
2F	154	99	70	10
3F	151	0	70	> 60
4F	160	0	65	0
5F	155	0	60	0
6F	150	0	50	0
7F	152	103	60	60
8F	138	111	63	> 60
9F	136	0	60	0
10F	146	0	63	0
11F	160	128	58	2
12F	148	126	65	10
1R	125	0	7	0
2R	138	70	> 60	>180
3R	127	71	> 60	>180
4R	133	0	60	0
5R	126	0	60	0
6R	125	0	30	0
7R	137	0	50	0
8R	154	0	60	0
11R	149	131	56	16
12R	150	107	60	5

Table 3. Contact angles and absorption times (F: front; R: reverse).

The untreated sample shows intrinsic water repellency (133° and 125° contact angles) due to the reagents used in the wet finishing, in particular in the greasing step. The absorption times of water drop are 30 and 7 min on front and reverse side, respectively, since the more relaxed structure of the latter justifies the faster absorption.

Instead, the untreated leather does not show oil repellency, as it can be deduced from the measurements with paraffin oil. The oil drop is immediately absorbed on both the sides with a persistent and hard to remove halo.

All the treatments on the front side improve the contact angle with water. Except the sample 8F, with the lowest amount of nanosol containing TEOS and Fluorolink, and 9F, with 10% Fluorolink alone, all the others show contact angles around 150°, threshold beyond which they can be defined superhydrophobic. The best results were obtained with the Nanosol-2, based on TEOS alone, and with the Nanosol-5, containing FOS alone; in fact, the samples 4F and 11F showed a contact angle of 160°.

The treatments on the reverse side generally yielded lower contact angles. This is ascribable to the less compact structure of the collagen fibers, which absorb the lyogel, preventing the formation of a homogeneous film. Nevertheless, the treatment with TEOS and Fluorolink and that based on FOS alone enabled to achieve superhydrophobicity in the case of samples 8R, 11R, and 12R.

The tests of oil repellency generally showed the same trend as those with water, but with the same treatment on the same side, the contact angles are obviously lower, due to the difference of surface tension of both test liquids. The Nanosol-1 and Nanosol-2, based on TEOS added by FOS and TEOS alone did not confer oil repellency. This was expectable with the latter nanosol that was lacking in oil repellent component, while with the former, the concentration of fluorocarbon resin added was not enough to give such property. The contact angles of oil on the samples 2R and 3R treated with Nanosol-1, based on TEOS added with FOS, did not attain the threshold of 90°, hence, these samples cannot be defined as oil repellent. The treatment with the Nanosol-5, based on FOS alone, was the most powerful, above all on the front side.

The results are confirmed by the images of drop shapes of water and oil on the Samples 11F and 11R reported in **Figures 4** and **5**.

3.2. Color fastness to rubbing

The resistance test to rubbing is fundamental to evaluate the property of a leather of not giving color release on textiles and at the same time, the color degradation of the leather itself. It is important for any use to simulate contact and friction between the leather and other fabrics, such as a bag on a shirt, an upper on trousers, and a shoe lining on a stocking. It is carried out by a continuous rubbing of wool felts loaded with 500 g, in the case of suede or nabuck, followed by a visual comparison with standard gray scales of the possible color release on the felts together with the color degradation of the leather. The number of cycles is varied according to the intended purpose, usually 100 cycles in dry conditions or 20 in wet.

Figure 4. Images of water and oil drops on the Sample 11F.

http://dx.doi.org/10.5772/67837

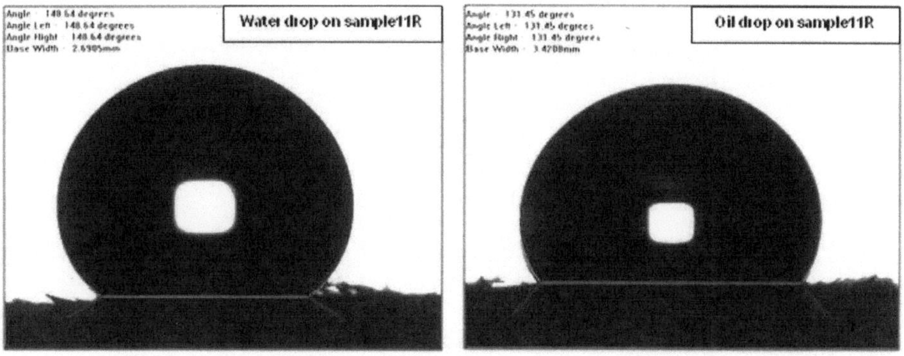

Figure 5. Images of water and oil drops on the Sample 11R.

The values of the standard gray scale for color release range from degree 1 for the maximum release to degree 5 for the minimum. The results of such evaluation on the all samples, in front and reverse sides, after rubbing tests in wet conditions are plotted in **Figure 6**. The color fastness to abrasion was generally improved by the finishing presence with respect to untreated leather (Sample 1), on which just a 1.5 value on gray scale was measured. The best results were obtained with Fluorolink alone (Samples 9 and 10); however, even FOS yielded a good fastness increase (Samples 2 and 11).

3.3. Esthetic aspect

The esthetic aspect should be critical in order to choose an optimal finish treatment for leather articles. In fact, for these products, the functional property, in this case water and oil repellency, is not the only aspect to consider.

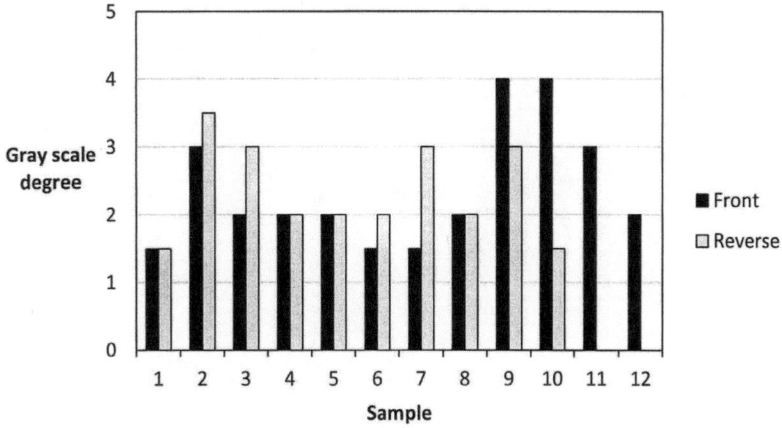

Figure 6. Color fastness evaluation after wet rubbing.

By comparison between treated samples and reference, four parameters were visually, or to the touch, evaluated by assigning a number decreasing as the difference with respect to the original sample is increasing:

- shade change of color: by analogy with the gray scale for fastness evaluation, the value 5 was assumed corresponding to the visual shade of the reference, while 1 was attributed to the maximum change toward darker shades;
- structural stiffness: the value 5 was assumed for the original texture, while 1 was attributed to the maximum stiffness;
- surface structure or "grain": the value 3 was assigned to the original sample, while 1 was attributed to maximum compaction of the fibers;
- writing effect, characteristic for nabuk: it was evaluated 3 if unvaried, 1 if absent.

The results are plotted in **Figures 7–10** for each aspect. With all the treatments, a worsening of the esthetic characteristics was generally observed, but not all can be considered equally important. An uniform darkening of color, if not excessive, is preferable to other changes, since in the previous dyeing a clearer shade can be conferred in order to obtain, after the sol-gel process, the desired shade. This is the case, for example, of the Sample 12F, treated with the lowest add-on of FOS alone, which showed only a color shade change of one degree lower with respect to 11F, one of the best samples, treated with the same nanosol, but with a higher add-on (10%).

Some treatments caused a compaction of surface fibers, with consequent worsening of hand properties and comfort; hence, they should be avoided.

The writing effect is considered a marginal aspect, and its change is not critical for a sample evaluation.

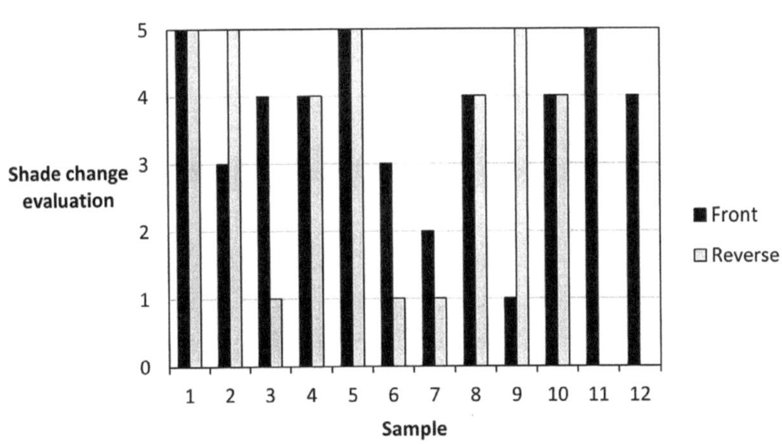

Figure 7. Shade change evaluation.

http://dx.doi.org/10.5772/67837

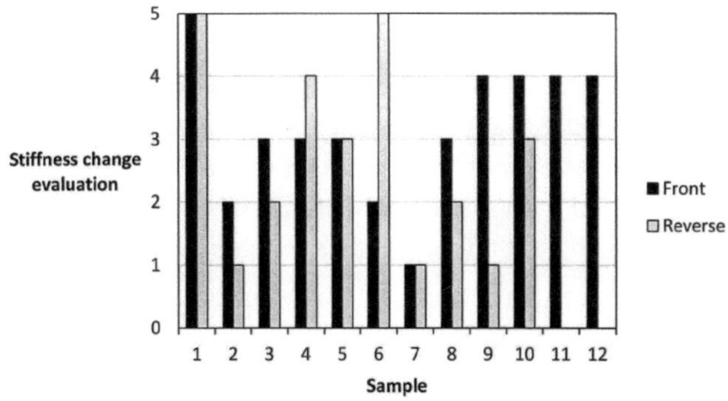

Figure 8. Stiffness change evaluation.

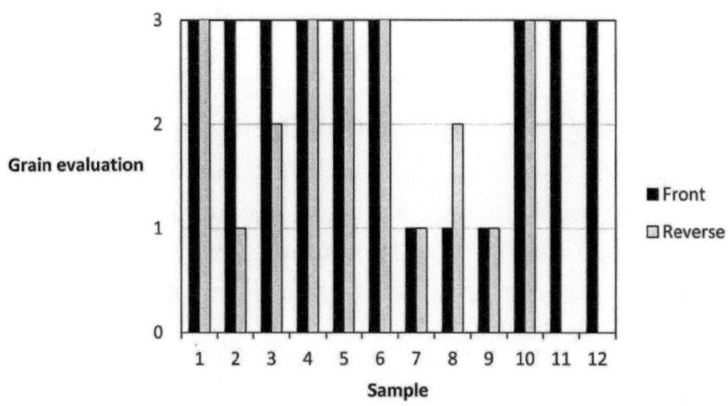

Figure 9. Grain evaluation.

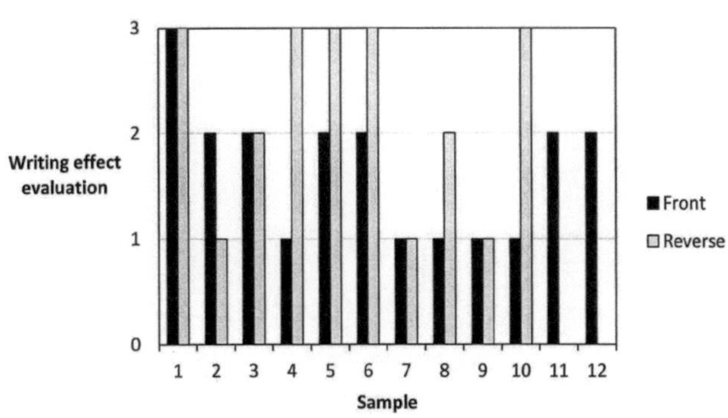

Figure 10. Writing effect evaluation.

3.4. FTIR-ATR analysis

All the samples were subjected to FTIR-ATR analysis to investigate the chemical composition of the leather substrate, highlighting the presence of characteristic groups of leather tanned and dyed as well as of finishing agents. In **Figure 11**, the spectra of the Sample 2F, treated with TEOS and FOS, are compared with that of untreated leather, and the peaks assignable to the characteristic groups according to literature data are evidenced.

The wavelength interval useful to reveal the presence of various finishing agents ranges from 1400 to 700 cm^{-1}. In particular, the peaks between 1250 and 1150 cm^{-1} shown by the finished leather are due to stretching of C-F bond in CF_2 e CF_3 groups and indicate the presence of fluorinated compounds on the leather surface. The peak at 1088 cm^{-1}, ascribable to Si-O-Si group, shows the presence of silicone nanoparticles on the surface. Such peak is present also with lower absorbance on the untreated sample since it is due to collagen component of the leather.

The spectrum of the Sample 9F, treated with the Nanosol-4, based on Fluorolink alone, shows a strong presence of fluorine and silicon compounds on the surface, as it can be observed in **Figure 12**. In the spectrum of the Sample 11F, treated with the Nanosol-5, based on FOS alone, a peak at about 900 cm^{-1}, ascribable to Si-C bond, also appears, as evidenced in **Figure 13**. Coupling it with the lowering, in comparison with the reference, of peaks related to OH and CH groups, at 3300 and 1600 cm^{-1}, it can be concluded that a grafting of the finishing agent on the leather surface could occur during the thermal treatment. The comparison between the spectra of the samples treated with nanosols containing TEOS and those with nanosols based on FOS or Fluorolink alone shows a decrease of surface concentration of fluorine groups confirming a dilution effect due to TEOS.

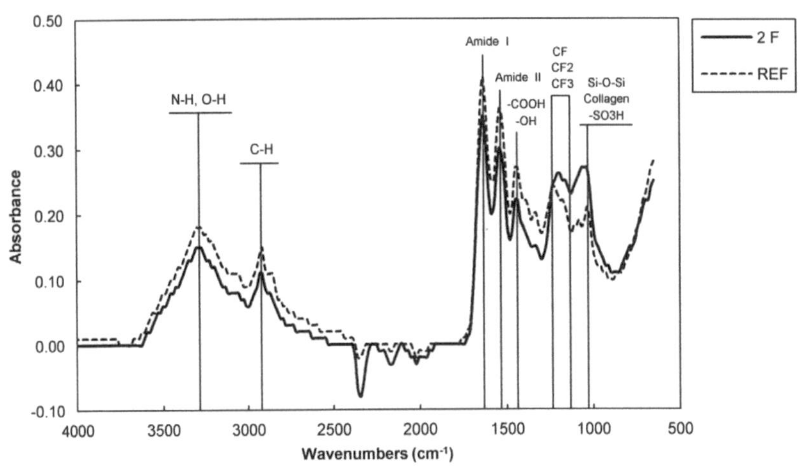

Figure 11. FTIR-ATR spectra of Sample 2F, treated with TEOS + FOS, and untreated reference (REF).

http://dx.doi.org/10.5772/67837

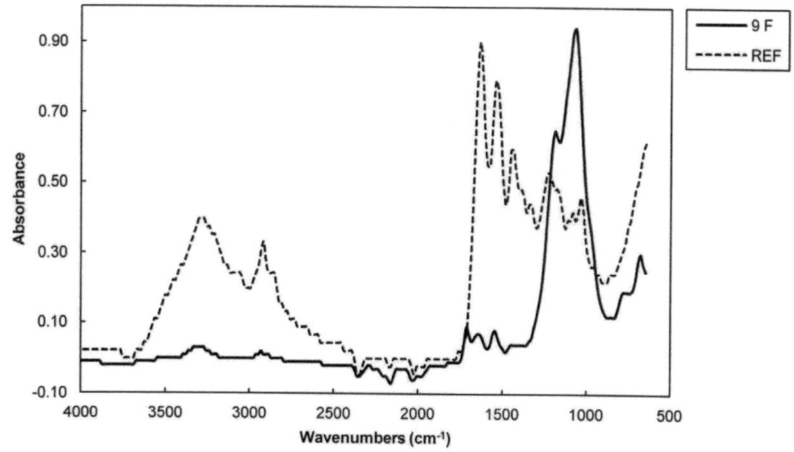

Figure 12. FTIR-ATR spectra of Sample 9F, treated with Fluorolink alone, and untreated reference (REF).

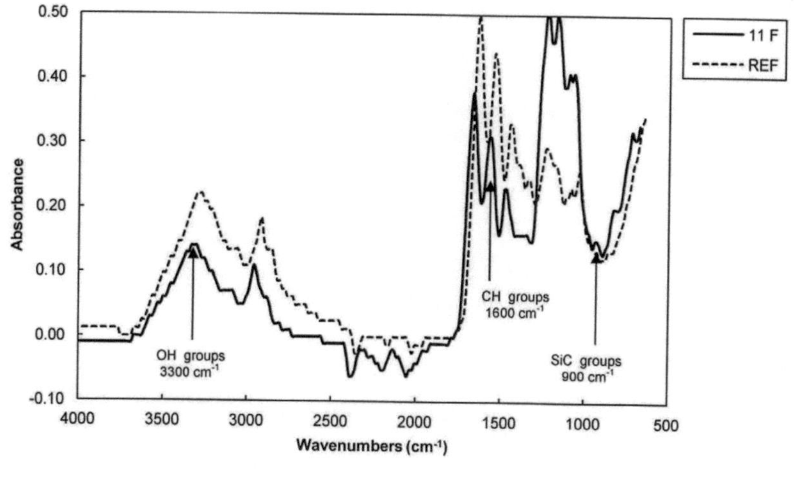

Figure 13. FTIR-ATR spectra of Sample 11F, treated with TEOS alone, and untreated reference (REF).

4. Conclusions

The research work on the leather was aimed to confer water and oil repellency and increase the color fastness to abrasion through a coating able to give the desired properties without unduly damaging the esthetic characteristics of the final product.

The application of a functional finishing on the leather by sol-gel conferred the desired properties, keeping the peculiar properties of the substrate unvaried. In particular for water and oil repellency, FOS-based nanosol at 10% add-on can be considered the best solution.

However, the nanosol based on Fluorolink alone enabled to obtain the best increase of color fastness to abrasion, but satisfactory results of this property were observed also with FOS-based nanosol at 10%, which can be considered as the optimal solution able to couple the two objectives. Moreover, the evaluation of the esthetic properties did not evidence a significant worsening of these with respect to the untreated leather.

Nanosols obtained by TEOS + FOS and TEOS + Fluorolink mixtures showed results generally worse both in terms of performance and esthetic appearance. This was attributed to the dilution effect due to TEOS.

The surface analysis carried out by FTIR-ATR has validated some previous knowledge in the textile field confirming the possibility to successfully apply the sol-gel treatment on a different substrate.

Thanks to the simple involved reactions and process, and the good obtained results, sol-gel can, in conclusion, be indicated as a valid finishing process for leather, opening the way for a future industrial scale up.

Acknowledgements

The authors are grateful to Foglizzo Leather Company for the technical support and to Ing. Danilo Picariello for the experimental work.

Author details

Franco Ferrero*, Monica Periolatto and Giuseppe Gozzelino

*Address all correspondence to: franco.ferrero@polito.it

Department of Applied Science and Technology, Polytechnic of Torino, Torino, Italy

References

[1] Gowri S., Amorim T., Carneiro N., Souto A.P., Esteves M.F. Polymer nanocomposites for multifunctional finishing of textiles—a review. Textile Research Journal 2010; 80: 1290–1306. DOI: 10.1177/0040517509357652.

[2] Mahltig B., Haufe H., Böttcher H. Functionalization of textiles by inorganic sol-gel coatings. Journal of Materials Chemistry 2005; 15: 4385–4398. DOI: 10.1039/B505177K.

[3] Textor T., Mahltig B. Nanosols and Textiles. Singapore: World Scientific Publishing Co. Pte. Ltd.: 2008. DOI: 10.1142/9789812833518.

[4] Mahltig B., Böttcher H. Modified silica sol coatings for water repellent textiles. Journal of Sol-Gel Science and Technology 2003; 27: 43–52. DOI: 10.1023/A:1022627926243.

[5] Satoh K., Nazakumi H., Morita M. Novel fluorinated inorganic–organic finishing materials for nylon carpeting. Textile Research Journal 2004; 74: 1079–1084. DOI: 10.1177/004051750407401209.

[6] Daoud W.A., Xin J.H., Tao X. Superhydrophobic silica nanocomposite coating by a low-temperature process. Journal of the American Ceramic Society 2004; 87: 1782–1784. DOI: 10.1111/j.1551-2916.2004.01782.x.

[7] Yu M., Gu G., Meng W.-D., Qing F.-L. Superhydrophobic cotton fabric coating based on a complex layer of silica nanoparticles and perfluorooctylated quaternary ammonium silane coupling agent. Applied Surface Science 2007; 253: 3669–3673. DOI: 10.1016/j.apsusc.2006.07.086.

[8] Tomšič B., Simončič B., Orel B., Černe L., Tavčer P.F., Zorko M., Jerman I., Vilčnik A., Kovač J. Sol-gel coating of cellulose fibres with antimicrobial and repellent properties. Journal of Sol-Gel Science and Technology 2008; 47: 44–57. DOI: 10.1007/s10971-008-1732-1.

[9] Bae G.Y., Min B.G., Jeong Y.G., Lee S.C., Jang J.H., Koo G.H. Superhydrophobicity of cotton fabrics treated with silica nanoparticles and water-repellent agent. Journal of Colloid and Interface Science 2009; 337: 170–175. DOI: 10.1016/j.jcis.2009.04.066.

[10] Erasmus E., Barkhuysen F.A. Superhydrophobic cotton by fluorosilane modification. Indian Journal of Fibre and Textile Research 2009; 34: 377–379. Available from: http://nopr.niscair.res.in/handle/123456789/6884 [Accessed 2017-01-31].

[11] Roe B., Zhang X. Durable hydrophobic textile fabric finishing using silica nanoparticles and mixed silanes. Textile Research Journal 2009; 79: 1115–1122. DOI: 10.1177/0040517508100184.

[12] Gao Q., Zhu Q., Guo Y., Yang C.Q. Formation of highly hydrophobic surfaces on cotton and polyester fabrics using silica sol nanoparticles and nonfluorinated alkylsilane. Industrial Engineering Chemistry Research 2009; 48: 9797–9803. DOI: 10.1021/ie9005518.

[13] Textor T., Mahltig B. A sol-gel based surface treatment for preparation of water-repellent antistatic textiles. Applied Surface Science 2010; 256 (6): 1668–1674. DOI: 10.1016/j.apsusc.2009.09.091.

[14] Liu J., Huang W., Xing Y., Li R., Dai J. Preparation of durable superhydrophobic surface by sol-gel method with water glass and citric acid. Journal of Sol-Gel Science and Technology 2011; 58: 18–23. DOI: 10.1007/s10971-010-2349-8

[15] Simončič B., Tomšič B., Cerne L., Orel B., Jerman I., Kovac J., Zerjav M., Simončič A. Multifunctional water and oil repellent and antimicrobial properties of finished cotton: influence of sol-gel finishing procedure. Journal of Sol-Gel Science and Technology 2012; 61: 340–354. DOI: 10.1007/s10971-011-2633-2.

[16] Shi Y., Wang Y., Feng X., Yue G., Yang W. Fabrication of superhydrophobicity on cotton fabric by sol-gel. Applied Surface Science 2012; 258: 8134–8138. DOI: 10.1016/j.apsusc.2012.05.008.

[17] Pan C., Shen L., Shang S., Xing Y. Preparation of superhydrophobic and UV blocking cotton fabric via sol-gel method and self-assembly. Applied Surface Science 2012; 259: 110–117. DOI: 10.1016/j.apsusc.2012.07.001.

[18] Vasiljević J., Gorjanc M., Tomšič B., Orel B., Jerman I., Mozetič M., Vesel A., Simončič B. The surface modification of cellulose fibres to create super-hydrophobic, oleophobic and self-cleaning properties. Cellulose 2013; 20: 277–289. DOI: 10.1007/s10570-012-9812-3.

[19] Periolatto M., Ferrero F., Mossotti R., Montarsolo A. Hydrorepellent finishing of cotton fabrics by chemically modified TEOS based nanosol. Cellulose 2013; 20: 355–364. DOI: 10.1007/s10570-012-9821-2.

[20] Ferrero F., Periolatto M. Application of fluorinated compounds to cotton fabrics via sol-gel. Applied Surface Science 2013; 275: 201–207. DOI: 10.1016/j.apsusc.2013.01.001.

[21] Montarsolo A., Periolatto M., Zerbola M., Mossotti R., Ferrero F. Hydrophobic sol-gel finishing for textiles: improvement by plasma pre-treatment. Textile Research Journal 2013; 83: 1190–1200. DOI: 10.1177/0040517512468823.

[22] Periolatto M., Ferrero F. Cotton and polyester surface modification by methacrylic silane and fluorinated alkoxysilane via sol-gel and UV-curing coupled process. Surface and Coating Technology 2015; 271: 165–173. DOI: 10.1016/j.surfcoat.2014.12.048.

[23] Kumar B.S. Self-cleaning finish on cotton textile using sol-gel derived TiO_2 nano finish. IOSR Journal of Polymer and Textile Engineering 2015; 2: 1–5. DOI: 10.9790/019X-0210105.

[24] Onar N., Mete G., Aksit A., Kutlu B., Celik E. Water- and oil-repellency properties of cotton fabric treated with Silane, Zr, Ti based nanosols. International Journal of Textile Science 2015; 4: 84–96. DOI: 10.5923/j.textile.20150404.03.

[25] Castelvetro V., Francini G., Ciardelli G., Ceccato M. Evaluating fluorinated acrylic latices as textile water and oil repellent finishes. Textile Research Journal 2001; 71: 399–406. DOI: 10.1177/004051750107100506.

[26] Ferrero F., Periolatto M., Udrescu C. Water and oil-repellent coatings of perfluoro-poly-acrylate resins on cotton fibers: UV curing in comparison with thermal polymerization. Fibers and Polymers 2012; 13: 191–198. DOI: 10.1007/s12221-012-0191-z.

[27] Grajeck E.J., Petersen W.H. Oil and water repellent fluorochemical finishes for cotton. Textile Research Journal 1962; 32: 320–331. DOI: 10.1177/004051756203200408.

[28] Liang J., Zhou Y., Jiang G., Wang R., Wang X., Hu R., Xi X. Transformation of hydrophilic cotton fabrics into superhydrophobic surfaces for oil/water separation. The Journal of the Textile Institute 2013; 104: 305–311. DOI: 10.1080/00405000.2012.721207.

[29] Feng Y., Liao X., Wang Y., Shi B. Improvement in leather surface hydrophobicity through low-pressure cold plasma polymerization. Journal American Leather Chemists Association 2014; 109: 89–95.

[30] Kwong C.H., Ng S.P., Kang C.W., Molina R. Inducing hydrophobic surface on polyurethane synthetic leather by atmospheric pressure plasma. Fibers and Polymers 2014; 15: 1596–1600. DOI: 10.1007/s12221-014-1596-7.

[31] Choi J.H., Lee E.S., Baik H.K., Lee S.-J., Song K.M., Kwang M.K., Huh C.S. Surface modification of natural leather using low-pressure parallel plate plasma. Surface and Coatings Technology 2003; 171: 257–263. DOI: 10.1016/S0257-8972(03)00282-2.

[32] Türk M., Ehrmann A., Mahltig B. Water-, oil-, and soil-repellent treatment of textiles, artificial leather and leather. Journal of the Textile Institute 2015; 106: 611–620. DOI: 10.1080/00405000.2014.931108.

[33] Serenko O., Nizamova Z., Kalinin M., Ostrovsky Y., Polukhina L., Muzafarov A. Effect of the morphology of leather surface on the hydrophobic-hydrophilic properties. Advances in Materials Physics and Chemistry 2014; 4: 13–19. DOI: 10.4236/ampc.2014.42003.

[34] Ma J., Zhang X., Bao Y., Liu J. A facile spraying method for fabricating superhydrophobic leather coating. Colloids and Surface A: Physicochemical and Engineering Aspects 2015; 472: 21–25. DOI: 10.1016/j.colsurfa.2015.02.019.

[35] Latthe S.S., Hirashima H., Rao A.V. TEOS-based water-repellent silica films obtained by a co-precursor sol-gel method. Smart Materials and Structures 2009; 18: 095017 (6 pp.). DOI: 10.1088/0964-1726/18/9/095017.

[36] Fabbri P., Messori M., Montecchi M., Pilati F., Taurino R., Tonelli C., Toselli M. Surface properties of fluorinated hybrid coatings. Journal of Applied Polymer Science 2006; 102: 1483–1488. DOI: 10.1002/app.24350.

[37] About Leather, What is Leather? [Internet]. 2012. Available from: http://leather-worker.blogspot.it/2012_09_01_archive.html [Accessed 2017-01-31].

[38] New Zealand Institute of Chemistry. Chemical Processes in New Zealand. The chemistry of the leather industry [Internet] Available from: http://nzic.org.nz/ChemProcesses/animal/5C.pdf [Accessed 2017-01-31].

[39] Kessman A.J., Cains D.R. Template-assisted encapsulation of fluorinated silanes in silica films for sustained hydrophobic–oleophobic functionality. Journal of Colloid and Interface Science 2011; 360: 785–792. DOI: 10.1016/j.jcis.2011.05.026.